Laboratory Manual to accom

Anatomy & Physiology

Laboratory Manual to accompany

Anatomy & Physiology

Gary A. Thibodeau, Ph.D.
Chancellor and Professor of Biology
University of Wisconsin—River Falls
River Falls, Wisconsin

Kevin T. Patton, Ph.D.
Professor, Department of Life Sciences
St. Charles County Community College
St. Peters, Missouri

Third Edition

St. Louis Baltimore Boston Carlsbad Chicago Naples New York Philadelphia Portland
London Madrid Mexico City Singapore Sydney Tokyo Toronto Wiesbaden

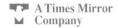

Vice President & Publisher: James M. Smith
Editor: Ronald E. Worthington, Ph.D.
Developmental Editor: Jean Sims Fornango
Project Manager: Carol Sullivan Weis
Production Editor: Florence Achenbach
Manufacturing Supervisor: Karen Lewis

Micrographs by Ed Reschke
Color Anatomy Plates by Douglas Eder,
    Shari Lewis Kaminsky, and John Bertram

Copyright © 1996 by Mosby–Year Book, Inc.

All rights reserved. No part of this publication may be reproduced, stored in a retrieval system, or transmitted, in any form or by any means, electronic, mechanical, photocopying, recording, or otherwise, without prior written permission from the publisher.

Permission to photocopy or reproduce solely for internal or personal use is permitted for libraries, or other users registered with the Copyright Clearance Center, provided that the base fee of $4.00 per chapter plus $.10 per page is paid directly to the Copyright Clearance Center, 27 Congress Street, Salem, MA 01970. This consent does not extend to other kinds of copying for general distribution, for advertising or promotional purposes, for creating new collected works, or for resale.

Printed in the United States of America.

Mosby–Year Book, Inc.
11830 Westline Industrial Drive
St Louis, Missouri 63146

ISBN: 0-8151-8826-9

Contents

Preface *p. vii*
Welcome to Anatomy and Physiology Laboratory *p. ix*
    The scientific method *p. ix*
    Measurement and data collection *p. ix*
    How to dissect *p. xi*
    Using this manual *p. xii*
    SAFETY FIRST! *p. xiii*
    How to study for lab quizzes *p. xv*
Laboratory Reference

The Basics

1 Organization of the Body *p. 1*
2 Dissection: The Whole Body *p. 11*
3 The Microscope *p. 23*
4 Cell Anatomy *p. 31*
5 Transport through Cell Membranes *p. 41*
6 The Cell's Life Cycle *p. 47*
7 Epithelial Tissue *p. 55*
8 Connective Tissue *p. 67*
9 Muscle and Nerve Tissue *p. 81*

Support and Movement

10 The Skin *p. 89*
11 Overview of the Skeleton *p. 99*
12 The Skull *p. 109*
13 The Vertebral Column and Thoracic Cage *p. 119*
14 The Upper Extremities *p. 129*
15 The Lower Extremities *p. 137*
16 Joints *p. 145*
17 Organization of the Muscular System *p. 157*
18 Skeletal Muscle Identification *p. 165*
19 Dissection: Skeletal Muscles *p. 179*
20 Skeletal Muscle Contractions *p. 201*

Integration and Control

21 Nerve Tissue *p. 211*
22 Nerve Reflexes *p. 219*
23 The Spinal Cord and Spinal Nerves *p. 227*
24 The Brain and Cranial Nerves *p. 233*
25 The Dissection: The Brain *p. 243*
26 Somatic Senses *p. 249*

27 Senses of Taste and Smell *p. 255*
28 The Ear *p. 259*
29 Hearing and Equilibrium *p. 265*
30 The Eye *p. 269*
31 Visual Function *p. 277*
32 Endocrine Glands *p. 283*
33 Hormones *p. 291*

Regulation and Maintenance

34 Blood *p. 299*
35 Structure of the Heart *p. 311*
36 Electrical Activity of the Heart *p. 319*
37 The Pulse and Blood Pressure *p. 329*
38 The Circulatory Pathway *p. 335*
39 The Lymphatic System *p. 345*
40 Dissection: Cardiovascular and
    Lymphatic Systems *p. 353*
41 Respiratory Structures *p. 363*
42 Dissection: Respiratory System *p. 369*
43 Pulmonary Volumes and Capacities *p. 375*
44 Digestive Structures *p. 383*
45 Dissection: Digestive System *p. 393*
46 Enzymes and Digestion *p. 399*
47 Urinary Structures *p. 405*
48 Dissection: Urinary System *p. 413*
49 Urinalysis *p. 419*

Reproduction

50 The Male Reproductive System *p. 425*
51 The Female Reproductive System *p. 431*
52 Dissection: Reproductive Systems *p. 437*
53 Development *p. 443*
54 Genetics and Heredity *p. 449*

Synthesis

55 The Whole Body *p. 455*
Front Cover (inside)
    Anatomical Directions
Back Cover (inside)
    Metric Measurement

Preface

Anatomy and physiology laboratory courses provide the essential hands-on learning opportunities required for a thorough understanding of the human body. This manual contains a series of 55 exercises that provide several guided explorations of human structure and function to accompany our textbook *Anatomy and Physiology (3rd edition)* or any standard human anatomy and physiology textbook. These activities include:

❑ **Labeling exercises** provide opportunities to identify important structures learned in the laboratory and lecture portions of the course. Once completed, they provide guidance to laboratory examinations of models and specimens. Students are encouraged to *write out* the labels so that the terms are more easily learned.

❑ **Coloring exercises** are becoming the most popular and effective way for many learners to grasp the essential spatial relationships of anatomical structures. This manual contains an accurate and comprehensive collection of human anatomy coloring plates, several of which are new to this edition.

❑ **Dissection of anatomical models** and examination of charts are an integral part of any beginning anatomy and physiology laboratory experience. This manual's instructions give valuable guidance for the effective use of models and charts.

❑ **Dissection of fresh and preserved specimens** of tissues, organs, and whole organisms enhances each student's appreciation of anatomical and functional relationships. Examination of tissues and organs is suggested in specific exercises throughout this manual. The rat, the cat, and the fetal pig are offered as options for many of the more comprehensive dissection activities. The last unit offers a guide to a live or taped demonstration with a prosected human cadaver.

❑ **Physiological experiments** emphasizing a variety of functional processes of the human body offer students immediate and dramatic examples of physiological concepts. When possible, these activities center around examination of the student's *own* physiological processes.

❑ **Optional computerized experiments** allow students to use the latest and most accurate methods for observing and measuring important physiological phenomena.

❑ **Content and concept review** questions and fill-in tables in each Lab Report and throughout the text of various exercises encourage students to reinforce and apply their knowledge of human structure and function.

❑ **Modern anatomical imaging** techniques such as computed tomography (CT), magnetic resonance imaging (MRI), and ultrasonography are introduced where appropriate. In each special presentation of imaging technology, students are challenged to interpret actual images of the human body.

❑ **Practical applications** to exercise and athletics, clinical situations, and everyday experiences increase student motivation and place important concepts in a useful context. Each practical example includes application questions that encourage students to think about how concepts apply to the situation described.

This laboratory manual also offers other features that enhance learning and ensure a safe and effective laboratory experience:

❑ **Special reference supplements** are included where appropriate to help students in their examination of laboratory specimens. The LABORATORY REFERENCE offers numerous full-color, labeled examples of commonly seen histology specimens and dissection specimens. The ANATOMICAL ATLAS OF THE RAT in Exercise 2 includes a complete series of labeled anatomical drawings of the laboratory rat.

❑ **Learning objectives** presented at the beginning of each exercise offer a framework for learning.

- **Complete lists of materials** for each exercise give the students and instructor a handy reference for efficient setup of laboratory activities.

- **Boxed hints** provide students with special tips on handling specimens, using equipment, and otherwise managing their laboratory activities.

- **Safety tips** are highlighted in special boxes to remind students of potential hazards, such as fire, chemical spills, cuts, or biological contamination.

- **Numerous illustrations** of proper procedures complement the text's complete description of laboratory activities.

The design of this laboratory package not only makes the laboratory course fun and effective for the student but also provides essential support for the instructor and lab preparation technician. Here are some examples of elements designed to aid instruction and preparation:

- **A comprehensive instruction and preparation guide** is provided to each adopting instructor. The guide contains a complete set of hints and special notes and instructions for each laboratory exercise. The guide provides a list of materials broken down by exercise, and a comprehensive list of all materials suggested for the course. Substitutions and special sources are given where appropriate. Lists of solution-preparation guidelines and other aids are found throughout the guide. Reproducible handouts to supplement certain exercises are also provided in the guide.

- **Modular organization of laboratory exercises** allows their use in virtually any order required by the needs of individual courses. Comprehensive cross-references in the instruction and preparation guide alert instructors about other Lab Exercises that may involve similar material. For example, the Hormone exercise may be more appropriately done near the reproductive system exercises in some courses.

- **Complete instructions for lab activities** are given to the student, freeing the instructor to interact with laboratory students on an individual basis rather than spending a great deal of time introducing the lab procedure to the whole class.

- **Easy-to-evaluate Lab Report formats** allow instructors to check student work or assign grades in an efficient manner. The Instruction and Preparation Guide provides correct answers to objective questions in each Lab Report.

Production of this lab manual was a team effort in many ways. Thanks especially to professors who reviewed our work and offered many helpful suggestions:

Cary Chaney
Northwest Arkansas Community College
Bentonville, Arkansas

Robert Eyres
Connors State College
Warner, Oklahoma

Zoe A. Fitzgerald
St. Charles County Community College
St. Peters, Missouri

G. M. Morris
Frank Phillips College
Borger, Texas

Illustrator Eileen Draper contributed many of the coloring plates and other figures, for which we are grateful. We also thank Mosby's editorial and production staff for their constant support and encouragement, as well as professional savvy. Our sincerest thanks to everyone involved.

Gary A. Thibodeau
Kevin T. Patton

Welcome to Anatomy and Physiology Laboratory

Anatomy and physiology laboratory challenges you to learn a great deal about human structure and function in a rather informal, practical atmosphere. Despite its informality as a learning situation, laboratory work requires some appreciation of the scientific method in general and laboratory policy and procedure in particular. Read this introductory section carefully. It will introduce you to laboratory science and will give you tips on how to complete this course with great success.

The scientific method

The **scientific method** is merely an approach to discovery. From its early days as a discipline, science has relied on this very simple, logical method for gaining an understanding of the universe. The basics of the scientific method can be summarized as a set of steps that are following in scientific discovery:

Hypothesis. First, one makes a tentative explanation, called a **hypothesis,** about some aspect of nature. A hypothesis is a reasonable guess based on previous informal observations or on previously tested explanations.

Initial experimentation. After a hypothesis has been proposed, it must be tested. The testing of a hypothesis is called **experimentation.** Scientific experiments are designed to be as simple as possible, to avoid the possibility of errors. Often, **experimental controls** are used to ensure that the test situation itself is not affecting the results. For example, if a new cancer drug is being tested, half the test subjects will get the drug and half the subjects will be given a harmless substitute. The group getting the drug is called the *test group,* and the group getting the placebo is called the *control group.* If both groups improve, or if only the control group improves, the drug's effectiveness hasn't been proven. If the test group improves, but the control group doesn't, the hypothesis that the drug works is tentatively accepted as true. Experimen-tation requires accurate measurement and recording of data.

Interpretation and conclusion. After an experiment, or series of experiments, the researcher analyzes all of the experimental data. If the results support the original hypothesis, it is tentatively accepted as true, and the researcher moves on to the next step. If the data does not support the hypothesis, the researcher tentatively rejects the hypothesis. If an experimental error is suspected, the hypothesis may not be rejected but retested. Knowing which hypotheses are untrue is almost as valuable as knowing which are true. Every rejected hypothesis brings the scientific community a little closer to the truth.

Replication. This step in the scientific method is the one least appreciated by nonscientists. After a hypothesis is tested and accepted, it is retested over and over to make sure that it is true. Because researchers often make minor mistakes in the design or execution of experiments, it is important that other scientists verify the original work. Usually, initial research experiments and their results are published in scientific journals so that others in the same field of research can benefit from them and verify them. If experimental results cannot be replicated (recreated) by other scientists, the hypothesis is not widely accepted. If a hypothesis withstands this rigorous retesting, the level of confidence in the hypothesis increases. A hypothesis that has gained a high level of confidence is called a **theory** or **law.**

The "facts" presented in this course are merely the latest hypotheses of how the body is built and how it functions. As methods of imaging the body and measuring functional processes improve, we find new data that causes us to replace old hypotheses with newer ones.

Measurement and data collection

Scientific experimentation is valid only when data are accurately measured and accurately recorded. In this lab course, you will be invited from time to time to execute various experiments. Your results will be meaningful only if you are careful to measure and record your results properly.

In this manual, as in nearly all scientific works, only the **metric system** of measurement is used. The metric system is useful for two important reasons: it is commonly used throughout the world, and metric units are easily converted because they are all based on units of ten.

> HINT → Do not worry about being able to convert metric units to United States (English) units or vice versa. You will rarely, if ever, be required to do that type of unit conversion in this course. It is more important that you learn to "think in metric," without regard to equivalent units in another system.

In this course, you will use metric units to measure time, temperature, length, volume, mass, and pressure. Each of these measurable characteristics has a basic metric unit that can be increased or decreased by factors of ten as needed. For example, the basic unit of measuring length is the **meter.** The meter is a useful unit in measuring the height of an elephant, 2 to 3 meters, but not in measuring the distance to the moon. Units a thousand times larger, **kilometers,** are used instead. Likewise, the diameter of a bacterial cell is measured in **micrometers,** which are a million times smaller than a meter. Different size units can be converted back and forth by multiplying or dividing by factors of ten.

Each of the tables in this section gives the basic units and common alternate units for time, temperature, length, volume, mass, and pressure.

> HINT → The abbreviations of metric units are given in parentheses. Note that they do not have a period (.) after them. Be careful to notice whether they are capital (uppercase) letters.

TIME

Basic unit: second (sec)

0.000001 second = microsecond (μsec)
0.001 second = millisecond (msec)
60 seconds = minute (min)
3600 seconds = hour (hr)

TEMPERATURE

Basic unit: degree Celsius (°C)

No alternate units are commonly used.

LENGTH

Basic unit: meter (m)

0.000000001 meter = nanometer (nm)
0.000001 meter = micrometer (μm)
0.001 meter = millimeter (mm)
0.01 meter = centimeter (cm)
1000 meters = kilometer (km)

VOLUME

Basic unit: liter (l or L)

0.001 liter = milliliter (ml)*
0.01 liter = centiliter (cl)
0.1 liter = deciliter (dl)

milliliters = cubic centimeters (cc)

MASS

Basic unit: gram (g)

0.001 gram = milligram (mg)
0.01 gram = centigram (cg)
1000 grams = kilogram (kg)

PRESSURE

Basic unit: millimeters of mercury (mm Hg)

No alternate units are commonly used.

Accurate measurement means using units of measurement correctly, but it also means using measuring devices accurately. If you are not familiar with reading the markings on metric rulers, balances, thermometers, and other common measuring devices, ask your instructor to demonstrate.

> **HINT** → A handy reference table of metric units and conversions, as well as the more commonly used prefixes, is printed on the inside back cover of this lab manual.

How to dissect

The term *anatomy* literally means "to cut apart," so it is no wonder that dissections are commonly done in anatomy lab courses. Dissection activities recommended in this manual are not proposed without recognition of humane concerns. The purpose of these dissections is to instruct in a way that no other method can duplicate. The specimens called for are usually from animals raised specifically to be euthanized and used as resources. For health professionals, dissection of anatomical preparations is a necessary prerequisite to working with living bodies.

The goal of any dissection exercise is the exploration of anatomical relationships. Proper dissection requires patience and skill. Some students slice and hack away at their specimens until they have a tray of ground meat. Others hardly touch their specimen. Good technique is somewhere in the middle of these two extremes. Organs should be separated from one another only enough to see surrounding structures. Rarely should structures be cut or removed. The instructions given in this manual state when, where, and how cuts should be made.

Each dissection activity in this manual offers safety advice concerning proper handling of the dissection specimens. Protective gloves, lab coat, and eyewear may be called for when using some specimens. Always be careful when using dissection tools. Severe injuries can result from their careless use.

Figure A shows some commonly used dissection instruments. A brief discussion of each is in order here:

1. **Scalpel or knife.** The scalpel is probably the most overused instrument in an anatomy lab course. This tool should be used *seldom,* only when you want to cut all the way through a specimen. A pathology **knife,** or butcher knife, is a much more useful tool.

2. **Scissors.** Scissors are probably the most underused dissection tool. Whenever you are tempted to use a scalpel for cutting, try the scissors first. Often, your results will be much better and you will not have damaged important underlying parts.

3. **Probes.** You may want to have several types of probes. Both dull and sharp probes (dissecting needles) are useful in separating tissues, exploring cavities, tracing blood vessels, and pointing to structures.

4. **Forceps.** Forceps are one of the handiest dissection tools. They can be used to grasp small objects, to separate structures, to point to structures, to explore cavities, and to pull on structures.

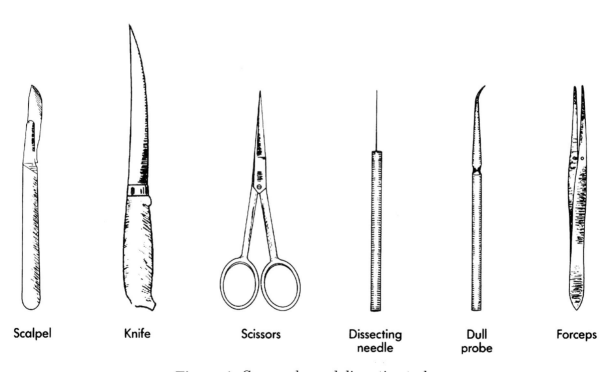

Figure A Commonly used dissection tools.

5. **Ruler.** Your metric ruler should be marked in both centimeters and millimeters. It is useful in measuring organs and in many nondissection lab activities.

6. **Dissection pins.** These are standard, heavy-duty straight pins. Dissection pins are useful in pinning membranes and other structures to a dissection board to keep them temporarily out of the way.

7. **Dissection tray.** There are many varieties and shapes of dissection trays, and your instructor will recommend the best for your situation. Dissection trays help organize the dissection activity by keeping everything together, and they protect the lab table surface. Some types can also be used for storing your specimen for later study.

Using this manual

The format of each Lab Exercise is self-explanatory. After some brief introductory remarks, each exercise begins with a section entitled "Before you begin." This section recommends that before starting any lab activity you should:

❑ Read the appropriate chapter(s) in *Anatomy and Physiology (3rd edition)* or whatever textbook you are using for this course.

❑ Set your learning goals so that you know what important concepts you should be trying to learn.

❑ Prepare your materials so that your lab activity will run smoothly.

❑ Read the entire activity before starting. In this way, you will not be surprised (and therefore unprepared) for any step in the procedure.

Each exercise contains one or more activities. Your instructor may suggest that you do all the activities in an exercise or only one. Each activity includes some or all of these helpful features:

❑ Large, bold step numbers help you keep your place as your eyes move back and forth between the manual and your lab setup. Each step number is preceded by a check-box (❑) that you can use to check off each step as it is completed.

❑ Boxed hints, safety tips, and landmark characteristics (for identifying specimens) highlight useful information for completing the activity safely and successfully.

❑ Anatomical dissection illustrations and photographs are identified with a colored tab for quick and easy reference.

❑ Labeling exercises encourage you to apply your knowledge of human anatomy in a practical test. Long label lines allow you room to write the name of the structure directly on the illustration. When each figure is completely labeled, the terms should be transferred to the Lab Report so that the instructor can check your work.

❑ Coloring exercises have been cited by educators and students as being an effective method of learning anatomy. By using multiple senses to trace and highlight shapes and relationships, you reinforce your knowledge of the human form. Each coloring plate is like a paint-by-number activity. Each label has a small number that corresponds to a number on the illustration itself. Use a colored pen or pencil (not a crayon) to write the term over the outline letters of the label. This will reinforce your familiarity with the term and its correct spelling and makes the label a "color code" so you can later find the structure by its color. After filling in the label, use the same color to shade in the matching structure in the illustration. Use contrasting colors so that parts that are next to one another can be easily distinguished. Do not attempt to color everything in realistic colors otherwise many parts will look the same. If you use pens, make sure they won't bleed through the paper. If you use pencils, you may want to insert a piece of plain paper into your manual over each completed coloring plate so that the colors do not rub off on the facing page. The materials list in the first exercise reminds you to bring colored pens or pencils to your work area, but it is assumed that you will have these for all the lab exercises.

Each exercise ends with a Lab Report. The report includes handy tables and sketching areas for you to record your laboratory observations and results. Where appropriate, the Lab Report also presents objective and subjective questions for you to answer. Some of these questions drill you on your knowledge of important terms, whereas others ask you to use your basic knowledge to interpret, organize, or otherwise process basic facts to ensure that you understand conceptual relationships. A number of highlighted practical application boxes in the exercise itself often ask you to apply your basic knowledge in this way. These questions, along with practice in making scientific observations, help you develop your scientific reasoning skills.

Safety First!

As already mentioned, numerous boxed safety tips appear throughout this manual. Entitled "SAFETY FIRST!", their large letters and urgent tone are meant to call your attention to potential hazards in the laboratory. Before beginning the course, it is essential that you learn some basic laboratory safety policies:

❏ Check the labels of all chemicals for safety warnings before using them. If the container does not have a safety label, consult the instructor.

❏ Wear protective gear if you are handling dangerous chemicals.

❏ Plan ahead about what to do if a spill occurs. If you spill a dangerous substance on your body, you may have to remove your clothing. If your lab has a safety shower, stand under it and pull the ring.

❏ Avoid using chemicals for which you have a known sensitivity.

❏ Locate first aid equipment in the laboratory and familiarize yourself with its use.

Chemical Safety

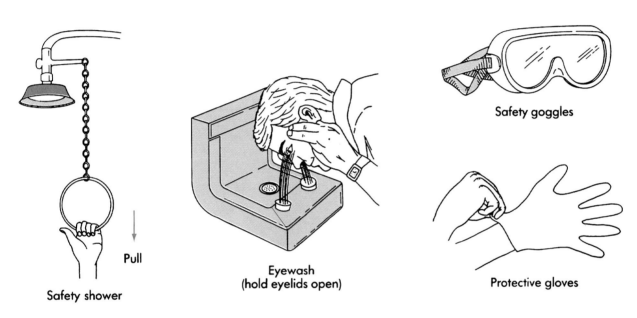

Fire Safety

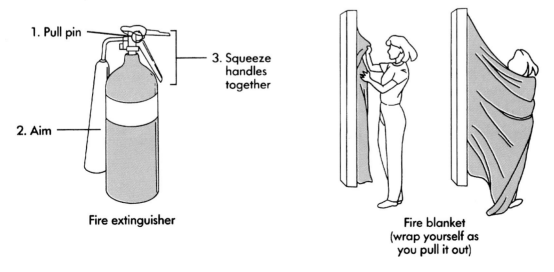

Figure B Commonly used chemical and fire safety equipment.

IMPORTANT SAFETY INFORMATION

(Use the spaces that follow to write important safety information.)

Location of first aid box:

Emergency medical help:

Fire extinguisher location:

Fire extinguisher type:

Fire alarm box location:

Primary escape route:

Alternate escape route:

Weather safety shelter:

Important notes:

- Locate the nearest medical help and identify the easiest and fastest way to access it (for example, phoning an ambulance).

- Locate the fire safety equipment in your lab and familiarize yourself with its use. Locate the nearest fire alarm box and identify the recommended primary and alternate fire exit routes.

- **Always** follow the directions provided with equipment and supplies, even if they are different than those given in this manual. Injuries often result from the misuse of equipment.

- Remember that absolutely no eating, drinking, or smoking is allowed in the laboratory.

- To avoid injury or contamination, properly dispose of or clean and store all lab equipment and supplies before vacating the laboratory. Wash your hands thoroughly on entering and leaving a biological laboratory.

- Always supervise experiments or demonstrations in progress. Never leave a laboratory experiment unattended.

How to study for lab quizzes

Tips for effective studying could fill several volumes. Instead of an entire work on studying, this section presents a few tips that many students have found to be particularly helpful in the anatomy and physiology laboratory course. As with any list of study tips, you will find some useful to you and others not so useful.

- Identify the type of quizzes you will encounter. Written quizzes may involve multiple choice, matching, or other objective questions. Written quizzes may also have subjective short answer or essay questions in which you must interpret or restate concepts. Figures may be offered for labeling. Practical tests may involve stations at which specimens have been placed. You may be asked to perform a procedure or answer a question about the specimen while at each station. You may have an individualized practical quiz in which the instructor will ask you questions or watch you perform a procedure individually. Ask your instructor for some sample questions if you do not understand how an upcoming quiz is constructed. Knowing the manner in which you will be evaluated is important to designing your study strategy.

- Organize a study group (which may be the same as your lab group) to meet several times before a quiz to help each other with the material. Quizzing each other is a useful technique.

- Spend many brief sessions reviewing specimens and other lab work, rather than one or two long sessions just before the quiz. Ask your instructor if there are open lab times or a learning center at which you can review models or specimens.

- Study actively. Do not just sit and look at your study materials but do some hands-on activities with them. Make flash cards to quiz yourself. Set up your own practical, then take it. Sketch specimens and models again. Draw a "concept map" in which you draw boxes or pictures connected in a logical way. This will help you organize concepts in your mind.

- Use all your resources. Campus libraries and learning centers often offer the services of specialists who can help you sharpen your study skills. They can show you examples of concept maps and flash cards. Your instructor has successfully completed laboratory courses and has helped numerous previous students in this course. Ask your instructor for study tips. Use the textbook and study aids in the textbook to help you learn the lab material.

LAB EXERCISE 1

Organization of the Body

Many explorers use maps, figures, and photographs to help orient themselves to the terrain to be explored. The anatomist uses maps, figures, and photos to explore the body and its parts. The use of maps and other aids to find a geographical position is called *orienteering* and is a useful analogy to human anatomical study. At the beginning of this exercise, you will learn how anatomical "maps" and models are read. Later in this exercise, you will become familiar with the major body systems and some of their organs so that you will be comfortable with the "lay of the land" in the human body.

Before you begin

❏ Read the appropriate chapter in your textbook.

❏ Set your learning goals. When you finish this exercise, you should be able to
- use anatomical terms correctly
- discuss the nature of an anatomical section
- describe the basic plan of the human body
- identify the major body cavities
- list the major systems of the body, their principal organs, and their primary functions

❏ Prepare your materials:
- dissectible human torso model (or comparable charts)
- models or figures showing different anatomical sections

❏ Read the directions and safety tips for this exercise carefully before starting any procedure.

A. Planes and sections

All terms describing the anatomy of organisms assume that the body is in the classic **anatomical position.** For the human, that means standing, facing the viewer. The hands are held down along the side of the trunk, with the palms facing forward.

> **HINT** → The anatomical position of a four-legged animal, such as a rat, cat, or fetal pig, is standing on all four limbs, head facing forward. See exercise 2 for an example.

It is often useful to show a figure of a *sectioned* human body or organ. A section refers to a part cut in a **plane.** A plane is a geometrical concept referring to an imagined flat surface. The term *cross section (c.s.),* for example, refers to a part cut crosswise. A *longitudinal section (l.s.)* is a cut made lengthwise. These terms are useful only in limited circumstances because they do not really identify whether the cuts are made top to bottom, front to back, or side to side. There are three anatomical planes used to describe sections of the body:

❏ 1 **Sagittal plane**—A sagittal plane extends from front to back and top to bottom, dividing the body into left and right portions. A *midsagittal plane* refers to a sagittal plane that divides the body into exactly equal left and right portions.

❏ 2 **Frontal plane**—The frontal plane, also called a *coronal plane,* divides the body into front and back portions.

❏ 3 **Horizontal plane**—Also called a *transverse plane,* the horizontal plane divides the body into top and bottom portions.

Using the models or figures provided, find at least three examples of each of the sections described. For each, ask yourself what perspective the section gives that a section cut along a different plane does not give.

B. Anatomical directions

To locate structures within a body, you must use *directional terms.* Actually, you use these kinds of terms all the time: left, right, up, down, north, south, for example.

❏ 1 Review the directional terms given in Table 1-1. Notice that they are grouped in relative pairs. Each member of a pair is the opposite, or complement, of the other member of the pair. For example, *right* is the opposite direction of *left*.

To make the reading of anatomical figures a little easier, an *anatomical compass* is used throughout this lab manual, just as it is in the textbook. On many figures, you will notice a small compass rosette similar to those on geographical maps. Rather than being labeled N, S, E, and W, the anatomical rosette is labeled with abbreviated anatomical directions.

Table 1-1

Directional term	Definition	Example of usage
Left	To the left of the body (not *your* left, the subject's)	The stomach is to the left of the liver.
Right	To the right of the body or structure being studied	The *right* kidney is damaged.
Lateral	Toward the side; away from the midsagittal plane	The eyes are *lateral* to the nose.
Medial	Toward the midsagittal plane; away from the side	The eyes are *medial* to the ears.
Anterior	Toward the front of the body	The nose is on the *anterior* of the head.
Posterior	Toward the back (rear)	The heel is *posterior* to the toes.
Superior	Toward the top of the body	The shoulders are *superior* to the hips.
Inferior	Toward the bottom of the body	The stomach is *inferior* to the heart.
Dorsal	Along (or toward) the vertebral surface of the body	Her scar is along the *dorsal* surface.
Ventral	Along (toward) the belly surface of the body	The navel is on the *ventral* surface.
Caudad (caudal)	Toward the tail	The neck is *caudad* to the skull.
Cephalad	Toward the head	The neck is *cephalad* to the tail.
Proximal	Toward the trunk (describes relative position in a limb or other appendage)	This joint is *proximal* to the toenail.
Distal	Away from the trunk or point of attachment	The hand is *distal* to the elbow.
Visceral	Toward an internal organ; away from the outer wall (describes positions inside a body cavity)	This organ is covered with the *visceral* layer of the membrane.
Parietal	Toward the wall; away from internal structures	The abdominal cavity is lined with the *parietal* peritoneal membrane.
Deep	Toward the inside of a part; away from the surface	The thigh muscles are *deep* to the skin.
Superficial	Toward the surface of a part; away from the inside	The skin is a *superficial* organ.
Medullary	Refers to an inner region, or *medulla*	The *medullary* portion of the organ contains nerve tissue.
Cortical	Refers to an outer region, or *cortex*	The *cortical* area produces hormones.

❏ 2 Review this list of directional terms and abbreviations:
A = Anterior
I = Inferior
L (opposite **R**) = Left
L (opposite **M**) = Lateral
M = Medial
P = Posterior
R = Right
S = Superior

❏ 3 Review the examples of how the anatomical compass is used in illustrations by looking at Figure 1-1, *A* through *D*. Test your knowledge of directions and the use of the anatomical compass by labeling the rosettes given in Figure 1-1, *E* through *J*.

> **HINT →** A handy summary of anatomical terms and their usage is found on the inside front cover of this manual, for easy location later in your studies.

C. Body cavities and regions

The inside of the human body contains the viscera, or internal organs. The viscera are found in any of a number of cavities (spaces) within the body. The two principal body cavities are the dorsal body cavity and ventral body cavity. Because these spaces are so large, they are subdivided into smaller units.

❏ 1 Using a dissectible torso model, find these divisions of the dorsal body (and organs within):
- **Cranial cavity**—Within the skull
   Organ: *brain*
- **Spinal cavity**—Within the vertebral column
   Organ: *spinal cord*

❏ 2 Using the torso model, find these divisions and organs of the ventral body cavity:
- **Thoracic cavity**—Within the rib cage
  - **Pleural cavities**—Left one third and right one third of the thoracic cavity
     Organ: *lung*
  - **Mediastinum**—Middle one third of thorax
     Organs: *heart, trachea, esophagus*
- **Abdominopelvic cavity**—From the diaphragm to the bottom of the trunk
  - **Abdominal cavity**—From the diaphragm to the rim of the pelvic bones
     Organs: *stomach, liver, most of the intestines, pancreas, spleen, kidneys*
  - **Pelvic cavity**—From the pelvic rim to the floor of the trunk
     Organs: *portions of the intestines, ovaries, uterus, urinary bladder*

❏ 3 Because the abdominopelvic cavity is so large, and contains so many different organs, it is often convenient to subdivide it into nine abdominopelvic regions. The regions are bounded by a grid made by imagining two horizontal planes (one just below the ribs, the other just above the hip bones) and two sagittal planes (each just medial to a nipple). This arrangement forms a 3-D, tic-tac-toe grid in the abdominopelvic cavity. Identify the approximate locations of each of the nine regions on a model of the human torso.
- **Right hypochondriac region**—Top right region (*hypochondriac* means "below (rib) cartilage")
- **Epigastric region**—Top middle region (*epigastric* means "near the stomach")
- **Left hypochondriac region**—Top left region
- **Right lumbar region**—Middle right region (*lumbar* refers to lumbar vertebrae in lower back)
- **Umbilical region**—Central region (*umbilical* refers to the umbilicus, or navel)
- **Left lumbar region**—Middle left region
- **Right iliac region**—Lower right region (*iliac* refers to ilium, the bowl-like part of the hip bone)
- **Hypogastric region**—Lower middle region (*hypogastric* means "below the stomach")
- **Left iliac region**—Lower left region

D. Surface regions

There are hundreds of terms that describe specific locations on the surface of the human body. These names are useful for identifying not only surface features but also underlying muscles, bones, nerves, and blood vessels. In this activity, locate regions named by a few of the more common terms.

❏ 1 Locate the following surface regions on the anterior aspect of a human model or figure:
- **Abdominal**—Area overlying the abdominal cavity
- **Antebrachial**—Forearm
- **Axillary**—Armpit
- **Brachial**—Upper arm
- **Buccal**—Cheek (side of mouth)
- **Carpal**—Wrist
- **Cervical**—Neck
- **Coxal**—Hip
- **Crural**—Anterior lower leg (shin)
- **Cubital**—Anterior of elbow
- **Femoral**—Upper leg (thigh)
- **Mental**—Chin
- **Orbital**—Eye
- **Patellar**—Anterior knee joint
- **Pubic**—Lower front of trunk, between legs
- **Tarsal**—Ankle
- **Thoracic**—Chest

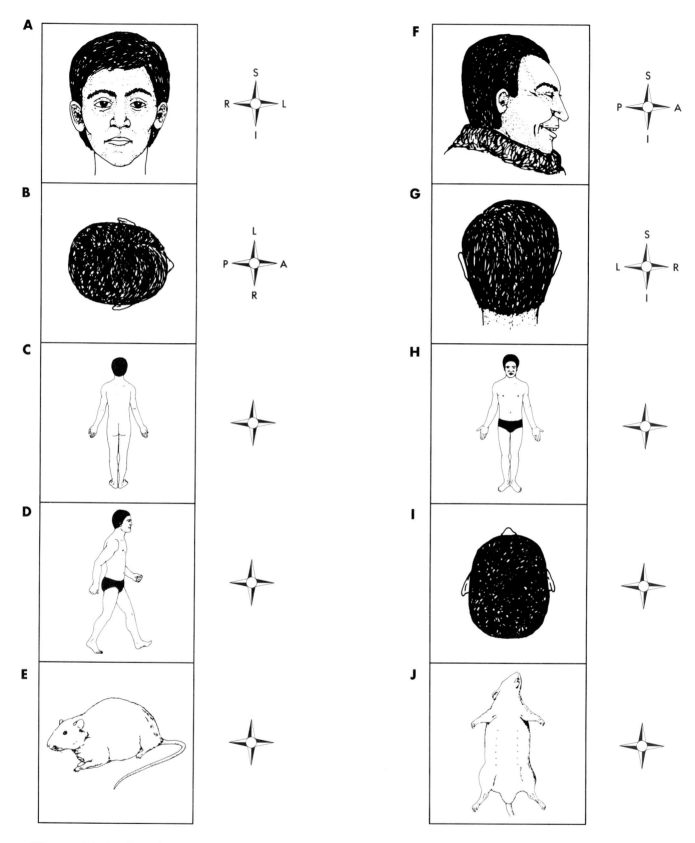

Figure 1-1 Look at the examples of labeling anatomical directions (**A** and **B**, **F** and **G**), then label the rosettes in **D** and **E, I** and **J** yourself.

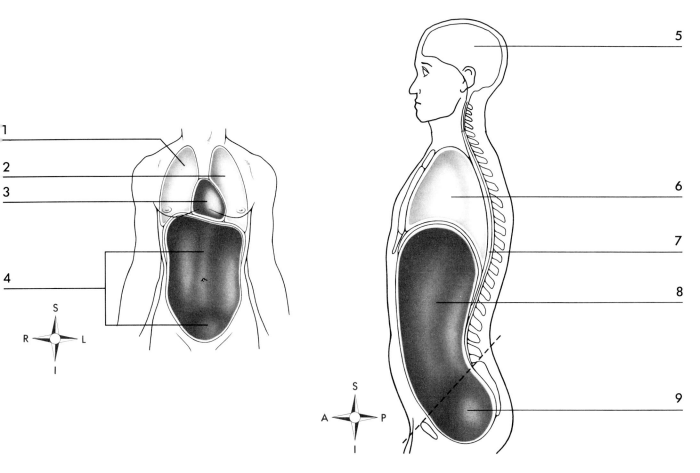

Figure 1-2 Label the names of the body cavities indicated.

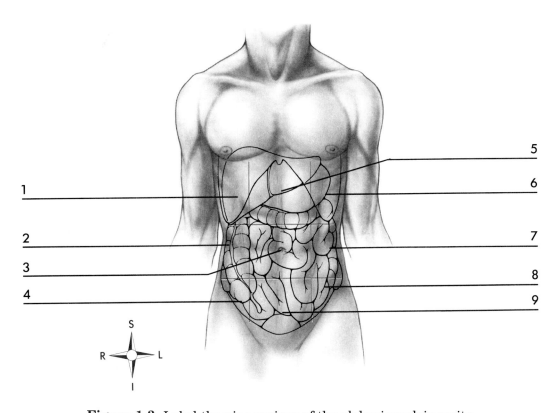

Figure 1-3 Label the nine regions of the abdominopelvic cavity.

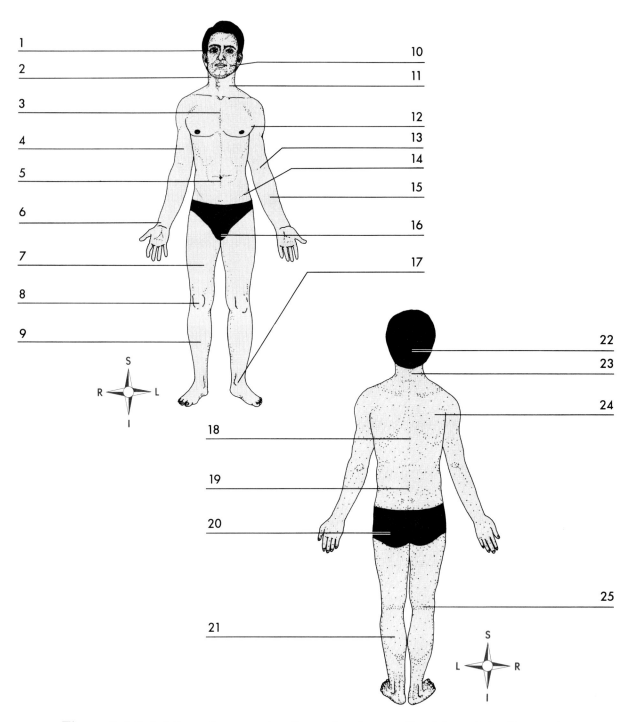

Figure 1-4 Label these figures using the regional terms listed in activity D, page 3.

❑ 2 Identify these regions on the posterior aspect of your subject:
- **Cervical**—Neck
- **Gluteal**—Buttocks
- **Lumbar**—Lower back
- **Occipital**—Posterior of head
- **Popliteal**—Posterior knee joint
- **Scapular**—Shoulder blade
- **Sural**—Calf
- **Thoracic**—Upper back

E. Body systems

As you know, the human organism is composed of organ groups called *systems*. The organs of a system work together in an organized manner to accomplish the function(s) of the system.

As an introduction to human body systems, study Table 1-2. Each of the systems will be discussed in more detail later in this course.

Table 1-2

Organ system	Principal organs	Primary function(s)
Integumentary	Skin	Protection, temperature regulation, sensation
Skeletal	Bones, ligaments	Support, protection, movement, mineral/fat storage, blood production
Muscular	Skeletal muscle, tendons	Movement, posture, heat production
Nervous	Brain, spinal cord, nerves, sensory organs	Control/regulation/coordination of other systems, sensation, memory
Endocrine	Pituitary gland, adrenals, pancreas, thyroid, parathyroids, other glands	Control/regulation of other systems
Cardiovascular	Heart, arteries, veins, capillaries	Exchange and transport of materials
Lymphatic	Lymph nodes, lymphatic vessels, spleen, thymus, tonsils	Immunity, fluid balance
Respiratory	Lungs, bronchial tree, trachea, larynx, nasal cavity	Gas exchange, acid-base balance
Digestive	Stomach, intestines, esophagus, liver, mouth, pancreas	Breakdown and absorption of nutrients, elimination of waste
Urinary	Kidneys, ureters, bladder, urethra	Excretion of waste, fluid and electrolyte balance, acid-base balance
Reproductive male	Testes, vas deferens, prostate, seminal vesicles, penis	Continuation of genes (reproduction)
Reproductive female	Ovaries, fallopian tubes, uterus, vagina	Reproduction, nurturing of offspring

Anatomical Planes and Directions

PLANES
SAGITTAL 1
MIDSAGITTAL 2
FRONTAL 3
HORIZONTAL 4

DIRECTIONS
LATERAL 5
MEDIAL 6
ANTERIOR 7
POSTERIOR 8
SUPERIOR 9
INFERIOR 10
PROXIMAL 11
DISTAL 12

COLORING EXERCISE Using colored pens or pencils, shade in the figure and accompanying labels in contrasting colors of your choice as indicated by the red numerals.

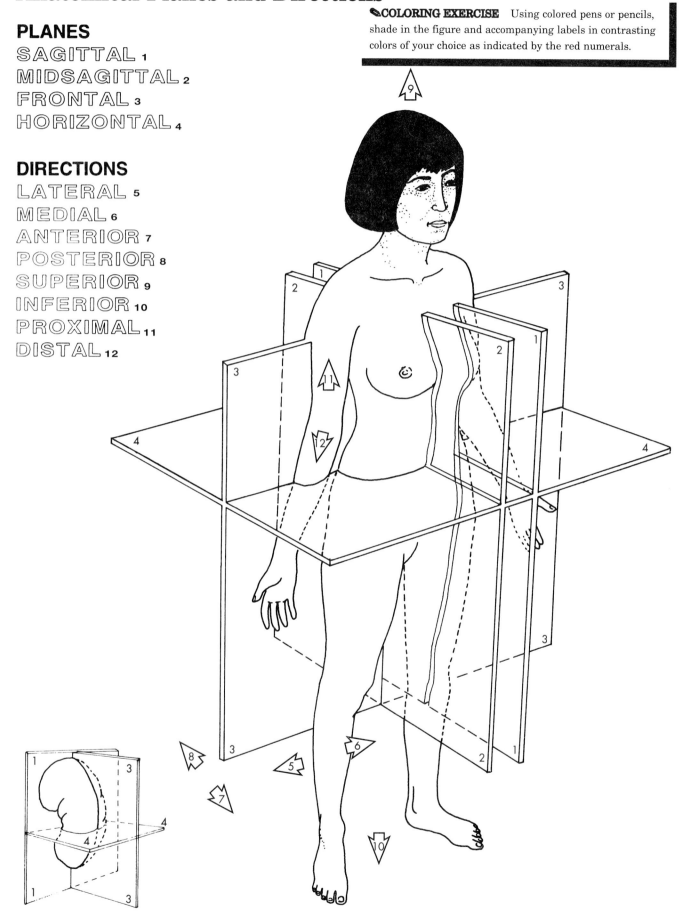

Figure 1-5

NAME _____ DATE _____ SECTION _____

LAB REPORT 1

Organization of the Body

Figure 1-2

1. _____
2. _____
3. _____
4. _____
5. _____
6. _____
7. _____
8. _____
9. _____

Figure 1-3

1. _____
2. _____
3. _____
4. _____
5. _____
6. _____
7. _____
8. _____
9. _____

Multiple choice

___ 1
___ 2
___ 3
___ 4
___ 5
___ 6

Multiple choice (only one response is correct in each item)

1. An anatomist cuts a cadaver (preserved body) with a large saw in a way that divides the cadaver into equal left and right halves. The cut is along a __?__ plane.
   a. sagittal
   b. midsagittal
   c. frontal
   d. horizontal
   e. a and b are correct

2. In many study skulls, the top of the skull can be removed so that inner features can be seen. Along which plane should one cut to open the top of a human study skull?
   a. sagittal
   b. coronal
   c. horizontal
   d. frontal
   e. b and d are correct

3. A surgeon makes an incision medially from the left axillary region, turning inferiorly at the midline and proceeding to the pubic region. The path of the cut can be mapped on the patient's chest as:

   a. ⌐   b. ╋   c. ⌊   d. ⌉   e. —   f. /

4. Which of these regions contains the spleen?
   a. epigastric
   b. hypogastric
   c. left hypochondriac
   d. right hypochondriac
   e. right lumbar

5. Soccer players often wear shin protectors, which shield the __?__ region of each leg.
   a. femoral
   b. tibial
   c. popliteal
   d. crural
   e. gluteal

6. Control and regulation of other systems are primary functions of the
   a. nervous system
   b. cardiovascular system
   c. endocrine system
   d. urinary system
   e. a and c are correct

Figure 1-4

_____ 1
_____ 2
_____ 3
_____ 4
_____ 5
_____ 6
_____ 7
_____ 8
_____ 9
_____ 10
_____ 11
_____ 12
_____ 13
_____ 14
_____ 15
_____ 16
_____ 17
_____ 18
_____ 19
_____ 20
_____ 21
_____ 22
_____ 23
_____ 24
_____ 25

Fill-in

_____ 1
_____ 2
_____ 3
_____ 4
_____ 5
_____ 6
_____ 7
_____ 8
_____ 9
_____ 10
_____ 11
_____ 12

Fill-in (give the correct anatomical term for each item below)

1. The head is __?__ to the feet.
2. The liver is part of the __?__ system.
3. A leg amputation is likely to involve a __?__ cut, or section, through bone.
4. My lower back, or __?__, is sore.
5. The first finger is __?__ to the hand, no matter which position it is in.
6. The popliteal vein is found in the __?__.
7. The heart is __?__ to the right lung.
8. The shoulder is __?__ to the elbow, no matter how one's arm is held.
9. The skin is __?__ relative to the skeleton.
10. Adipose tissue is often just __?__ to the skin.
11. An occipital scar is on the back of the __?__.
12. The thoracic wall is lined with the __?__ layer of the double-layered pleural membrane.

Sketch (make a rough sketch of a human figure in the position indicated by the compass rosette and label)

LAB EXERCISE 2

Dissection: The Whole Body

This exercise may be used as either an introduction to vertebrate anatomy or as a synthesis activity to conclude the laboratory course. Succeeding exercises in this manual concentrate on organs or systems one by one. This exercise begins the study of the body with an overview of everything together in an exploration of the entire vertebrate body. Because the vertebrate body plan is similar among species, another laboratory animal specimen (e.g., a rabbit) could be substituted.

> **HINT** → In a short course, a well-preserved specimen can be used from time to time throughout your studies. If you will be looking at your dissected specimen for the next several months, make sure that it is kept in the appropriate container under conditions suggested by your instructor.

Before you begin

❏ Set your learning goals. When you finish this exercise, you should be able to
- perform a whole-body dissection of a vertebrate animal
- identify the major anatomical features of the vertebrate body in a dissected specimen

❏ Prepare your materials:
- preserved (plain or double-injected) or freshly killed laboratory rat (or similar vertebrate)
- dissection tools and trays
- mounted rat skeleton (optional)
- storage container (if preserved specimen is to be reused)

> **SAFETY FIRST!** Observe the usual precautions when working with a preserved or fresh specimen. Heed the safety advice accompanying preservatives used with your specimen. Use protective gloves and perhaps safety goggles while handling your specimen. Avoid injury with dissection tools. Dispose of your specimen as instructed.

❏ Read the directions and safety tips for this exercise **carefully** before starting any procedure.

A. The external aspect

Examine the external aspect of your specimen.

❏ 1 Determine the anatomical orientation of the specimen. Which direction is anterior? posterior? Which direction is ventral? dorsal? Identify sagittal, transverse, and frontal planes in your specimen.

❏ 2 Identify these externally visible features:
- **Pinna (auricle)**
- **External nares (nostrils)**
- **Vibrissae (whiskers)**
- **Incisors**
- **Integument**
- **Forelimbs**
- **Hindlimbs**
- **Thoracic region**
- **Abdominal region**
- **Nipples**
- **Anus**
- **Tail**

❏ 3 Determine the sex of your specimen by examining the external genitals:
- **Female**—Immediately anterior to the anus, on the ventral surface, is the **vulva** with an opening to the **vagina.** Anterior to the vulva is the **clitoris** with the **urethral** opening.
- **Male**—The **scrotum** containing the **testis** is immediately posterior to the anus, perhaps even hiding the anus from view. Near the anterior edge of the scrotum is the **penis** with its **prepuce,** or skin-fold covering. Locate the opening of the **urethra** in the penis.

> **HINT** → A complete set of anatomical sketches of the rat are found in the ANATOMICAL ATLAS OF THE RAT found in this exercise. A complete set of full-color photographs of the rat dissection, including both male and female specimens, can be found in the LABORATORY REFERENCE, Plates 1 through 5.

B. Skin, bones, and muscles

☐ 1 Remove most of the skin from your specimen by following this procedure:
- Place the animal in the tray with its ventral surface facing you.
- Pull up on the skin over the sternum and puncture it with the tip of a scissors. Slide the bottom tip of the scissors into the **subcutaneous** area under the skin.
- Begin cutting along the lines indicated in Figure 2-2. Be careful not to cut into the skeletal muscles under the skin.
- With your forceps, pull the two flaps of skin over the neck away from the animal's body. Notice the **areolar tissue** under the skin that is pulled apart as you remove the skin. Pull the flaps of skin over the abdomen and over the **groin** area away in a similar fashion.
- Turn the animal over, so that the dorsum is facing you.
- If you made all the cuts properly, you should be able to pull the skin off the back in one complete piece. Sometimes it helps if you scrape at the loose connective tissue under the skin with your scalpel as you peel the skin away. The skinning process is difficult unless you have patience and proceed slowly.

☐ 2 Examine the skin, identifying these features:
- **Dermis**—The thick inner layer of the skin
- **Epidermis**—The thinner outer layer of the skin with *hair* (fur)

☐ 3 Explore the shape of the skinned rat body. How many **bones** of the rat's skeleton can you see or palpate (feel)? If you have a mounted rat skeleton available, identify as many of the bones of the skeleton as you can. If you become stumped, refer to Figure 2-3, *A* in the rat atlas. Notice the similarity between the rat's skeletal plan and that of the human.

☐ 4 Observe the rat's musculature. Some of the external muscles of the torso can be separated from each other for easier viewing. Slide a probe into the loose connective tissue joining adjacent muscles and run the probe along their margins. Using Figures 2-3 and 2-4 as guides, try to identify the major superficial muscles of the rat's body.

C. Cardiovascular structures

☐ 1 Open the **ventral body cavity** by cutting into its muscular wall in a manner similar to your earlier cut into the skin. Cut flaps in the neck, abdomen, and groin areas as you did with the skin, but do not cut all the way around to the back. Be careful not to damage any **visceral organs** with your scissors as you cut. Fold back the flaps and anchor them with pins or remove them.

☐ 2 Locate the **heart** near the middle of the **thoracic cavity,** in the **mediastinum.** Can you identify the four chambers?

☐ 3 If you have a double-injected preserved specimen, the **arteries** are filled with red latex and the **veins** are filled with blue latex. If not, the arteries can usually be distinguished from veins because they are stiffer and lighter in color than veins. Locate the **aorta,** the large artery leaving the heart and arching posteriorly. Trace the branches of the aorta, naming them if you can. Use Figure 2-5, *A* if you need help.

☐ 4 Locate the **anterior vena cava** where it drains into the heart. This is analogous to the superior vena cava in the human. Follow its tributary veins and identify them with the help of Figure 2-5, *B*. Locate the **posterior vena cava** and trace its tributaries.

> **HINT** → Once you have cut into the ventral body cavity, you may be tempted to cut and remove organs. It is important that you keep everything as intact as possible. You may pull organs to one side or another to view deeper structures, but avoid making cuts.

D. The viscera

The viscera, or major internal organs, can be seen within the ventral body cavity. Use Figures 2-6 and 2-7 to guide you in locating the following:

☐ 1 Locate some of these features of the lower **respiratory system:**
- **Larynx**
- **Trachea**
- **Primary bronchi**
- **Lungs** (Can you distinguish the parietal and visceral *pleurae*?)
- **Diaphragm**

☐ 2 Locate these structures of the **digestive system:**
- **Submandibular salivary glands**
- **Esophagus**
- **Stomach**
- **Liver**

- **Pancreas**
- **Small intestine**
- **Mesentery**
- **Cecum** (This structure is very large in the rat and does not have an appendix as does the human cecum.)
- **Colon**

❏ 3 Locate these **lymphatic** organs:
- **Spleen**
- **Thymus** (The thymus may be small in older animals.)

❏ 4 Locate these features of the **urinary system**:
- **Kidney**
- **Ureter**
- **Urinary bladder**
- **Urethra**

❏ 5 Try to locate these **endocrine glands** in your specimen:
- **Thyroid gland**
- **Thymus gland**
- **Pancreas**
- **Adrenal glands**
- **Testes**
- **Ovaries**

❏ 6 Identify these structures associated with the **male reproductive system**:
- **Testes**
- **Epididymis**
- **Ductus deferens**
- **Seminal vesicle**
- **Prostate gland**
- **Penis**

❏ 7 Find these **female reproductive system** structures:
- **Ovaries**
- **Uterine tubes**
- **Uterus** (notice that the rat uterus has a Y shape, with a right and left *uterine horn*)
- **Vagina**

> **HINT** → Unless your lab group has both a male and a female specimen, you may want to temporarily trade specimens with a group that has a rat of a different gender than yours. By doing so, you will be able to find the features of both reproductive systems.

❏ 8 If your specimen is a pregnant female, open the uterus *very carefully* and identify these structures:
- **Amniotic sac**—This is a fluid-filled cushion for each embryo.
- **Embryo**—Examine the structure of the embryo. Do you think that it is in a late stage or an early stage of development?
- **Umbilical cord**—Can you distinguish the umbilical vessels?
- **Placenta**—It is attached to the uterine lining.

ANATOMICAL ATLAS OF THE RAT

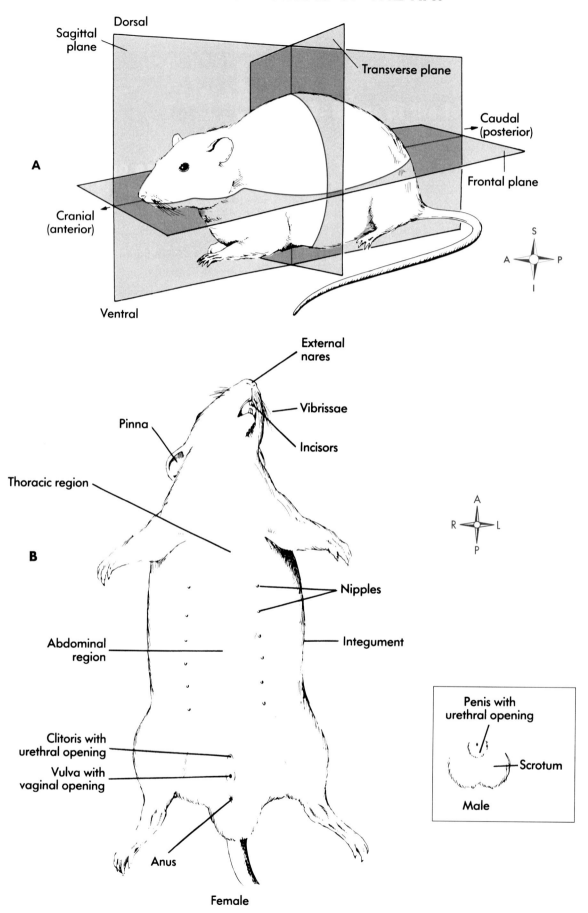

Figure 2-1 External aspect of the rat. **A,** Anatomical planes and directions. **B,** The ventral surface.

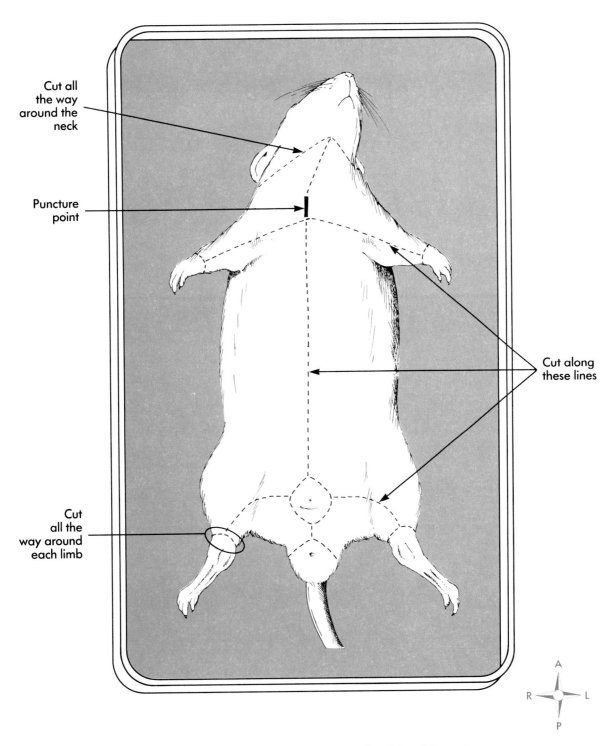

Figure 2-2 Directions for cutting the skin of the rat.

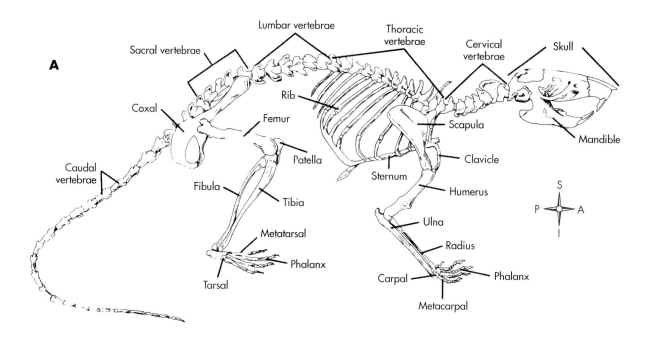
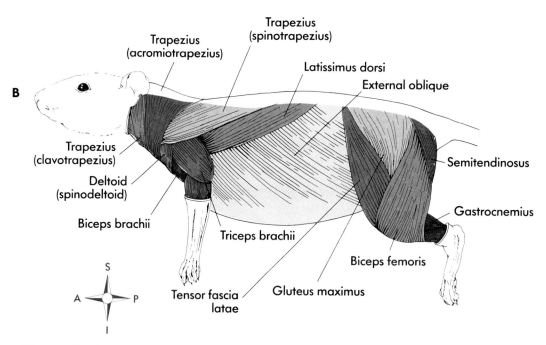

Figure 2-3 Lateral views of the rat. **A,** The skeleton. **B,** Superficial skeletal muscles.

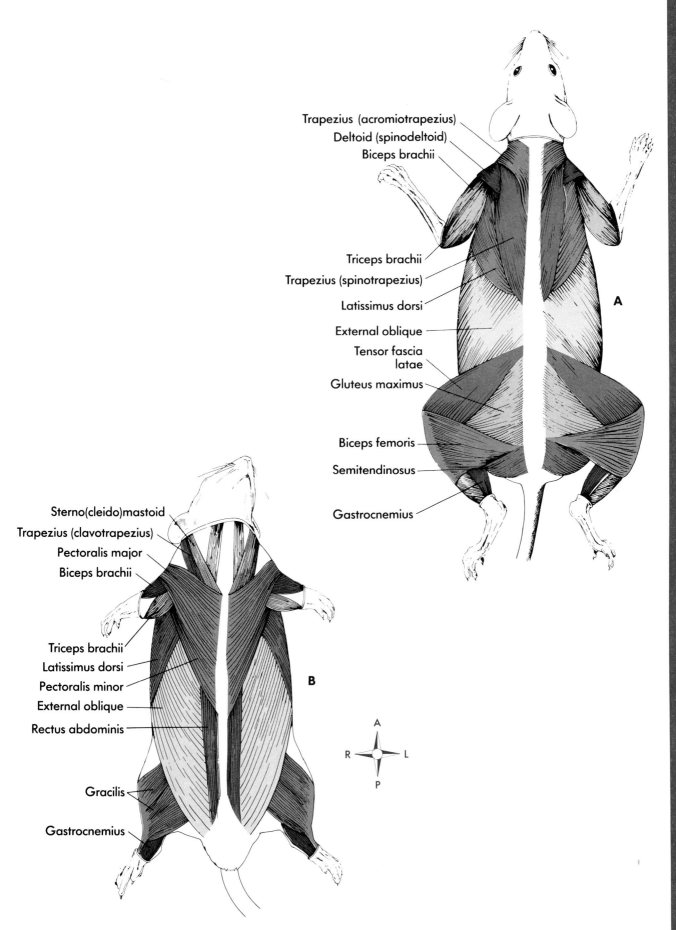

Figure 2-4 Superficial skeletal muscles of the rat. **A,** Dorsal aspect. **B,** Ventral aspect.

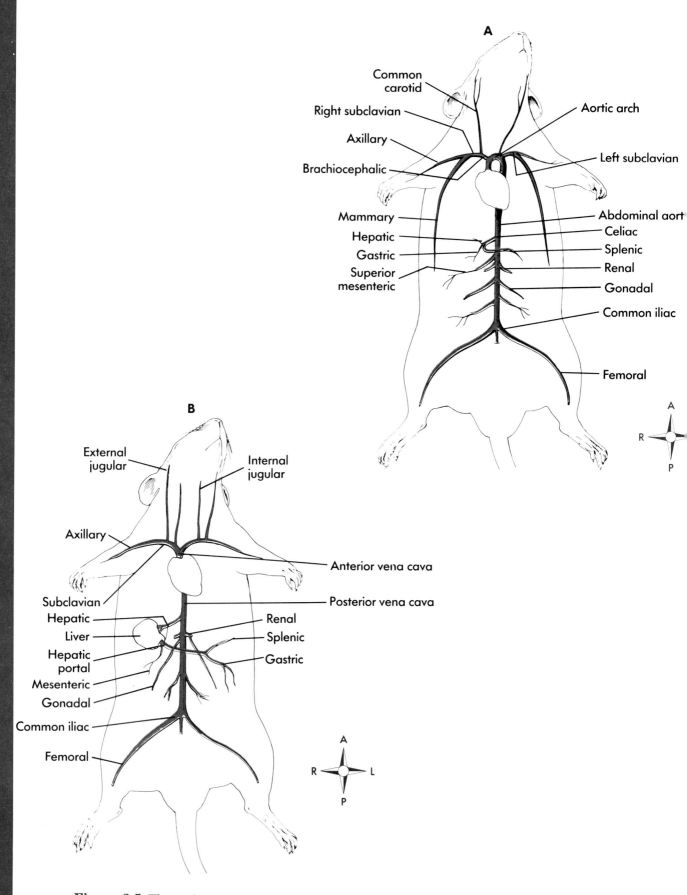

Figure 2-5 The rat's systemic circulation. **A,** Major systemic arteries. **B,** Major systemic veins.

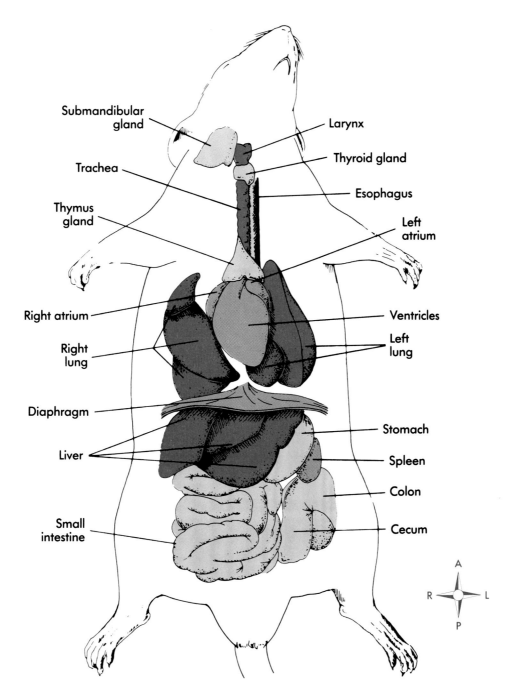

Figure 2-6 The ventral body cavity of the rat.

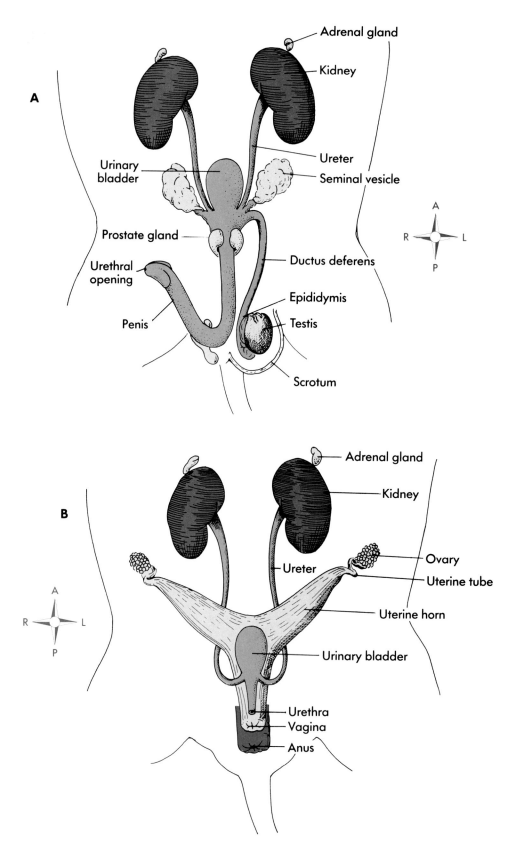

Figure 2-7 The pelvic cavity of the rat. **A,** Male. **B,** Female.

NAME _____ DATE _____ SECTION _____

LAB REPORT 2

Dissection: The Whole Body

Rat dissection checklist

■ **External aspect**
- pinna (auricle)
- external naris (nostril)
- vibrissae (whiskers)
- incisors
- integument
- forelimbs
- hindlimbs
- thoracic region
- abdominal region
- nipples
- anus
- tail

Gender: ❏ female
- vulva
- vagina
- clitoris
- urethral opening

Gender: ❏ male
- scrotum (testes within)
- penis
- prepuce (foreskin)
- urethral opening

■ **Skin, bones, and muscles**
- subcutaneous areolar tissue
- dermis
- epidermis, fur
- axial skeleton
  - cranial bones
  - facial bones
  - cervical vertebrae
  - thoracic vertebrae
  - lumbar vertebrae
  - sacral vertebrae
  - caudal vertebrae
  - ribs
  - sternum
- appendicular skeleton
  - scapula
  - clavicle
  - humerus
  - radius
  - ulna
  - carpal bones
  - metacarpal bones
  - phalanges of the forelimb
  - coxal bone
  - femur
  - patella
  - tibia
  - fibula
  - tarsal bones
  - metatarsal bones
  - phalanges of the hindlimb
- head, neck, and shoulder muscles
  - sternomastoid (human: sternocleidomastoid)
  - clavotrapezius (human: trapezius)
  - acromiotrapezius (human: trapezius)
  - spinotrapezius (human: trapezius)
  - spinodeltoid (human: deltoid)
  - biceps brachii
  - triceps brachii
  - latissimus dorsi
- abdominal and hindlimb muscles
  - external oblique
  - rectus abdominis
  - tensor fasciae latae
  - gluteus maximus
  - biceps femoris
  - semitendinosus
  - gastrocnemius
  - gracilis

■ **Cardiovascular structures**
- heart
  - left and right atria
  - left and right ventricles
- aorta
- major aortic branches
  - common carotid arteries
  - brachiocephalic artery
  - left subclavian artery
  - celiac artery (trunk)
  - superior mesenteric artery
  - renal arteries
  - gonadal arteries

- ❏ common iliac arteries
- ❏ anterior vena cava (human: superior vena cava)
- ❏ subclavian veins
- ❏ posterior vena cava (human: inferior vena cava)
- ❏ hepatic vein
- ❏ renal veins
- ❏ common iliac veins
- ❏ hepatic portal vein

■ **The viscera**

- ❏ respiratory structures
  - ❏ larynx
  - ❏ trachea
  - ❏ primary bronchi
  - ❏ lungs
  - ❏ pleurae
  - ❏ diaphragm
- ❏ digestive structures
  - ❏ submandibular salivary glands
  - ❏ esophagus
  - ❏ stomach
  - ❏ liver
  - ❏ pancreas
  - ❏ small intestine
  - ❏ mesentery
  - ❏ cecum
  - ❏ colon
- ❏ lymphatic organs
  - ❏ spleen
  - ❏ thymus
- ❏ urinary structures
  - ❏ kidney
  - ❏ ureter
  - ❏ urinary bladder
  - ❏ urethra
- ❏ endocrine glands
  - ❏ thyroid gland
  - ❏ thymus gland
  - ❏ pancreas
  - ❏ adrenal glands
  - ❏ testes
  - ❏ ovaries
- ❏ male reproductive organs
  - ❏ testes
  - ❏ epididymis
  - ❏ ductus (vas) deferens
  - ❏ seminal vesicles
  - ❏ prostate gland
  - ❏ penis
- ❏ female reproductive organs
  - ❏ ovaries
  - ❏ uterine tubes (oviducts)
  - ❏ uterus
  - ❏ vagina
- ❏ pregnant uterus
  - ❏ amniotic sac
  - ❏ embryo (or fetus)
  - ❏ umbilical cord
  - ❏ placenta

Notes or sketches:

LAB EXERCISE 3

The Microscope

In the seventeenth century, the amateur Dutch scientist Anton van Leeuwenhoek used one of the first microscopes to discover a whole new world of living organisms (Figure 3-1, *A*). Using a single lens, or **simple microscope,** he observed tiny organisms in pond water and other substances. Robert Hooke, an English scientist, discovered that larger organisms had small microscopic subunits that he called *cells*. Ever since this early era of discovery, biological microscopy has been essential in the study of living organisms.

The microscopes used in this course are **compound microscopes,** made of a set of lenses. They are more powerful, and more complex, than those used by van Leeuwenhoek and Hooke. This exercise will introduce you to the use and care of the standard compound microscope.

Before you begin

❑ Read the appropriate chapter in your textbook.

❑ Set your learning goals. When you finish this exercise, you should be able to
- identify each major part of a compound light microscope and describe its function
- determine total magnification at different settings
- use a microscope to observe prepared specimens
- prepare a wet-mount slide
- stain microscopic specimens
- record microscopic observations accurately

❑ Prepare your materials:
- compound light microscope
- prepared microslides: *newsprint "e"*
  3 colored threads
- clean slides
- coverslips
- paper wipes
- methylene blue stain
- flat toothpicks

❑ Read the directions and safety tips for this exercise **carefully** before starting any procedure.

A. Parts of the microscope

❑ 1   Obtain a compound light microscope and identify each of the structures described as follows.

> **HINT** → Because not all microscope models are alike, some of the structures described as follows may be different on your scope. Refer to Figure 3-1, *B* for help in this exercise.

❑ 2   The set of lenses closest to your eye is the **ocular,** or eyepiece. The ocular magnifies an image by the factor indicated on the ocular's barrel, usually 10×. If the factor is 10×, the image is magnified ten times. If the factor is 5×, the image is magnified five times.

❑ 3   The **body tube** holds the ocular in place. Although called a *tube*, it may be more like a box in some models.

❑ 4   At the bottom of the body tube is the **revolving nosepiece.** A turretlike circular mechanism rotates so that different lenses can be selected. *Always rotate the nosepiece by holding the outside of the revolving disk—never push on the lens barrels.*

❑ 5   Each of the lens sets attached to the revolving nosepiece is an **objective.** As with the ocular, each objective is marked with its magnification factor. Microscopes may have any or all of the following objectives:

> 4× (scanning objective) is used for initial location of the specimen. Some scopes have 3.2× or 3.5× scanning objectives. List the factor on your scope's scanning objective here ____.
>
> 10× (low power objective) may also be used for initial location of the specimen. It is also used for observing specimens that do not need greater magnification.

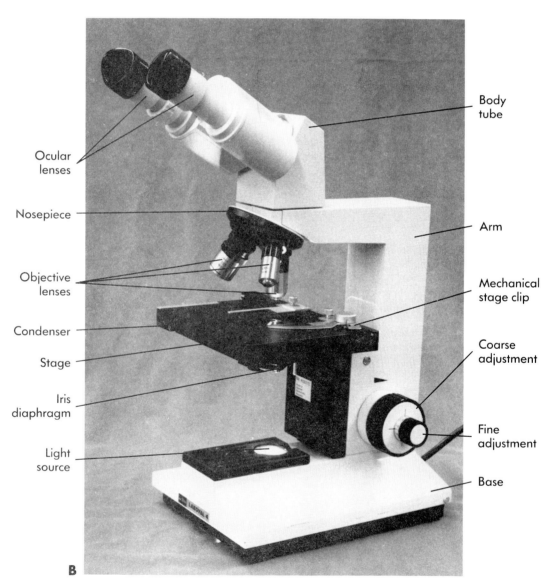

Figure 3-1 A, van Leeuwenhoek's simple microscope. **B,** Important parts of a typical, modern compound light microscope.

40× (high-dry objective) is used for specimens requiring greater magnification. This objective is called *dry* because it does not require the use of oil, as other high-power objectives do. Some scopes have 43× or 45× high-dry objectives. List the factor on your scope's high-dry objective here ____.

100× (high-oil objective) is used for magnification of extremely small specimens, such as bacterial cells. It must be immersed in oil, so it is called the high-oil objective. This objective will not be used in this course.

In this manual, *low power* will refer to use of the 10× objective and *high power* will refer to use of the 40× objective.

❏ 6 **Total magnification** is determined by multiplying the power of the ocular by the power of the objective in use. Thus, when using a 10× ocular and the low power objective, total magnification is 100× (10 × 10 = 100). List all ocular and objective combinations, then determine all the total magnifications possible on your scope and record them in the Lab Report.

❏ 7 The specimen is usually mounted on a glass or plastic **microscope slide** that rests on the **stage,** a platform just below the objective. The stage has a hole so that light can pass through the specimen from below. If the stage has an adjustable bracket that moves the slide around mechanically, the stage is called a **mechanical stage.** If not, the slide is held by **stage clips** and must be moved by hand.

❏ 8 Below the stage is a high-intensity **lamp.** Light rays from the lamp travel through a hole in the stage, through the specimen mounted on a slide, then through the objective and ocular, to the eye.

❏ 9 A **condenser,** a lens that concentrates light, may be found between the lamp and the stage.

❏ 10 Sometimes the light from the lamp is too strong to see the specimen clearly. The light level may be reduced by adjusting the lamp intensity (if possible). Light intensity may also be adjusted by adjusting the **diaphragm** just below the stage (Figure 3-2). A *disk diaphragm* is a rotating disk with holes of different diameters. An *iris diaphragm* is made of overlapping slivers of metal in a pattern resembling the iris flower, as does the iris of the human eye. Like the iris of the eye, the iris diaphragm can dilate or constrict its opening. You can change the amount of light passing to the specimen by rotating the edge

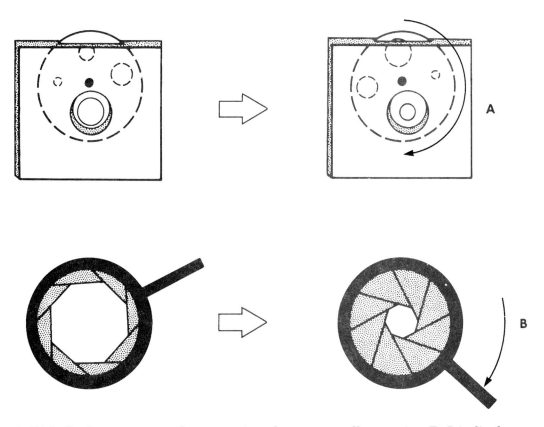

Figure 3-2 A, Disk diaphragm set at a large opening, then at a smaller opening. **B,** Iris diaphragm open, then partially closed.

of the disk diaphragm or by rotating the lever projecting from the iris diaphragm.

❏ 11 The entire upper assembly of the microscope is held in an upright position by a bar called the **arm.** The scope is supported by a square or horseshoe-shaped **base.** The arm may be connected to the base by a **pivot,** which allows the upper assembly to move into a more comfortable viewing position.

❏ 12 The **coarse-focus knobs** and **fine-focus knobs** are on the arm. These knobs adjust the distance between the stage and objective, thus focusing an image of the specimen. The fine-focus knob changes the distance very little, whereas the coarse-focus knob changes the distance greatly.

B. Using the microscope

Skill in using the microscope is necessary for many of the exercises in this lab manual. Fortunately, learning to use the microscope is both easy and fun. Before beginning, acquaint yourself with these basic rules.

❏ Always carry the scope with two hands: one under the base and the other grasping the arm. Carry it in an upright position.

❏ Unwind the lamp cord carefully. Avoid damaging the parts around which it is coiled. Plug the cord into an outlet in a safe manner.

> **SAFETY FIRST!** Be careful when plugging the power cord into the outlet. Always plug it into a receptacle at *your* station, taking care not to string it over a chair, across an aisle, or in any other dangerous position.

❏ Make sure the stage and objective are at their farthest distance apart and that the *lowest power objective is in position*. Start each new observation at low power.

❏ 1 Obtain a microscope slide with a newsprint letter *e* mounted on it. This is your practice specimen.

❏ 2 Place your slide carefully on the stage and secure it with stage clips or the brackets of the mechanical stage. Move the slide so that the *e* is centered in the stage's hole.

❏ 3 While looking from the side, use the coarse-focus knob to move the objective (still on low power) as close as possible to the slide.

❏ 4 Look through the ocular and use the coarse-focus knob to slowly move the objective and slide apart. When the image becomes clear, switch to the fine-focus knob to make the image even sharper. To avoid damaging the scope and slide, *never move the objective and stage toward each other while looking through the ocular.*

> **HINT** → To avoid eyestrain and a possible headache, always keep both eyes open when viewing a specimen. With binocular scopes this is easy, once you have adjusted the distance between oculars. However, it may take some practice if you are using a monoocular scope. If you have trouble keeping both eyes open, try covering the unused eye with your hand.

❏ 5 Adjust the light intensity by using the lamp controller or diaphragm until the detail of the image is at its clearest.

❏ 6 Sketch the entire field of view in the space provided in the Lab Report.

> **HINT** → Here are some rules for recording microscopic observations properly:
>
> ■ Label each sketch with the name of the specimen and what type of section it is.
> ■ Label each sketch with the *total magnification*.
> ■ Label as many parts of the sketch as you can. Labels should be orderly, never crossing lines with each other.
> ■ Sketches should be done with care, not with haste.

❏ 7 Center the part of the specimen that you wish to see more clearly and switch to a higher-power objective. Most scopes are **parfocal,** which means that the image remains focused when you change objectives. However, some minor adjustment with the fine focus is usually necessary. *Never adjust the coarse focus when using high power.*

❏ 8 Sketch the entire field that you see using the high-power (40× to 45×) objective.

❏ 9 When you remove the slide, make sure that the low-power objective is in viewing position and that the objective and stage are as far apart as possible.

❑ 10 Practice focusing at different depths by using a prepared slide of three crossed strands of colored thread. The strands (one red, one yellow, and one blue) cross at the same point. Determine which is on top, which is in the middle, and which is on the bottom. Report your results in the Lab Report.

❑ 11 When you have finished for the day, return the scope to its original configuration. The low-power objective should be in position. Make sure the slide has been removed.

> **HINT →** If you can't see an image in your scope . . .
>
> - Make sure the lamp is on.
> - Make sure the diaphragm is open. The amount of light should be set low on low power and high on high power.
> - Check for obstructions in the light path.
> - Make sure the objective is seated properly.
> - Check to see if the specimen is centered.
> - Clean the lenses with lens paper and lens solution, not facial tissues or lab wipes.
>
> If the image seems to fade in and out . . .
>
> - Watch to see if the body tube or stage is shifting or dropping.
> - Make sure the scope doesn't require a rubber eyecup for proper viewing.
> - Make sure the specimen is not in a medium that obstructs viewing.
> - Check the lamp or cord for short circuits.
>
> If the image doesn't look like the figure in the book . . .
>
> - Get a different book (it may have a better figure).
> - Use your imagination.
> - Join the club (nobody else's specimen is identical to that in the book either).

C. Preparing microscopic specimens

We will use prepared slides of human and other tissues often in this course. However, occasionally we may be making our own specimens.

A **wet-mount** slide is a slide on which a wet specimen is placed, then covered with a **coverslip.** **Stains** are used to make a specimen, or some of its parts, more visible. Some require special techniques, but most stains can simply be added to the specimen and viewed. The usual method for preparing a stained wet-mount slide is described in Figure 3-3 and in the following steps.

❑ 1 Obtain some skin cells by scraping the inner surface of your cheek with a clean, flat toothpick.

❑ 2 Wipe the scrapings on a clean microscope slide and put a small drop of *methylene blue* stain directly on the smear.

> **SAFETY FIRST!** Use only fresh toothpicks and dispose of used toothpicks *immediately* by placing them in the waste container indicated by your instructor. Take precautions to avoid the spread of disease through your saliva.

❑ 3 Place one edge of a coverslip on the slide next to the specimen, then let it drop slowly on the specimen. This method avoids forming air bubbles.

❑ 4 Absorb any excess fluid around the edges of the coverslip with the edge of a paper wipe.

❑ 5 Locate some cells with the low-power objective, shift to high power, and sketch your observations in the Lab Report.

> **LANDMARK CHARACTERISTICS**
>
> You are looking for scattered epithelial cells. Each cell will appear as a lightly stained, flat polygon or circle with a dark center. The dark center is the nucleus of the cell. Some cells may be separate, whereas others are clumped together.
>
> You may see a variety of *other things* in your specimen: bacteria, vegetable or meat fibers, and other matter. Don't worry; this is normal. Distinct dark circles with hollow centers that may be in motion are air bubbles.

> **HINT →** If your instructor wants you to learn how to use an oil-immersion objective, follow these directions. After focusing on high-dry, rotate the nosepiece *half-way* to the high-oil objective. Put a drop of immersion oil on the slide, then rotate the high-oil objective into the oil drop. You may need to increase the lighting. When finished, clean the objective and other surfaces as your instructor directs.

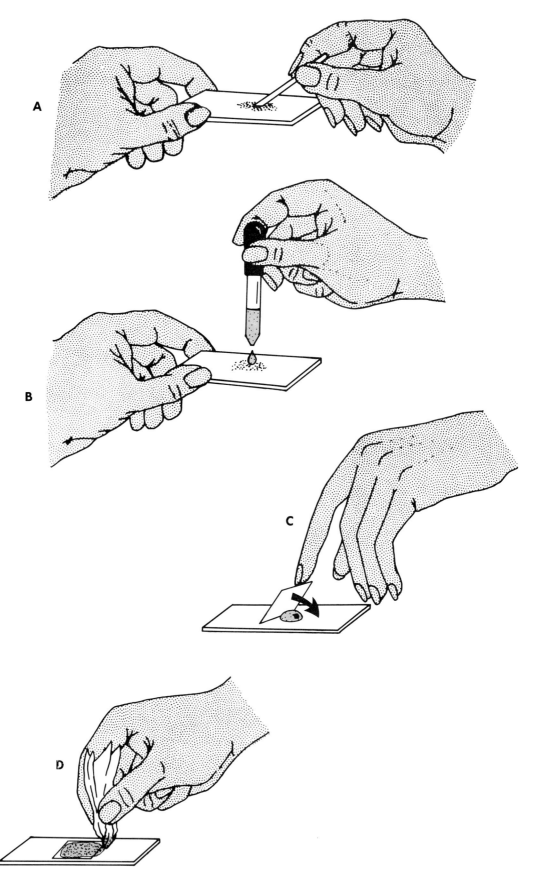

Figure 3-3 Staining a wet mount specimen. **A,** Place the specimen on the slide. **B,** Add a small drop of stain. **C,** Drop the cover slip slowly to avoid air bubbles. **D,** Absorb excess fluid with a paper wipe.

NAME _____ DATE _____ SECTION _____

LAB REPORT 3

The Microscope

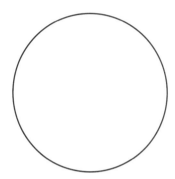

Specimen: *newsprint e*   Total Magnification: _____

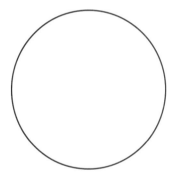

Specimen: *newsprint e*   Total Magnification: _____

Does the *e* in the newsprint appear to be oriented the same way it is on the stage? Explain your observations.

Is the *diameter of field* the same when observing the specimen under high power and low power?

Which of the three colored threads in the prepared slide is on the bottom?

Which thread is on the top?

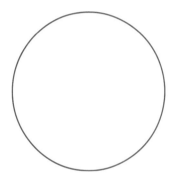

Specimen: *human cheek cells*

Total Magnification: _____

What is the darkly stained region of each cell?

What advantage is gained by staining this specimen?

List all the possible ocular-objective combinations on your microscope and calculate the total magnification for each.

Power of ocular	Power of objective	Total magnification
_____	_____	_____
_____	_____	_____
_____	_____	_____
_____	_____	_____

Fill-in

1. _____
2. _____
3. _____
4. _____
5. _____
6. _____
7. _____
8. _____

Multiple choice

___ 1
___ 2
___ 3
___ 4
___ 5
___ 6
___ 7

Fill-in

1. A __?__ is a glass or plastic rectangle on which specimens are mounted.
2. A __?__ is at the base of the body tube, allowing selection of different objectives.
3. Below the stage is a high-intensity __?__ or other light source.
4. Different lens sets, or __?__, are attached to the revolving nosepiece.
5. If the stage has a bracket that moves the slide, it is called a(n) __?__ stage.
6. The __?__ allows the user to limit the amount of light from a steady light source.
7. The set of lenses closest to the viewer's eye is the __?__.
8. The __?__ is a lens under the stage that concentrates light.

Multiple choice (only one choice is correct)

1. A microscope __?__ the image of a specimen.
    a. magnifies
    b. reverses
    c. reduces
    d. erases
    e. a and b are correct

2. The use of __?__ may make the specimen (or some of its parts) more visible.
    a. stain
    b. the diaphragm
    c. fine focus
    d. all of the above
    e. none of the above

3. If you select a 40× objective to use with a 5× ocular, the total magnification would be:
    a. 200×
    b. 45×
    c. 400×
    d. 405×
    e. none of the above

4. When you move a specimen to the right, its image appears to
    a. move to the right
    b. move to the left
    c. remain stationary

5. Which is the correct path of light in a compound microscope?
    a. lamp, objective, diaphragm, specimen, ocular
    b. lamp, diaphragm, specimen, objective, ocular
    c. diaphragm, lamp, specimen, ocular, objective
    d. lamp, diaphragm, specimen, ocular, objective

6. If your scope is not working properly, consult
    a. the lab instructor
    b. your lab partner
    c. the owner's manual
    d. the manufacturer

7. When carrying your microscope, use __?__ hands.
    a. one
    b. two

LAB EXERCISE 4

Cell Anatomy

The cell is often considered the basic unit of living organisms. Actually, cells are only one of several *levels of organization* in the human organism:

ATOMS
MOLECULES
ORGANELLES
CELLS
TISSUES
ORGANS
ORGAN SYSTEMS
ORGANISM

Thus, cells are parts of larger units (tissues and organs) and are composed of smaller units (organelles).

For now, we will concentrate on the structure and function of the cell and its organelles. We will study some of the other levels of organization in later units.

Before you begin

❑ Read the appropriate chapter in your textbook.

❑ Set your learning goals. When you finish this exercise, you should be able to
- place the cell and its organelles within the scheme of organizational levels
- identify the major organelles on a model, chart, specimen, or micrograph
- describe primary function(s) of typical organelles
- discuss the anatomical relationship among various cell parts

❑ Prepare your materials:
- cell model or chart
- compound light microscope
- clean slides
- coverslips
- paper wipes
- methylene blue stain
- colored pencils or pens
- prepared slides (instructor's option)

❑ Read the directions and safety tips for this exercise **carefully** before starting any procedure.

A. Parts of the cell

Cells are microscopic units composed of a bubble of fatty material filled with a water-based mixture of molecules and tiny particles. Parts of the cell are called **organelles** (meaning *small organs*). Begin your study of the cell by examining a *generalized* cell, such as that in the coloring exercise in Figure 4-1. A generalized cell is one with many cell features that are not all found in a single natural cell. Locate each of the listed organelles on a generalized cell model or chart.

❑ 1 The outer boundary of the cell is the **plasma membrane.** It is composed of a double layer (or *bilayer*) of **phospholipid** molecules imbedded with other molecules. Each phospholipid molecule has a polar, hydrophilic *head* made up of phosphate and glycerol and a nonpolar, hydrophobic *tail* made up of two fatty acid chains. Imbedded in the membrane are **integral proteins,** which may have additional protein molecules called **peripheral proteins** attached to them on one side of the membrane or the other. Figure 4-2 shows some of the types of molecules often associated with the plasma membrane. The plasma membrane has many functions, most involving transport and communication between the inside and outside of the cell.

> **HINT** → Each membranous organelle described as follows is made of a lipid membrane similar to the cell membrane. Because they are so thin, cell membranes may seem invisible when observed with a light microscope.

❑ 2 The double-walled **nucleus** is a large bubble containing the cell's genetic code. The code is in the form of **deoxyribonucleic acid (DNA)** strands called **chromatin.** Portions of chromatin accept stains readily, giving the nucleus a very dark appearance.

The Cell

COLORING EXERCISE Using colored pens or pencils, shade in the figure and accompanying labels in contrasting colors of your choice as indicated by the red numerals.

PLASMA MEMBRANE 1
CENTRIOLE 2
CENTROSOME 3
CHROMATIN 4
CYTOSOL 5
GOLGI APPARATUS 6
LYSOSOME 7
MICROVILLI 8
MITOCHONDRION 9
NUCLEAR MEMBRANE 10
NUCLEOLUS 11
NUCLEAR PORE 12
RIBOSOME 13
ROUGH ER 14
SMOOTH ER 15
VESICLE 16

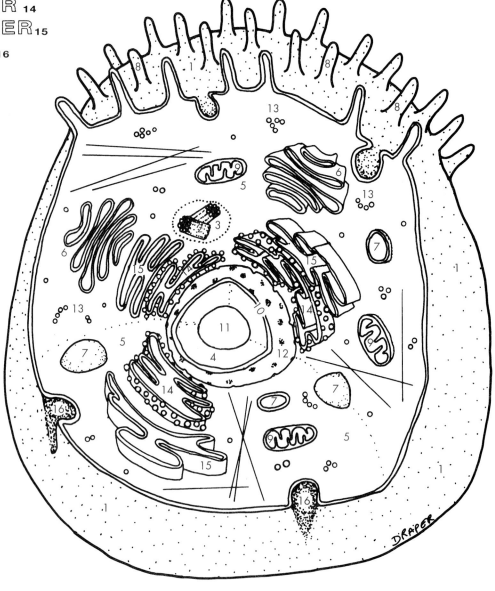

Figure 4-1

The Plasma Membrane

CARBOHYDRATE₁
CARRIER PROTEIN₂
CELL MARKER₃
CHANNEL PROTEIN₄
CHOLESTEROL₅
ENZYME₆
INTEGRAL PROTEIN₇
PERIPHERAL PROTEIN₈
PHOSPHOLIPID, NONPOLAR TAIL₉
PHOSPHOLIPID, POLAR HEAD₁₀
RECEPTOR PROTEIN₁₁

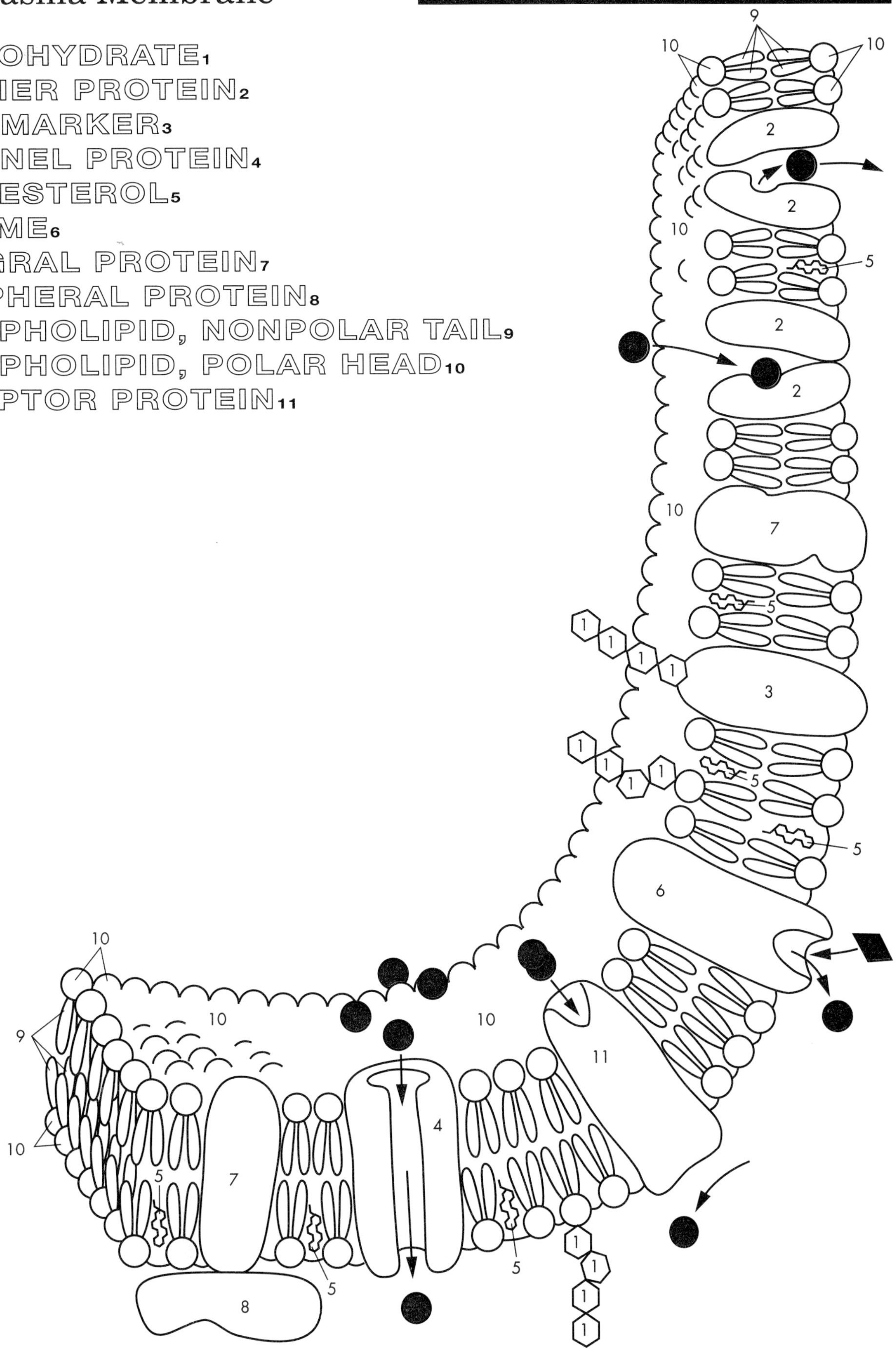

COLORING EXERCISE Using colored pens or pencils, shade in the figure and accompanying labels in contrasting colors of your choice as indicated by the red numerals.

Figure 4-2

❑ 3 A **nucleolus,** literally *tiny nucleus,* is a small area within the nucleus for the synthesis of ribosomal **ribonucleic acid (RNA).** There may be more than one nucleolus (plural *nucleoli*) in a nucleus.

❑ 4 The **endoplasmic reticulum (ER)** is a network of membranous tubes and canals winding through the interior of the cell. A *rough ER* is specked with tiny granules (ribosomes); a *smooth ER* is not. The ER transports proteins synthesized by ribosomes and other molecules synthesized by its membrane. The material within the plasma membrane is the **cytoplasm** (literally, *cell stuff*) and includes both the organelles and the liquid, or **cytosol,** surrounding the organelles.

❑ 5 **Ribosomes** are tiny bodies that serve as a site for protein synthesis. Some ribosomes are found on the outer surface of the ER, and some are found scattered elsewhere within the cell.

❑ 6 The **Golgi apparatus** or *Golgi body* appears as a stack of flattened sacs. The apparatus receives material from the ER, processes it, then packages it in tiny *vesicles* (bubbles) for possible export from the cell.

❑ 7 **Mitochondria** (singular *mitochondrion*) are tiny bodies similar to bacteria that serve as sites for ATP synthesis (energy conversion). Mitochondria have an outer membrane, forming a round or oblong capsule, and a folded inner membrane. The folds of the inner membrane are called *cristae*.

❑ 8 **Lysosomes** are vesicles containing digestive enzymes that digest foreign particles and worn cell parts.

❑ 9 **Microtubules** are very tiny, hollow beams that form part of the supporting cell skeleton, or *cytoskeleton*. They also form parts of other cell organelles, such as *flagella, cilia, centrioles,* and *spindle fibers*. Other components of the cytoskeleton include *microfilaments* and *intermediate filaments*.

❑ 10 The **centrosome,** or *microtubule organizing center,* is a dense area of cell fluid near the nucleus. The centrosome contains a pair of **centrioles,** cylinders formed by parallel microtubules. A network of microtubules called *spindle fibers* extend from the centrosome during cell division. Spindle fibers distribute DNA equally to the resulting daughter cells.

❑ 11 The cell may have any number of other assorted organelles. **Microvilli** are tiny, fingerlike projections of the cell that increase the membrane's surface area for more efficient absorption. **Cilia** are numerous short, hairlike organelles that propel material along a cell's surface. **Flagella** are single long, hairlike organelles found in sperm cells to propel them through the female reproductive tract toward the egg. **Vesicles** are membranous bubbles that may be formed by the Golgi apparatus, or by the pinching inward of the cell membrane to engulf external substances.

B. Microscopic cell specimen

Prepare a stained wet-mount specimen of human cheek cells (see Lab Exercise 3 for instructions) or obtain another specimen provided by the instructor; try to identify as many cell parts as possible. Use high-power magnification (after first finding the specimen with low-power magnification). Sketch and label your observations in the Lab Report.

> **HINT →** Many organelles are very small and can only be seen when properly prepared and examined with a more powerful microscope.

C. Interpreting micrographs

An **electron microscope** is an instrument that uses a beam of electrons, rather than a beam of light, to form the image of a tiny specimen. *Transmission* electron microscopes send an electron beam through the specimen, similar to the manner in which a light microscope sends a light beam through a specimen. However, the magnifying power and resolution are much greater in the electron microscope. **Resolution** is the ability to distinguish detail. *Scanning* electron microscopes, on the other hand, reflect an electron beam off the specimen. The shadows produced by a scanning electron beam lend a three-dimensional effect.

A **TEM** is a transmission electron micrograph (photograph taken with a transmission electron microscope). Figure 4-3 shows TEM representations of major cell parts. Use them to help you identify all of the cell parts in the labeling exercises in Figure 4-4, which shows artists' renderings of TEM images, and in Figure 4-5, which shows actual TEMs of human cells.

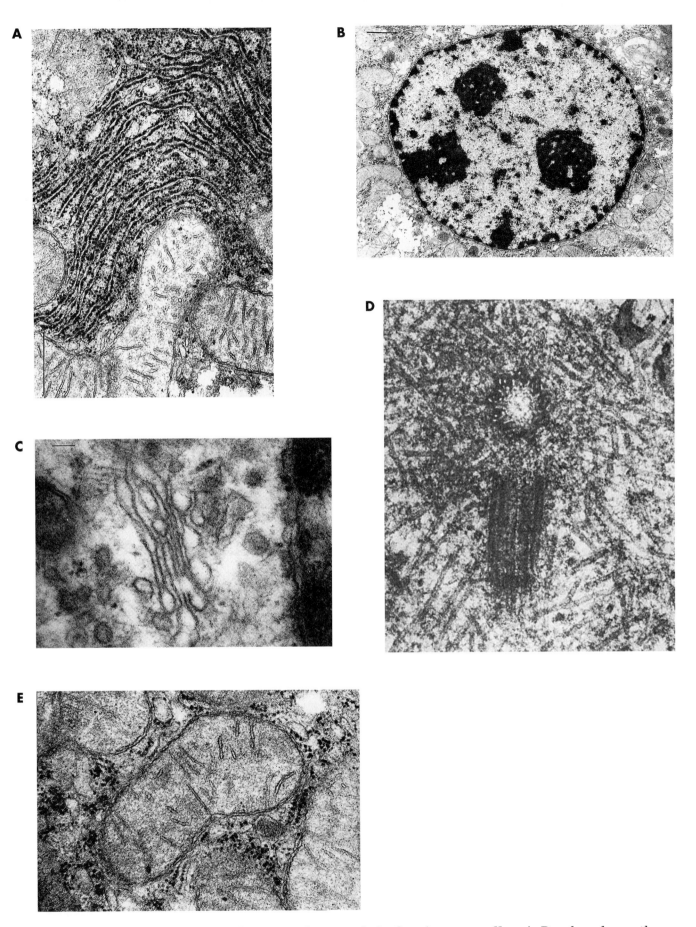

Figure 4-3 TEMs (transmission electron micrographs) of major organelles. A, Rough and smooth endoplasmic reticulum (portions of several mitochondria are also visible). **B,** Nucleus with several nucleoli. **C,** Golgi apparatus. **D,** Centrosome with both centrioles visible. **E,** Mitochondrion with cristae visible.

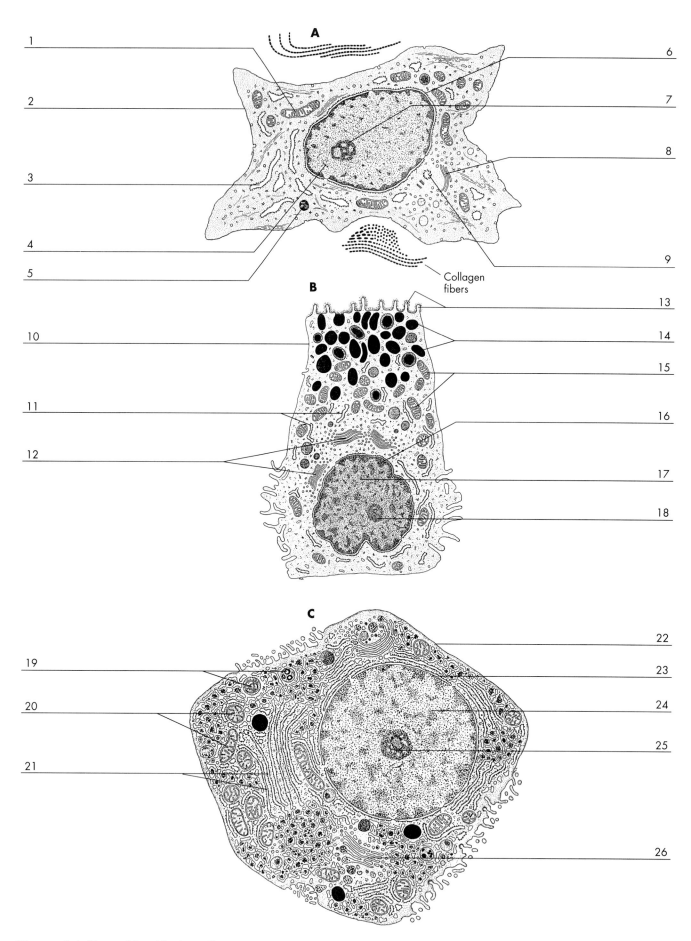

Figure 4-4 Try to identify the cell organelles labeled in these artists' renderings of TEMs (transmission electron micrographs). **A,** A fibroblast (cell that produces collagen fibers). **B,** A mucus-producing cell. **C,** A liver cell.

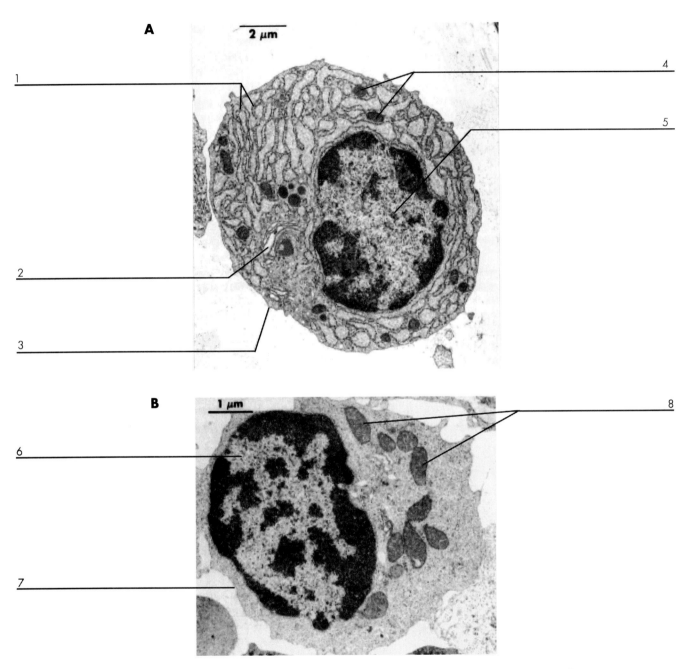

Figure 4-5 Try to identify the cell organelles labeled in these actual TEMs (transmission electron micrographs) of human cells.

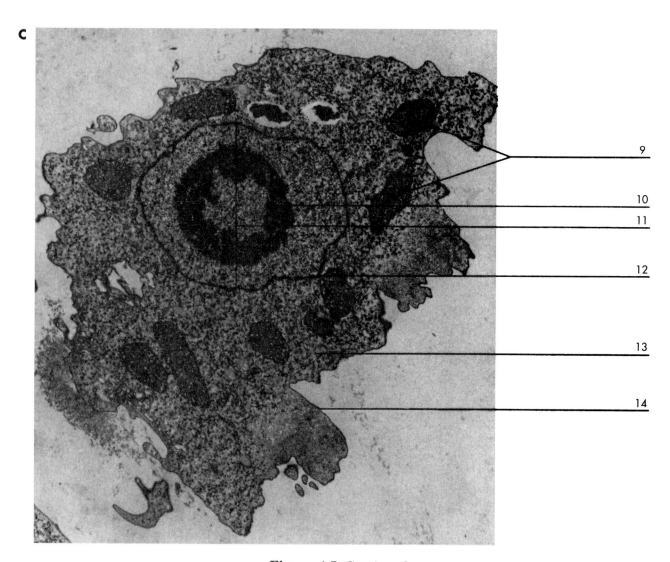

Figure 4-5, Continued.

NAME _____ DATE _____ SECTION _____

LAB REPORT 4

Cell Anatomy

Checklist for cell model
- ❏ plasma membrane
- ❏ cytosol
- ❏ nucleus
  - ❏ chromatin
- ❏ nucleolus
- ❏ endoplasmic reticulum
  - ❏ rough ER
  - ❏ smooth ER
- ❏ ribosomes
  - ❏ free
  - ❏ associated with ER
- ❏ Golgi apparatus
  - ❏ secretory vesicles
- ❏ mitochondria
- ❏ lysosomes
- ❏ microtubules
- ❏ centrosome
  - ❏ centrioles
- ❏ microvilli
- ❏ cilia
- ❏ flagella
- ❏ _____
- ❏ _____
- ❏ _____

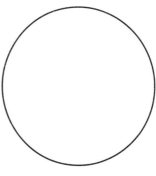

Specimen: _____
Total Magnification: ____

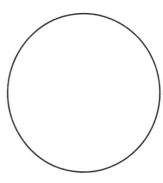

Specimen: _____
Total Magnification: ____

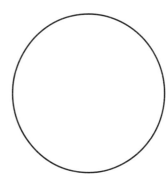

Specimen: _____
Total Magnification: ____

Figure 4-4

_____ 1
_____ 2
_____ 3
_____ 4
_____ 5
_____ 6
_____ 7
_____ 8
_____ 9
_____ 10
_____ 11
_____ 12
_____ 13
_____ 14
_____ 15
_____ 16
_____ 17
_____ 18
_____ 19
_____ 20
_____ 21
_____ 22
_____ 23
_____ 24
_____ 25
_____ 26

Matching (may be used more than once or not at all)

a. plasma membrane
b. centriole
c. endoplasmic reticulum
d. Golgi apparatus
e. lysosome
f. mitochondrion
g. nucleolus
h. nucleus
i. ribosome

1. A double-walled structure containing the cell's genetic code
2. A network of membranous tubes and canals that transports proteins
3. A stack of flattened sacs that process and package proteins
4. Site of manufacture of ribosomal RNA
5. A cylinder formed by parallel microtubules
6. An organelle that serves as the site of protein synthesis
7. A bubble containing digestive enzymes
8. May be *rough* (with ribosomes) or *smooth* (ribosome-free)
9. Allows communication between the internal and external cell environment
10. Forms secretory vesicles

Fill-in table (write out the names of the organelles listed, *a* to *i*, in the section above in the appropriate column of the table)

Membranous	Nonmembranous

Figure 4-5

_____ 1
_____ 2
_____ 3
_____ 4
_____ 5
_____ 6
_____ 7
_____ 8
_____ 9
_____ 10
_____ 11
_____ 12
_____ 13
_____ 14

Matching

_____ 1   _____ 6
_____ 2   _____ 7
_____ 3   _____ 8
_____ 4   _____ 9
_____ 5   _____ 10

LAB EXERCISE 5

Transport through Cell Membranes

Cells make use of both **passive** and **active** processes to transport substances across their membranes. Passive processes are those that require no metabolic energy from the cell but rely solely on the physical properties of the substances themselves. Active transport processes require energy expenditure by the cell to move substances whose physical properties prevent their independent motion. In this exercise, we will observe large-scale examples of passive transport processes.

Before you begin

❏ Read the appropriate chapter in your textbook.

❏ Set your learning goals. When you finish this exercise, you should be able to
- discuss the nature of diffusion and osmosis
- use the terms *hypertonic, hypotonic,* and *isotonic* to compare solutions
- compare and contrast diffusion, osmosis, and filtration

❏ Prepare your materials:
- India ink (dropper bottle)
- microscope
- microscope slides and coverslips
- potassium permanganate crystals
- Petri dishes
- hot (100°C) water and cold (0°C) water
- forceps
- dialysis tubing (8 to 10 cm, presoaked)
- string (or tubing clamps)
- sucrose solutions (15% and 40%)
- syringe (or pipette) to fill dialysis bags
- laboratory balance
- jar or beaker (75 to 200 ml)
- distilled water
- filter paper (circular, large pore)
- glass funnel
- flask (to hold funnel and filtrate)
- copper sulfate (saturated solution)
- boiled starch solution (10%)
- charcoal (powdered)
- Lugol's reagent

❏ Read the directions and safety tips for this exercise.

A. Physical basis of passive transport

Naturally occurring **Brownian motion** drives passive transport processes. Discovered by Scottish scientist Robert Brown, Brownian motion is the constant movement of all particles of matter.

❏ 1 Prepare a wet-mount slide with a drop of India ink.

❏ 2 Observe the slide under high-power magnification. You should be able to see the dye particles moving short, irregular distances. This is Brownian motion.

Because small particles of matter are always bouncing around, Brownian motion causes matter to spread **(diffuse)** to areas where there is more room to bounce around. That is, particles tend to move toward an area of lower particle concentration. You can observe this movement by following steps 3 and 4.

❏ 3 Using forceps, place a small crystal of potassium permanganate in the center of a Petri dish of calm, hot water. See Figure 5-1. Place a second crystal of the same size in a dish of calm, cold water. Be careful not to disturb the dishes by touching them or causing them to shake.

❏ 4 Observe the two dishes over a period of 30 minutes, carefully noting any changes in the crystals. Answer the questions in the Lab Report.

B. Diffusion and osmosis

Diffusion can be defined as the net movement of particles from an area of high concentration to an area of low concentration. Some particles move down such a *concentration gradient* when they are allowed to cross a cell membrane. If the particles in question happen to be water molecules and there is at least one impermeant solute present, we call the process **osmosis.** Osmosis, then, is a particular type of diffusion. In this activity, you will observe the osmosis of water across a nonliving membrane called *dialysis tubing.* Dialysis tubing, like the cell membrane, is *semipermeable* (allowing only certain substances to pass through its

Figure 5-1 Place the potassium permanganate crystal slowly, taking care not to disturb the water.

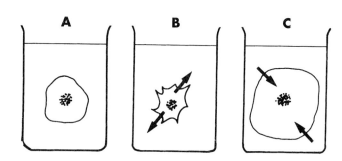

Figure 5-2 A cell in **A,** isotonic solution, **B,** hypertonic solution, and **C,** hypotonic solution. The arrows show direction of net osmosis.

pores). The bag that you construct from a piece of dialysis tubing serves as a model of the cell.

One of the three model cells you construct will be immersed in an **isotonic** solution. *Isotonic* is a comparative term that refers to a solution that has the same *potential osmotic pressure* as another solution. In other words, isotonic solutions have the same relative water concentrations. With the cell model, the solution inside is isotonic to the solution outside (see Figure 5-2).

Another cell model will be immersed in **hypotonic** solution. A hypotonic solution is one with a lower potential osmotic pressure (that is, a higher water concentration). Thus, the outside solution is hypotonic to the solution inside the cell model. Another cell model will be immersed in **hypertonic** solution, a solution with a higher potential osmotic pressure (lower water concentration). When the *tonicity* of two solutions on either side of a membrane is known, the net direction of osmosis can be predicted.

As you set up each of the three situations described as follows, predict the direction of net osmosis.

> **HINT** → These hints may help you:
>
> - Either presoak the dialysis tubing before using or hold it under water and rub the end of the tube between your fingers until it opens.
> - Set up the jars described in step 4 *before* beginning this activity.
> - Because this activity calls for an extended waiting period, you may want to do the other activities of this exercise while you wait.

❑ 1 Obtain three 8 to 10 cm pieces of dialysis tubing. Tie one end shut with string, forming a water-tight bag. Leave about 10 cm of string dangling freely.

❑ 2 With a syringe, carefully fill each of three bags with 15% sucrose solution. Try to avoid spilling any solution on the outside of the bag. Tie off the open end of each tube with another piece of string.

❑ 3 Quickly measure the mass of each tube on a laboratory balance.

❑ 4 Place bag 1 in a jar or beaker half filled with distilled water. Place bag 2 in a jar half filled with 15% sucrose solution. Place bag 3 in a jar half filled with 40% sucrose solution. Leave a piece of string dangling out of the jar, as you would with the string on a tea bag (see Figure 5-3).

❑ 5 After 45 to 60 minutes, remove the tubes from the jars, drying and weighing each as you do so. Record and interpret your result in Lab Report 5.

C. Filtration

The movement of particles across a membrane from an area of high pressure to an area of low pressure (down a *hydrostatic pressure gradient*) is called **filtration.** In this experiment, gravity creates the pressure needed to push some particles across a paper membrane. In the human body, forces such as blood pressure push particles through living membranes.

❑ 1 Place a cone made of filter paper inside a glass funnel.

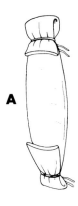
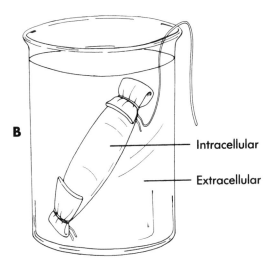

Figure 5-3 A, Folding the end of the dialysis tube before tying it prevents leaking. **B,** Your model cell contains a sucrose solution that represents *intracellular* fluid. The bath solution represents *extracellular* fluid.

> **HINT** → To make the cone, fold a circle of paper in half, then in half again. You should now have a triangle four layers thick. Pull one of the layers away from the rest. You now have a paper cone. Place it in the funnel and wet it with distilled water so that it sticks to the side of the funnel.

What substances were not allowed through the filter? Record and explain your results in Lab Report 5.

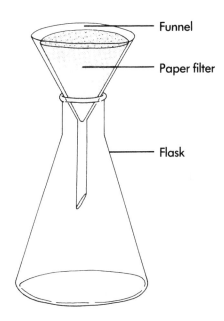

Figure 5-4 Set up your filtration apparatus as shown.

☐ **2** Position the funnel over an empty flask, as shown in Figure 5-4.

☐ **3** Mix together 3 ml of copper sulfate solution, 4 ml of boiled starch solution, a pinch of charcoal powder, and 5 to 8 ml of distilled water. Note the physical characteristics of each as you mix them.

☐ **4** Pour the mixture into the funnel.

☐ **5** Observe the mixture being filtered by the paper. Does the process become progressively faster, progressively slower, or remain steady? Record and explain your results in Lab Report 5.

☐ **6** Examine the *filtrate* (filtered material). Test for the presence of each component of the original mixture:
a. *Charcoal*—observe visually
b. *Copper sulfate*—observe visually
c. *Starch*—add a few drops of Lugol's reagent (a color change to black means that starch is present)

CLINICAL APPLICATION

Results similar to those in your osmosis experiment occur when red blood cells come into contact with various solutions that are injected into the blood supply, as an *intravenous (IV)* therapy. When blood cells are bathed in a solution that is isotonic to them, they remain unchanged. If a hypotonic solution is introduced, the cells experience an inflow of water and usually burst. Bursting of red blood cells caused by osmosis is called **hemolysis.** If hypertonic solution is introduced, the cells lose water and shrivel. Shriveling that results from osmotic loss of water is called **crenation.** Therefore, the concentration of injected materials is critical to the survival of the patient.

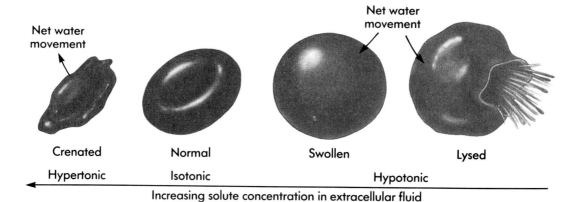

Figure 5-5 The effects on red blood cells when changing the tonicity of the extracellular fluid (plasma).

Predict the expected results in these situations:

❑ 1 Randy is a nursing student. He has been asked to intravenously inject his patient with 10 cc of an isotonic preparation. Mistakenly, he injects his patient with 10 cc of pure water. What is likely to happen to the red blood cells near the site of injection?

❑ 2 Jennifer is Randy's patient. She is carefully watching him fill an IV bottle with a mixture specially ordered by the physician. She notices that he accidentally fills the bottle with concentrated (10%) salt solution instead of the mixture from the pharmacy. Why should she refuse to allow Randy to attach the IV bottle to her system?

❑ 3 Randy is now assigned to the surgical unit. He is given a piece of living tissue and asked to put it in fluid before taking the sample to the pathology lab. What essential characteristic must such fluid have so that it will not damage any cells in the tissue?

NAME _____ DATE _____ SECTION _____

LAB REPORT 5

Transport through Cell Membranes

1. Describe the movement of India ink particles. Was it uniform or random? Were the particles moving great distances?

Observation	Elapsed time	Extent of diffusion in COLD WATER	Entent of diffusion in HOT WATER
1	0 minutes	No diffusion of dye particles	No diffusion of dye particles
2			
3			
4	30 minutes		
Results			

2. Using potassium permanganate as an example, explain in your own words how Brownian motion relates to diffusion.

3. Based on your results in the potassium permanganate experiment, would you state that increased temperature increases or decreases Brownian motion?

Bag	Predicted result	Initial mass	Final mass	Gain or loss of mass	Explanation of results
1					
2					
3					

4. Did the filtration rate of your mixture become progressively faster or slower or was it steady? Explain.

Substance	Test result (present or absent)	Explanation of result
Charcoal		
Copper sulfate		
Starch		

Matching

____ 1   ____ 6
____ 2   ____ 7
____ 3   ____ 8
____ 4   ____ 9
____ 5   ____ 10

Identify

_____ 1
_____ 2
_____ 3
_____ 4
_____ 5

Matching (may be used more than once or not at all)

a. Brownian motion
b. diffusion
c. filtration
d. hydrostatic pressure
e. hypertonic
f. hypotonic
g. isotonic
h. osmosis
i. semipermeable

1. The tendency of matter to spread to areas of lower concentration
2. Movement through a membrane driven by a hydrostatic pressure gradient
3. Term that specifically describes the diffusion of water across a membrane
4. Term that describes a membrane that allows only some types of particles to pass through it
5. The natural vibration of particles, it drives diffusion
6. In the kidney, blood pressure forces some water and solute particles from a blood vessel and into a kidney tubule. What is this type of transport called?
7. A cell is bathed in solution X. The cell quickly shrivels. What term describes solution X (when compared to the cell's fluid)?
8. Solution Q has a higher water content than solution Z. Therefore, solution Q is __?__ to solution Z.
9. A saline (salt) solution is to be injected into a patient. The salt/water ratio should be adjusted so that the saline solution is __?__ to the patient's cells.
10. Particles of substance Y move into a cell because there are fewer particles inside the cell than outside. This is an example of __?__.

Identify (state whether each item is an example of *diffusion, filtration,* or *active transport*)

1. Movement of water from an area of high concentration to an area of low concentration.
2. Dye particles spread evenly through water.
3. Starch particles pass through a paper membrane.
4. A cell uses energy to "pump" sugar molecules from its external environment.
5. Osmosis.

LAB EXERCISE 6

The Cell's Life Cycle

Cells in many parts of the human body divide to produce more cells of the same type. The hereditary information contained within the nucleus of a resting *parent* cell must first be **replicated** (copied), then evenly distributed between the two cells that result from division. The process of distributing genetic material is termed **mitosis.** Mitosis was named in the late nineteenth century by Walther Flemming, who noticed threadlike structures in cells during cell division (*mitos-* "threads," *-osis* "condition of").

The pinching in of the plasma membrane, and eventual split of the membrane and its contents into two *daughter cells,* is termed **cytokinesis.** Cytokinesis occurs about the same time (or just after) the last phases of mitosis.

Life cycles are circular patterns of organisms' life histories. For example, the life cycle of humans would list conception, development, adulthood, reproduction, then list conception, and so forth, again for the offspring. Likewise, individual body cells are formed, they reproduce, and their daughter cells continue the cycle of life. In this exercise, we will explore the major events of the human cell's life cycle.

Before you begin

❏ Read the appropriate chapter in your textbook.

❏ Set your learning goals. When you finish this exercise, you should be able to
- list the major phases of a cell's life cycle
- describe the principal events of mitosis
- identify cell parts involved in mitosis
- explain the importance of mitosis
- define the term *cytokinesis*

❏ Prepare your materials
- chart or model: *animal mitosis series*
- colored pencils or pens
- microscope
- prepared microslide: *whitefish blastula*

❏ Read the directions and safety tips for this exercise **carefully** before starting any procedure.

A. Introduction to the cell life cycle

The process of mitosis can be described step by step to make it a little easier to picture. We will divide the whole life cycle into five phases: **interphase, prophase, metaphase, anaphase,** and **telophase.**

Note in Figure 6-1 the major events of each life cycle phase listed. Try to identify the physical representation of those events in a chart or model of cells at various life cycle stages.

❏ 1 **Interphase** is not a phase of mitosis but is the period between cell divisions. It is not an inactive time, however, because the chromatin replicates during interphase (forming two sister **chromatids** joined at a **centromere**). In anticipation of division, additional cell fluid and organelles are formed during interphase. As Figure 6-1 shows, an initial growth phase (G_1) is followed by the DNA replication or synthesis (S) phase, which is in turn followed by a second growth (G_2) phase.

> **HINT** → The cell life cycle is a continuous process. The cell does not suddenly jump from one phase to another but gradually changes. Therefore, cell models represent a snapshot of each phase at a point where it is most distinct from the phases before and after it.

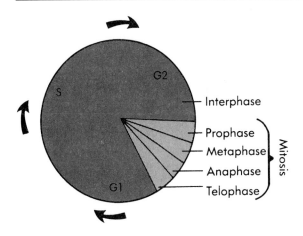

Figure 6-1 Cell life cycle and its major phases.

❏ 2 **Prophase** is the first phase of mitosis. During prophase, the nuclear membrane disappears, freeing the chromatin (which first shortens into tiny bodies called **chromosomes**). Also, centriole pairs move to opposite poles of the cell as **spindle fibers** begin to project from them. The spindle fibers extend toward the equator of the cell and may overlap with spindle fibers projecting from the opposite centriole pair. A spindle fiber may also attach to one side of a chromosome's centromere.

❏ 3 **Metaphase** is the period during which the chromosomes (each a pair of replicate chromatids joined at a centromere) line up along the cell's equator (imaginary center plane). Each chromosome now has a pair of spindle fibers attached to it, one from each centriole pair.

❏ 4 **Anaphase** is the phase during which the chromatids split at the centromere, each moving toward an opposite pole along the path of a spindle fiber. At the end of anaphase, each pole of the cell has a full group of 46 single chromosomes. The chromosomes on one side of the cell are replicates of the set of chromosomes on the other side.

❏ 5 **Telophase** is the time during which each side of the cell changes everything to the way it should be during interphase:
- A nuclear membrane forms around each group of chromosomes
- The chromosomes uncoil to form long chromatin strands
- Remnants of the spindle fibers disintegrate

❏ 6 During anaphase of mitosis, the separate (but concurrent) process of *cytokinesis* begins. By the end of anaphase, cleavage, or pinching in, of the parent cell is evident. By the end of telophase, complete splitting of the parent cell into two similar daughter cells is complete. Each daughter cell has a nucleus and roughly half of the cytoplasm and organelles of the parent cell.

❏ 7 Review the major events of the cell life cycle by completing the coloring exercise *Cell Life Cycle*.

The length of time between divisions, and the time required for division to take place, vary considerably from cell to cell. Even the relative length of different phases varies among individual cells. Cell division can range from 20 minutes to several hours.

B. Microscopic observations

In this activity, you will observe cells of the *whitefish blastula* in a prepared slide. The blastula is a developmental stage in the growth of many animals. The original cell formed by the joining of the egg and sperm undergoes mitosis many times, at one stage forming a ball of cells called a *blastula*. The cells of the blastula are still undergoing rapid cell division, making it a specimen likely to have many cells at various stages of the cell life cycle.

SAFETY FIRST! Do not forget to check the microscope's power cord for frays and for proper placement. Take care that you don't crack a slide or lens by zooming in with coarse focus without looking from the side. Accidents do not happen when you are careful.

❏ 1 Obtain a prepared microscope slide (microslide) of whitefish blastula cross sections (c.s.).

❏ 2 Locate a group of cells using low-power magnification. There are usually several different cell groups from which to choose on each slide.

❏ 3 Switch to high power and try to locate individual cells in each of the five phases of the cell life cycle. You may need to occasionally switch slides to find all the phases.

LANDMARK CHARACTERISTICS

Many prepared whitefish blastula specimens are stained so that the cytoplasm appears pinkish to brown-red. See Figure 6-3. The DNA (chromatin, chromosomes) is stained black. Therefore, look for fine, black formations when looking for the major events of mitosis. The plasma membranes and centrioles are so small that they will not appear distinctly in your specimen.

In a single microscopic field, you will see many cells, all at different points in the cell life cycle. There is no pattern to the way the cells in different phases are arranged (they are not laid out in order of mitotic phases, for instance). If you are fortunate, you may find at least one example of each phase in a single field.

Cell Life Cycle

MITOSIS₁
CYTOKINESIS₂
INTERPHASE₃
PROPHASE₄
METAPHASE₅
ANAPHASE₆
TELOPHASE₇

CELL MEMBRANE₈
NUCLEAR MEMBRANE₉
NUCLEOLUS₁₀
CENTRIOLES₁₁
CHROMATIN/CHROMOSOME₁₂
CENTROMERE₁₃

COLORING EXERCISE Using colored pens or pencils, shade in the figure and accompanying labels in contrasting colors of your choice as indicated by the red numerals.

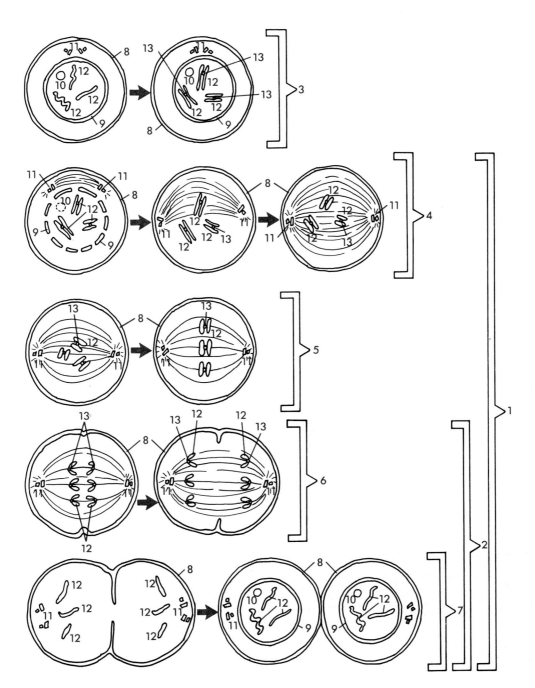

Figure 6-2

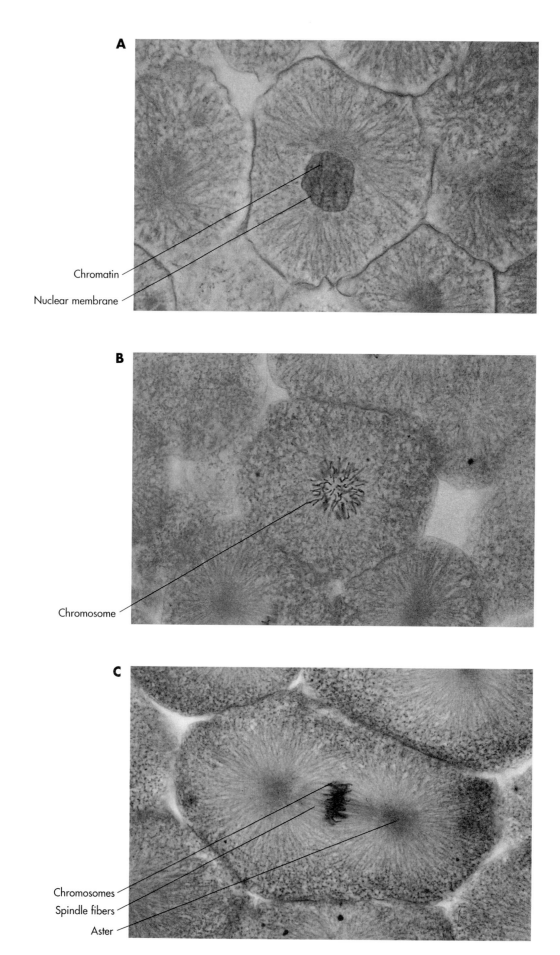

Figure 6-3 Micrographs of whitefish cells during the cell life cycle. **A,** Interphase. **B,** Prophase. **C,** Metaphase. **D,** Anaphase. **E,** Telophase.

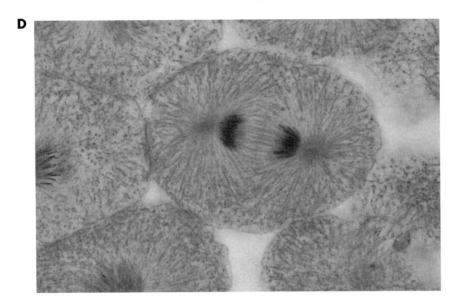

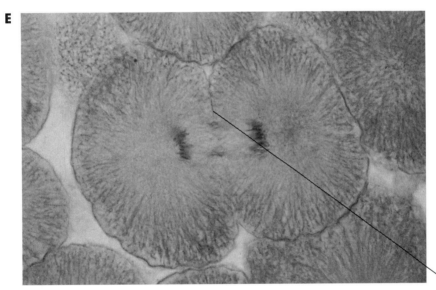

Cleavage furrow at equator

Figure 6-3, Continued.

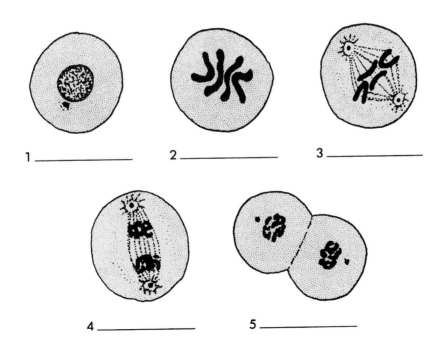

1 _____ 2 _____ 3 _____

4 _____ 5 _____

Figure 6-4 Examples of whitefish cells. Label each with the name of the appropriate phase.

ANOTHER KIND OF CELL DIVISION

Meiosis is a process distinct from mitosis. Meiosis is associated with a type of cell division that occurs during the formation of reproductive cells (*sperm* and *eggs*). Mitosis occurs in the division of all other cell types.

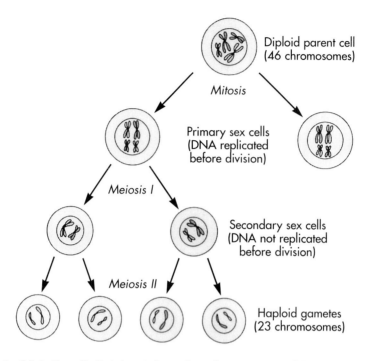

Figure 6-5 Meiosis. Meiotic cell division takes place in two steps: Meiosis I and Meiosis II. Meiosis is called *reduction division* because the number of chromosomes is reduced by half (from the diploid number to the haploid number).

Meiosis results in daughter cells that have only half the number of chromosomes that other cells, including the parent cells, have. That is, during meiosis, a parent cell with 46 chromosomes (the diploid number) produces daughter cells that have 23 chromosomes each (the haploid number). This must occur so that, when the sperm and egg unite during conception, the newly formed cell has 46 chromosomes (23 from the sperm plus 23 from the egg). Thus, the offspring has equal amounts of hereditary information from each parent.

Theoretically, what would happen if meiosis did not occur and sperm and egg cells could only form using mitosis?

What would happen if *all* your body cells divided using meiosis instead of mitosis?

NAME _____ DATE _____ SECTION _____

LAB REPORT 6

The Cell's Life Cycle

Sketch your observations of particularly clear examples of mitotic phases in whitefish cells.

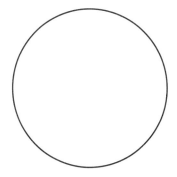

Specimen: *whitefish, phase* ____

Total Magnification: ____

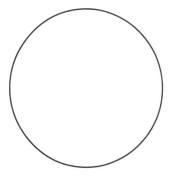

Specimen: *whitefish, phase* ____

Total Magnification: ____

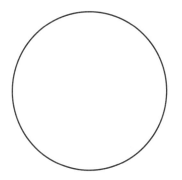

Specimen: *whitefish, phase* ____

Total Magnification: ____

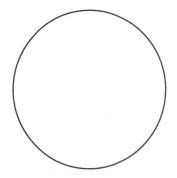

Specimen: *whitefish, phase* ____

Total Magnification: ____

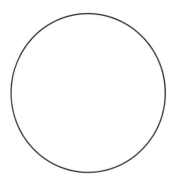

Specimen: *whitefish, phase* ____

Total Magnification: ____

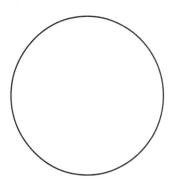

Specimen: *whitefish, phase* ____

Total Magnification: ____

Figure 6-3

_____ 1
_____ 2
_____ 3
_____ 4
_____ 5

Fill-in

_____ 1
_____ 2
_____ 3
_____ 4
_____ 5
_____ 6
_____ 7
_____ 8
_____ 9
_____ 10

Multiple choice

____ 1
____ 2
____ 3
____ 4
____ 5

Fill-in (write the name of the mitotic phase identified in each item)

1. The centrioles move toward opposite poles during __?__.
2. During __?__ the nuclear membrane disintegrates.
3. The nuclear membrane reappears during __?__.
4. The last phase of mitosis is __?__.
5. During __?__ the chromosomes align at the cell's equator.
6. Cytokinesis usually begins during __?__ of mitosis.
7. The first phase of mitosis is __?__.
8. During __?__ the chromatids separate and move toward opposite poles.
9. Spindle fibers appear during __?__.
10. Mitosis ends with __?__.

Multiple choice (only one choice is correct)

1. The function of mitosis is
    a. to distribute the cell's DNA equally between the daughters
    b. to divide the cytoplasm equally between the daughter cells
    c. to distribute the parent cell's organelles evenly
    d. all of the above
    e. none of the above

2. The chromosomes of daughter cells formed during mitosis
    a. are double the number of those in the parent cell
    b. are identical to the chromosomes of the other daughter
    c. are half the number of those in the parent cell
    d. b and c are correct
    e. all are correct

3. Which of these cells is likely to have been formed using mitosis?
    a. skin cell
    b. sperm cell
    c. egg cell
    d. heart muscle cell
    e. a and d are correct

4. A certain stain colors DNA violet. When stained, which of these should appear violet?
    a. chromatin
    b. chromosomes
    c. chromatids
    d. a and b are correct
    e. all of the above

5. Two DNA molecules, joined at a centromere, have coiled to form tiny bodies. This connected pair of DNA molecules should properly be called
    a. a centromere
    b. a centriole
    c. a chromosome
    d. a chromatid
    e. a chromatin strand

Epithelial Tissue

LAB EXERCISE 7

As we continue our study of the human body, we progress from the cellular level of organization to the tissue level. Tissues are masses of similar cells (and their extracellular *matrix*) that combine with other tissues to form *membranes* or *organs*. The systematic study of tissue types is called **histology.**

There are four basic tissue types in the human:

> EPITHELIAL TISSUE
> CONNECTIVE TISSUE
> MUSCLE TISSUE
> NERVE TISSUE

Each of these categories include several subcategories, based on differences in structure and function. We will consider some of the major subcategories in this and the next several exercises. We will begin our exploration with a brief look at epithelial tissue types.

Before you begin

❏ Read the appropriate chapter in your textbook.

❏ Set your learning goals. When you finish this exercise, you should be able to
- define the term *epithelium*
- classify epithelial tissue types
- identify and give examples of the major types of epithelia in figures and specimens

❏ Prepare your materials:
- unlabeled photos of epithelial types (optional)
- microscope
- prepared microslides:
  Simple squamous (surface)
  Stratified squamous (nonkeratinized) c.s.
  Stratified squamous (keratinized) c.s.
  Simple cuboidal c.s.
  Simple columnar c.s.
  Pseudostratified (ciliated) c.s.
  Transitional c.s.

❏ Read the directions and safety tips for this exercise **carefully** before starting any procedure.

A. How to classify epithelial tissues

Epithelial tissue can have either of two basic roles: covering/lining and glandular. The first kind, covering/lining epithelium, is found in sheets that cover body structures (as in the outer layer of the skin) or line body spaces (as in the lining of the stomach). A key characteristic of covering/lining epithelium is that it always has one side exposed, or free to face outward (covering) or inward (lining). Because the cells form a continuous sheet, they are held together very tightly and have very little matrix (extracellular material). The nonfree face of the epithelial sheet is attached to underlying connective tissue by the **basement membrane.** The basement membrane is a thin, gluelike layer that holds the epithelium in place while remaining highly permeable to water and other substances. This is important because epithelia do not have their own blood supply. Water and other important substances must diffuse between the epithelial cells and the underlying tissue through the basement membrane.

The second kind of epithelium is the type that forms glands. Glandular epithelium forms the functional portions of *exocrine* glands (glands that secrete substances into ducts that empty onto epithelial surfaces). It also forms *endocrine* glands (glands that secrete substances that diffuse into the blood stream).

Epithelial tissue that forms membranes is usually categorized by its structure. The most common scheme classifies types by identifying the number of cell layers and the shape of the cells in the outer layer. We will practice classifying membranous epithelia by using photomicrographs or drawings provided by your instructor, or the samples in Figure 7-3.

❏ 1 Randomly choose a sample from those provided.

❏ 2 The first step in classifying an epithelial tissue is determining how many layers of cells compose the sheet. The sample falls into one of three categories based on number of layers:
- **Simple**—This category includes all tissues that have exactly one layer of cells.
- **Stratified**—This group includes epithelia with more than one layer (or *stratum*) of cells.

Epithelial Tissue Classification

COLORING EXERCISE Using colored pens or pencils, shade in the figure and accompanying labels in contrasting colors of your choice as indicated by the red numerals.

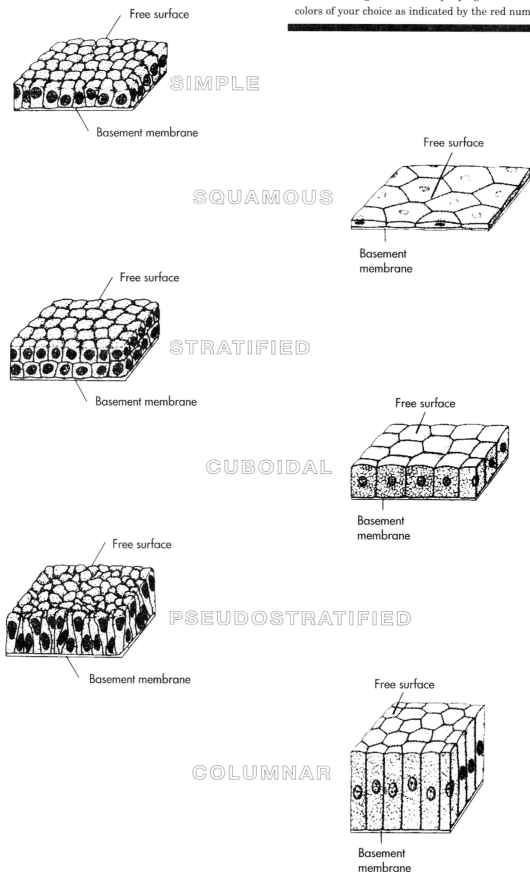

Figure 7-1 The figures on the left show how epithelia are classified according to number of layers. Figures on the right show how epithelia are classified according to shape.

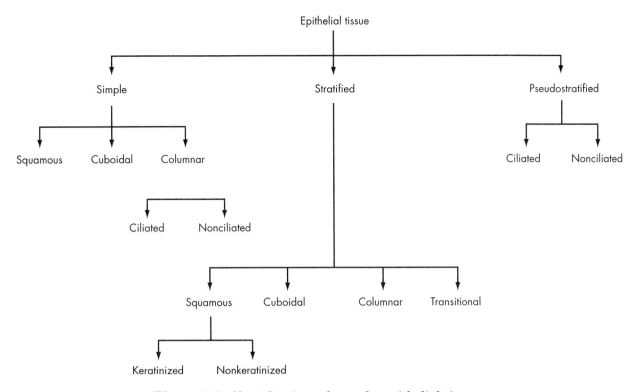

Figure 7-2 Classification scheme for epithelial tissues.

- **Pseudostratified**—The name of this category means "false stratified." This term is based on the fact that pseudostratified epithelium looks as if it's stratified but is really only one layer deep. This illusion is caused by the way the cells seem to be pushed together, with some nuclei pushed into the upper portion of the cell and nuclei of other cells located in the bottom part. The nuclei then seem to form an upper and lower row, giving the illusion of two rows of cells. Remember, cell membranes are often too thin for you to be able to see the boundary between epithelial cells.

❏ 3 If you have determined that your sample is simple or stratified, skip to step 4. If your sample is pseudostratified, you must determine whether it is **ciliated** or **nonciliated.** Cilia are short, hairlike projections from each cell's free surface that propel material along the epithelial surface. You have now completely classified your sample and can try it again with a new sample.

❏ 4 Look at the layer of cells along the sample's free surface. Determine which shape these cells have:
- **Squamous**—Shaped like fish scales, these cells are much wider than they are tall when viewed in a cross section. Nuclei may be absent in cross sections.
- **Cuboidal**—As their name implies, these cells are roughly cube-shaped—about as tall as they are wide. Do not expect squared corners, as in a cube. Although they sometimes appear that way in a cross section, they more often resemble rounded squares or even circles.
- **Columnar**—These columnlike cells are taller than they are wide when viewed in a cross section.
- **Transitional**—The term *transitional* refers to the fact that these cells change shape as the need arises, being stretched or compressed into any of the three shapes already listed. Transitional cells are usually found as a hodgepodge of different cell shapes in membranes subject to a great degree of stretching and recoil. Transitional epithelia are always stratified.

If your sample is simple squamous, simple or stratified cuboidal, or stratified columnar, you are finished with this sample and can try another one. If you have simple columnar, move to step 5. If you have stratified squamous, skip to step 6.

❏ 5 Determine whether the sample is ciliated or nonciliated. See step 3 for clarification. You are now finished classifying this sample and can try another sample.

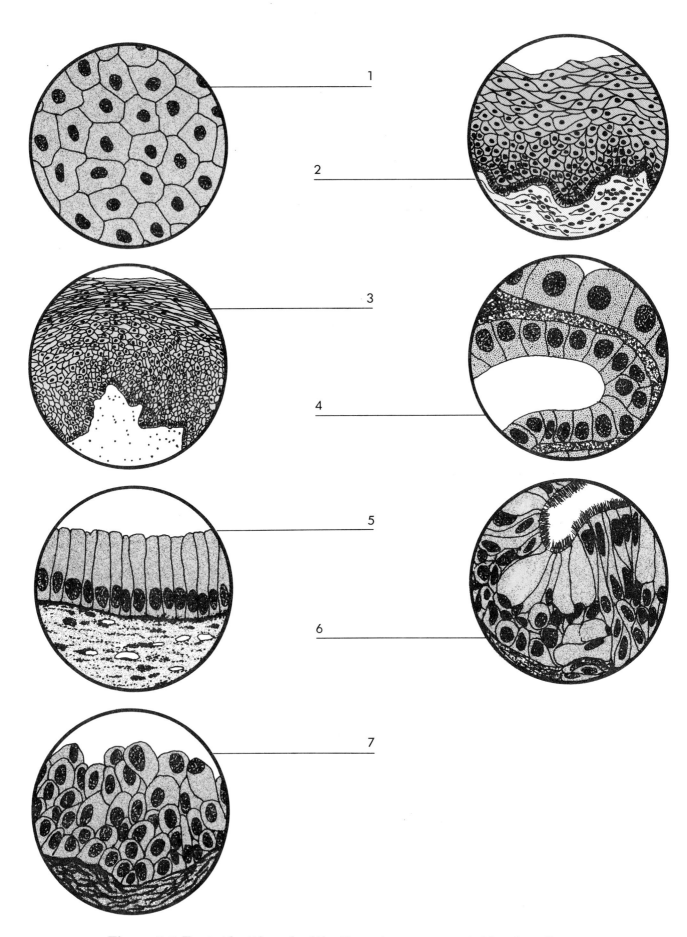

Figure 7-3 Try to identify each of the tissue types represented by these figures.

Table 7-1 Examples of Major Membranous Epithelial Tissues

Tissue	Location	Function
Simple squamous	Alveoli of lungs	Absorption by diffusion of respiratory gases between alveolar air and blood
	Lining of blood and lymphatic vessels (called endothelium; classed as connective tissue by some histologists)	Absorption by diffusion, filtration, osmosis
	Surface layer of pleura, pericardium, peritoneum (called mesothelium; classed as connective tissue by some histologists)	Absorption by diffusion and osmosis; also, secretion
Stratified squamous		
Nonkeratinized	Surface of mucous membrane lining mouth, esophagus, and vagina	Protection
Keratinized	Surface of skin (epidermis)	Protection
Simple cuboidal	In many types of glands and their ducts; also found in ducts and tubules of other organs, such as the kidney	Secretion; absorption
Pseudostratified	Surface of mucous membrane lining trachea, large bronchi, nasal mucosa, and parts of male reproductive tract (epididymis and vas deferens); lines large ducts of some glands (e.g., parotid)	Protection
Simple columnar	Surface layer of mucous lining of stomach, intestines, and part of respiratory tract	Protection; secretion; absorption; moving of mucus (by ciliated columnar epithelium)
Transitional	Surface of mucous membrane lining urinary bladder and ureters	Permits stretching

❏ 6 Determine whether your stratified squamous tissue is **keratinized** or **nonkeratinized.** In some stratified squamous tissues, the upper layers of cells are dead (evidenced by lack of nuclei) and filled with a tough, waterproof material called *keratin*. If your sample has a thick sheet of material with no visible nuclei along its free surface, it is probably keratinized. Nonkeratinized samples have nucleated squamous cells along the free edge.

❏ 7 Once you have determined the specific tissue type of each specimen, refer to Table 7-1 to discover its typical locations and functions.

B. Microscopic specimens

In this activity, we will examine prepared slides of human epithelial tissues. Before beginning, review the microscopy techniques outlined in Lab Exercise 3.

> **SAFETY FIRST!** Avoid the hazards associated with frayed power cords and broken glass slides. Do not reach for anything while looking into the ocular, or you may knock over something (or someone).

For each of the epithelial tissues listed, examine at least one example in a prepared slide. Try to practice looking for the key characteristics by which each type can be classified. Sketch some representative examples in Lab Report 7.

> **HINT** → In many (not all) prepared epithelia, the cytoplasm will appear either as a clear/cloudy area or a very pale pink. The cell membranes are usually faint but may be visible as medium-pink lines. The nuclei are often stained dark pink to violet or bluish.

❑ 1 **Simple squamous epithelium**—As the name implies, simple squamous epithelium is a single layer of flattened cells. This type of epithelium forms the very thin lining found in the blood vessels, alveoli (air sacs) of the lungs, and other areas where thin membranes are required. Because it is so thin, simple squamous is well adapted for diffusion or filtration of water, gases, and other substances.

> **LANDMARK CHARACTERISTICS**
> Viewed from the side, as in a cross section, simple squamous epithelium looks like a thin line of cells—often with distinguishable nuclei. See Figure 7-4 and LABORATORY REFERENCE, Plate 7. Viewed as a sheet from above, simple squamous epithelium looks like a two-dimensional layout of polygonal or rounded "tiles," each with a central nucleus. See LABORATORY REFERENCE, Plate 6.

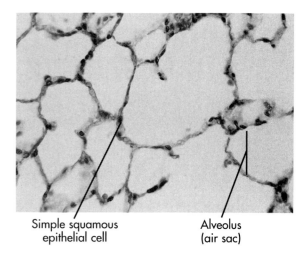

Figure 7-4 Simple squamous epithelium.

❑ 2 **Stratified squamous epithelium**—This tissue type is composed of multiple layers of cells: columnar cells along the basement membrane, topped by cuboidal cells, then by squamous cells. Cells divide in the columnar layer and are pushed upward, where they are distorted into cuboidal, then squamous, cells. The cells on the surface slough off but are continually replaced by cells moving up from the bottom layer. Because of its thickness and its constant renewal, it is well adapted for protection. For example, stratified squamous epithelium is found in the outer part of the skin and the mucous linings of the mouth, vagina, and esophagus.

> **LANDMARK CHARACTERISTICS**
> ■ *Nonkeratinized stratified squamous* epithelium typically has a dense concentration of nuclei (of columnar and cuboidal cells) near the basement membrane, becoming less dense toward the free surface. Most of the squamous cells near the free surface should have identifiable nuclei. See LABORATORY REFERENCE, Plate 8.
> ■ Keratinized stratified squamous samples will be from the epidermis (outer layer) of the skin. They will look similar to nonkeratinized specimens, except that they have distinct additional layers overlying the top layers of squamous cells. This layer has no distinguishable nuclei. See LABORATORY REFERENCE, Plate 9.

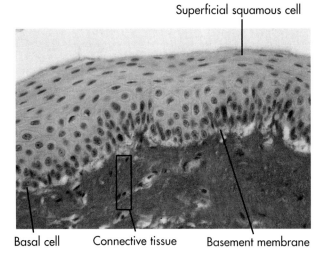

Figure 7-5 Stratified squamous (nonkeratinized) epithelium.

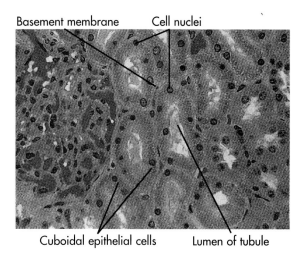

Figure 7-7 Simple cuboidal epithelium.

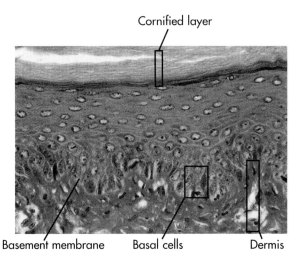

Figure 7-6 Stratified squamous (keratinized) epithelium.

❑ 3 **Simple cuboidal epithelium**—Composed of a single layer of almost cubic cells, this type is found in secreting organs, such as glands. Simple cuboidal epithelium also forms the kidney tubules, where it is specialized for water reabsorption and ion movement.

LANDMARK CHARACTERISTICS
The cuboid shape of cells is easily seen in cross sections of kidney tubules. See LABORATORY REFERENCE, Plate 10. Because the sample is formed by many tubules cut at an angle, you will see many circles and loops made of simple cuboidal epithelium. Similar specimens are seen in thyroid tissue or other glands and glandular ducts.

❑ 4 **Simple columnar epithelium**—Forming linings specialized for absorption and secretion, simple columnar epithelium is found in many parts of the body. For example, this type of epithelium lines portions of the reproductive tract, digestive tract, excretory ducts, and respiratory tract. A special cell that is often interspersed among the other columnar cells is the **goblet cell.** The goblet cell resembles its namesake, the wine goblet, in that it has a large cuplike vesicle that may open onto the free surface. The cup has no wine but rather **mucus,** which the goblet cells produce and secrete in great quantity. Mucus has many functions, including the lubrication and protection of the epithelial lining.

LANDMARK CHARACTERISTICS
Simple columnar sheets often line cavities with deeply folded or grooved walls. The specimen, then, will appear to zigzag when viewed in a cross section. One surface of the sheet is always free, however, even though the free surfaces may fold back and touch one another. A goblet cell is easily recognizable by the very large bubble (vesicle) in the center or near the top. See LABORATORY REFERENCE, Plate 11. Because the vesicle contains clear, unstained mucus, it will appear more lightly colored than the surrounding material.

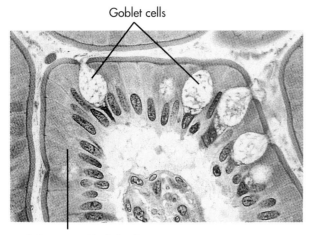

Figure 7-8 Simple columnar epithelium.

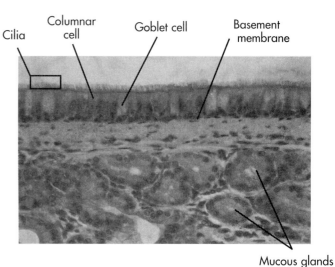

Figure 7-9 Pseudostratified ciliated epithelium.

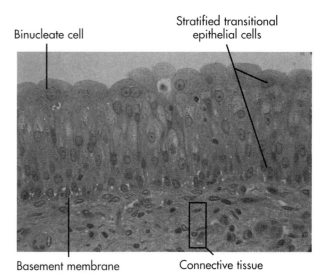

Figure 7-10 Transitional epithelium.

❑ 5 **Ciliated pseudostratified epithelium**—This type of epithelium is a single layer of columnar cells that all attach to the basement membrane. However, some cells are short and do not quite reach the free surface. Nuclei of short cells are nearer the basement membrane than the nuclei of tall cells, so there appear to be two rows of cells. Like simple columnar epithelium, pseudostratified epithelium is found in both ciliated and nonciliated forms. It is found in many of the same general areas of the body as simple columnar epithelium: the upper throat, upper respiratory tract, and parts of the male urinary and reproductive tracts.

LANDMARK CHARACTERISTICS
On initial examination, pseudostratified epithelium resembles simple columnar epithelium. See LABORATORY REFERENCE, Plate 12. However, the telltale double row of nuclei gives it away. As in simple columnar epithelium, goblet cells may be present. Cilia look like distinct, tiny hairs along the free surface. In some specimens, some of the cilia may have some areas of matted cilia or even "bald" spots. Under low magnification, the cilia look like "fuzz" on the free edge.

HINT → The terms *mucous* and *mucus* are often confused. *Mucus* is a noun, naming a glycoprotein-water solution. *Mucous* is an adjective describing something covered with mucus. Thus, mucous membranes are covered with mucus.

❑ 6 **Transitional epithelium**—Transitional tissue is adapted for stretching, so it is found in areas subject to a great deal of elastic stress. You will most likely see specimens from the lining of the urinary bladder, which must stretch a great deal from time to time throughout the day.

LANDMARK CHARACTERISTICS
Transitional epithelium resembles nonkeratinized stratified squamous epithelium at first glance. Transitional epithelium, however, often has rounded cuboidal cells in the top layer (rather than only squamous cells). Depending on the individual specimen, you may see a great variety of shapes scattered throughout the tissue. Some of them may have a sort of distorted "teardrop" shape. This gives transitional tissue a rather unorganized appearance, compared to the other epithelial types. See LABORATORY REFERENCE, Plate 13.

CLINICAL APPLICATION

Researchers are always looking for new ways to deliver therapeutic drugs to the blood stream for distribution throughout the body. Currently, biomedical research teams are using epithelial tissue cultures to investigate a relatively new approach to drug introduction. For some time, science has known that certain compounds are easily absorbed by epithelial membranes and picked up by the blood. For example, patches of material containing drugs that inhibit motion sickness have been used by tourists on cruises. Patients with heart disease or burns often receive therapy by means of similar patches. Today, however, researchers are looking for wider and more varied uses of epithelial absorption as a drug-delivery method.

Based on what you know of the different epithelial tissue types, which are the best candidates for this type of therapy? Explain your choices.

How could a clinician actually apply epithelial absorption techniques? Try to think of other strategies other than the skin-patch approach.

Explain some of the harmful side effects of smoking tobacco, cocaine, and marijuana in terms of epithelial absorption.

NAME_____ DATE_____ SECTION_____

LAB REPORT 7

Epithelial Tissue

Activity A:

Trial number	Identifying code*	Tissue type	Notes
1			
2			
3			
4			
5			
6			
7			

* The IDENTIFYING CODE is the code number (letter) assigned to your sample by the instructor. If you are using figures from this or another book, put the plate and/or figure number in this column.

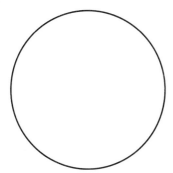

Specimen: *simple squamous epithelium (surface view)*

Total Magnification:____

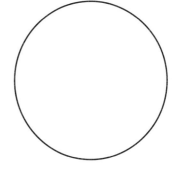

Specimen: *stratified squamous epithelium (keratinized)*

Total Magnification:____

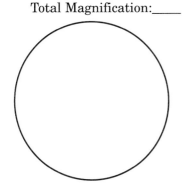

Specimen: *stratified squamous epithelium (nonkeratinized)*

Total Magnification:____

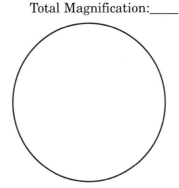

Specimen: *simple cuboidal epithelium*

Total Magnification:____

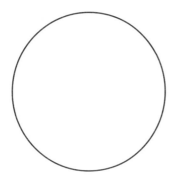

Specimen: *simple columnar epithelium*

Total Magnification:____

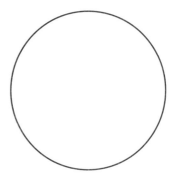

Specimen: *transitional epithelium*

Total Magnification:____

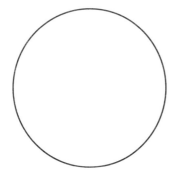

Specimen: *ciliated pseudostratified epithelium*

Total Magnification:____

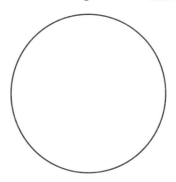

Specimen:_____

Total Magnification:____

Figure 7-3

_____ 1
_____ 2
_____ 3
_____ 4
_____ 5
_____ 6
_____ 7

Identify

_____ 1
_____ 2
_____ 3
_____ 4
_____ 5
_____ 6
_____ 7
_____ 8

Identify (identify the term associated with each description)

1. A type of cell found in simple columnar and pseudostratified tissue that secretes mucus
2. One of two kinds of glands, it secretes substances onto epithelial surfaces
3. A cellular organelle, it is a projection that moves substances along the surface of a cell
4. A type of epithelial tissue especially well adapted to excessive stretching
5. Mucous membranes secrete this water-based protein solution
6. A tough, waterproof material found in the upper layers of some examples of stratified squamous epithelium
7. The type of epithelial tissue likely to be found forming glands
8. The type of epithelial tissue likely to be found on the palm of the hand

LAB EXERCISE 8

Connective Tissue

Some tissues are called **connective tissues** because they act as connections among various other tissues. **Bone, cartilage,** and **fibrous** connective tissues actually hold parts together or support them in some way. **Blood** tissue connects other tissues in the sense of transporting materials between them.

An additional function of some connective tissues is storage. For example, bone tissue is an important storage site for calcium and phosphorus, minerals vital to proper function in many parts of the body.

In this exercise, we will learn to identify some of the more common connective tissue types found in the human body.

Before you begin

❑ Read the appropriate chapter in your textbook.

❑ Set your learning goals. When you finish this exercise, you should be able to
- describe the general pattern of structure in connective tissues
- classify major connective tissue types
- identify examples of connective tissue types in figures and specimens

❑ Prepare your materials:
- microscope
- prepared microslides:
  Dense fibrous (regular) connective tissue
  Dense fibrous (irregular) connective tissue
  Loose fibrous (areolar) connective tissue
  Adipose tissue
  Reticular tissue
  Hyaline cartilage
  Fibrocartilage
  Elastic cartilage
  Compact bone
  Cancellous bone
  Blood smear
- fresh animal connective tissue samples (as available)

❑ Read the directions and safety tips for this exercise **carefully** before starting any procedure.

A. Classifying connective tissues

A structural feature common to all connective tissues is the dominance of the matrix, or extracellular material. Recall that epithelial tissues have almost no extracellular material separating individual cells. Connective tissue cells are often widely separated by one of three basic types of matrix:

❑ 1 **Protein matrix** is extracellular material composed of many substances but with a dominance of protein fibers. **Collagen** is a common protein, forming bundles of tough, flexible fibers. Because they have a whitish color, collagen fibers are often called *white fibers*. **Elastin,** a stretchy, fibrous protein, forms thick, single fibers in connective tissue matrices. Elastin fibers are sometimes called *yellow fibers*. Identify these categories of protein-matrix connective tissues in Figure 8-1:
- **Fibrous**—Fibrous tissues are categorized as either **dense fibrous** or **loose fibrous,** depending on the density of protein fibers in the matrix. Dense fibrous tissue can either be **regular dense fibrous** (having regular parallel bundles of fibers) or **irregular dense fibrous** (having an irregular hodgepodge of fibers). Loose fibrous tissue is also called **loose, ordinary areolar tissue.**
- **Adipose**—Adipose tissue is often simply called *fat tissue* because its primary function is the storage of fat (for later use or for body support). Adipose tissue is actually a modified form of areolar tissue, with fat storage cells having been filled with stored lipids and expanded into the extracellular spaces.

❑ 2 **Protein/ground substance matrix** is extracellular material that has some protein fibers in it but also a great deal of nonfibrous protein and other substances. Identify these categories of protein/ground substance tissues in Figure 8-1:
- **Cartilage**—The matrix of cartilage is a combination of fibers and ground substance that gives it a rubbery quality. **Hyaline cartilage** has a moderate amount of collagen fiber in its matrix. **Fibrocartilage,** as its name implies, has a large amount of collagen in its matrix. **Elastic cartilage** is

67

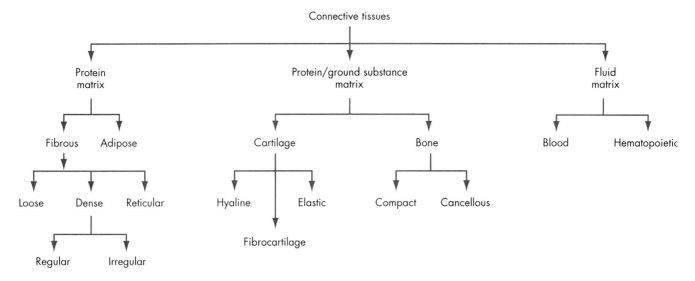

Figure 8-1 A classification scheme for connective tissues.

distinguished by the presence of elastin fibers, giving it a stretchy quality.

- **Bone**—There are two broad categories of bone: **compact bone** and **cancellous (spongy) bone.** Both have a matrix of collagen fibers encrusted with mineral crystals that give it a solid consistency. Compact bone forms rather large, dense pieces of bone matrix. Cancellous bone forms thin, narrow beams of hard bone matrix in which **red bone marrow** can be supported.

❑ 3 **Fluid matrix** is composed of a water-based solution with a fluid consistency. **Blood** is the major type of fluid matrix connective tissue. Blood cells are suspended within the fluid *plasma* and can slide past one another freely. Another type of fluid—matrix tissue—is **hematopoietic tissue,** which produces blood cells. Hematopoietic tissue is also called **myeloid tissue,** or simply *red bone marrow*. Identify fluid matrix tissue in the general scheme presented in Figure 8-1.

> **HINT** → Many cells found in connective tissues are named according to their action. Cells that produce matrix often have the suffix *-blast* ("make"). Cells that destroy matrix during remodeling have the suffix *-clast* ("break"). Cells that are in a relatively inactive mode have a *-cyte* ("cell") suffix. The first part of the name tells the specific kind of matrix involved: *fibro-* ("fiber"), *chondro-* ("cartilage"), or *osteo-* ("bone"). Thus *fibroblasts* make fibers, *chondroblasts* manufacture cartilage matrix, and *osteoblasts* lay down bone matrix.

Histologists classify tissues for convenience in naming them and for better identification of their nature. However, different histologists use slightly different schemes, depending on the context in which they are working. The histology schemes used in this manual are fairly universal.

Table 8-1 summarizes the locations and functions of some of the major types of connective tissue.

B. Microscopic specimens

> **SAFETY FIRST!** Avoid electrical hazards while using the microscope. Be sure to exercise care in dealing with broken glass slides. Do not reach for slides or other objects while you are looking into the ocular, or you may knock over something (or someone).

This activity asks you to observe representative connective tissue types in prepared specimens. For each type listed in the following steps, locate an example in a slide and sketch it in Lab Report 8.

❑ 1 **Loose, ordinary fibrous (areolar) connective tissue**—Areolar (meaning "spacious") tissue forms loose bonds between other tissues. For example, under the skin, it allows the skin to be slid around over, or pulled from, the underlying muscle to some degree. Both collagen and elastin fibers are found in this tissue but are widely spaced. Blood vessels, as well as nerves, course through it. A variety of cell types may be found here: fibrocytes, adipose cells, and white blood cells.

Table 8-1 Examples of major connective tissues.

Tissue	Location	Function
Fibrous		
Loose, ordinary (areolar)	Between other tissues and organs Superficial fascia	Connection Connection
Adipose (fat)	Under skin Padding at various points	Protection Insulation Support Reserve food
Reticular	Inner framework of spleen, lymph nodes, bone marrow	Support Filtration
Dense fibrous		
Regular	Tendons Ligaments Aponeuroses	Flexible but strong connection
Irregular	Deep fascia Dermis Scars Capsule of kidney, etc.	Connection Support
Bone		
Compact	Skeleton	Support Protection Calcium reservoir
Cancellous	Skeleton	Supports bone marrow
Cartilage		
Hyaline	Part of nasal septum Covering articular surfaces of bones Larynx Rings in trachea and bronchi	Firm but flexible support
Fibrocartilage	Disks between vertebrae Symphysis pubis	
Elastic	External ear Auditory (Eustachian) tube	
Fluid		
Blood	In blood vessels	Transportation Protection
Hematopoietic	Skeleton (in cavities of cancellous bone)	Blood tissue production

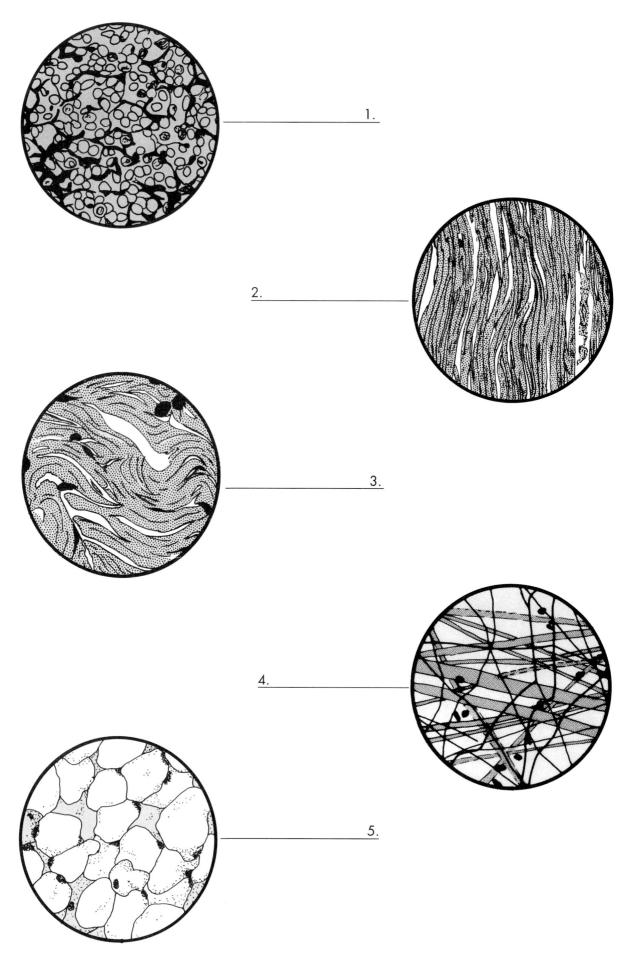

1. _____
2. _____
3. _____
4. _____
5. _____

Try to identify the connective tissues represented in these drawings.

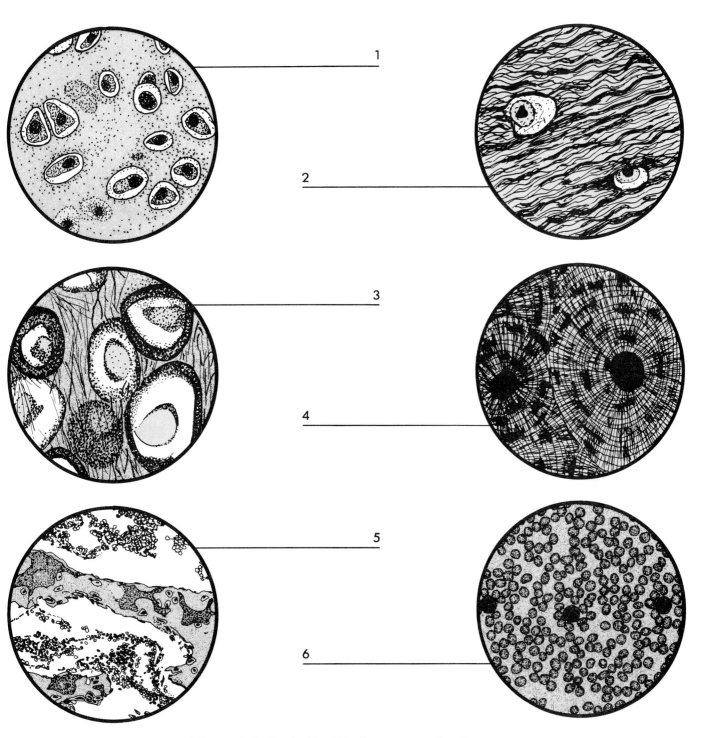

Figure 8-3 Try to identify these connective tissues.

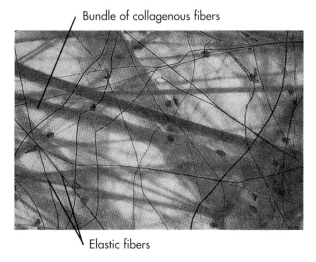

Figure 8-4 Loose, ordinary (areolar) connective tissue.

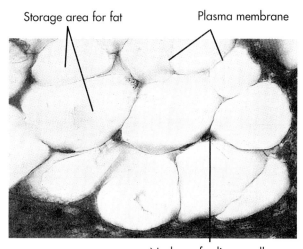

Figure 8-5 Adipose tissue.

> **LANDMARK CHARACTERISTICS**
> Areolar tissue's widely spaced fibers and variety of cell types make it easy tissue to identify. The elastin fibers appear as dark, thick, jagged strands. Collagen fibers form bundles that often appear as hazy pink or light purple crisscrossing lines. See LABORATORY REFERENCE, Plate 14.

❏ 2 **Adipose tissue**—Adipose cells are specialized to store lipids in large vesicles. The vesicle can be so large that it pushes the nucleus and other organelles up to the cell membrane, which enlarges to accommodate the large cell volume. Some histologists classify adipose tissue as a subcategory of areolar tissue. If the adipose cells in areolar tissue enlarge (because of increased storage of fat), they crowd the fibers and other cells, eventually forming adipose tissue. Thus, adipose tissue has very little matrix compared to other connective tissue types.

Adipose tissue is found wherever areolar tissue is found but is most often seen around the heart and kidney—and under the skin. It not only stores lipids for later use, it also serves as support (as in the breasts), as insulation (under the skin), and as a cushion (you are sitting on it).

> **LANDMARK CHARACTERISTICS**
> Unlike other connective tissue cells, adipose cells are very close to one another. The large fat globule inside each cell pushes the cytoplasm and organelles into a thin, dark ring around the inside of the cell membrane. The nucleus often appears as a bulge in the ring. The most obvious characteristic of this tissue is the presence of large fat vesicles, which generally look clear, yellowish, or light pink. The overall appearance is that of a host of large bubbles. See LABORATORY REFERENCE, Plate 15.

❏ 3 **Reticular tissue**—Reticular tissue is named for a word that means "network," referring to this tissue's characteristic three-dimensional web of fine reticular fibers. The fine fibers that characterize this tissue are made of a special type of collagen. Reticular cells are found overlying the fine fibers that form the reticular meshwork. Branches of the cytoplasm of reticular cells follow the branching reticular fibers. Reticular fiber tissue forms the framework of the spleen, lymph nodes, and bone marrow cavities. It functions as part of the body's defensive system.

> **LANDMARK CHARACTERISTICS**
> Reticular tissue is characterized by a network of extremely thin reticular fibers that are either stained darkly so they can be seen easily or not stained, which makes them practically invisible under ordinary magnification. Reticular cells can be seen against the fibers, with cell shapes that seem to conform to the branches of the reticular meshwork. See LABORATORY REFERENCE, Plate 16.

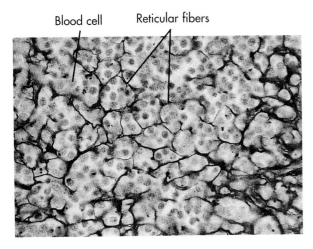

Figure 8-6 Reticular tissue.

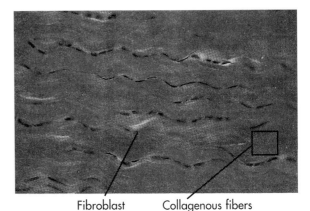

Figure 8-7 Dense fibrous (regular) connective tissue.

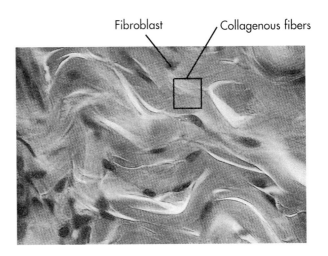

Figure 8-8 Dense fibrous (irregular) connective tissue.

☐ 4 **Dense fibrous connective tissue**—As its name indicates, this tissue is a dense arrangement of fibers. They may be collagen fibers or elastic fibers. Their arrangement may be regular (roughly parallel) or irregular (swirling or random). Interspersed among the fibers of a mature tissue are **fibrocytes.** Irregular dense fibrous tissue forms the lower layer of the skin (dermis), much of the body's fascia, and the capsules of many organs. Regular dense fibrous tissue is used for structures that require a better-engineered connection between parts that are pulled with great force. For example, this tissue forms *tendons* (connecting muscle to bone) and *ligaments* (connecting bone to bone).

LANDMARK CHARACTERISTICS

■ Regular dense fibrous tissue usually has a roughly parallel arrangement of either collagen or elastic fibers. Collagen appears in wavy bundles stained pink or bluish. Elastin appears as thick fibers stained dark violet or blue. Darkly stained fibrocytes are scattered between fibers or fiber bundles, often in groups.

■ Irregular dense fibrous tissue is virtually the same as regular tissue in composition. Irregular tissue is different because the fibers appear as chaotic swirls rather than parallel lines. See LABORATORY REFERENCE, Plates 17 and 18.

☐ 5 **Hyaline cartilage**—All cartilage is distinctive in the consistency of its semisolid rubbery matrix of protein fibers mixed with other substances. **Chondrocytes** are scattered throughout the matrix in little pockets called **lacunae** (meaning "lakes"). Hyaline cartilage has a moderate amount of collagen, giving it a great deal of toughness along with its cushiony quality. This cartilage type forms the bulk of the fetal skeleton (before it is replaced by bone) and continues to be the most abundant type of cartilage throughout life. It forms the thin, rubbery layer over the ends of long bones and is found in parts of the larynx, nose, and trachea. Specimens of cartilage may exhibit a fibrous **perichondrium** surrounding the cartilage tissue.

LANDMARK CHARACTERISTICS

The collagen fibers of hyaline cartilage are not distinct in the matrix, which has a smooth, pinkish, or lavender appearance in many preparations. The chondrocytes are usually pink to violet and appear to have shriveled within their respective lacunae. Because of chondrocyte shrinkage, a clear ring appears around the inside of many lacunae. See LABORATORY REFERENCE, Plate 19.

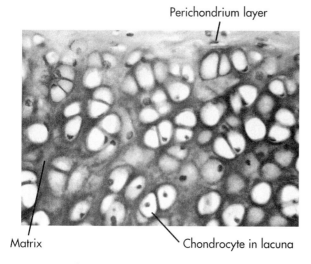

Figure 8-9 Hyaline cartilage.

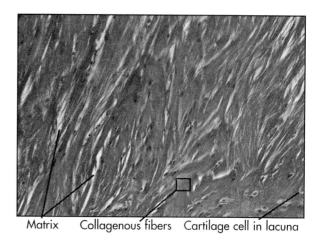

Figure 8-10 Fibrocartilage.

❏ 6 **Fibrocartilage**—The name of this tissue indicates its high concentration of collagen fibers. These fibers give the tissue a distinctive fibrous appearance. Fibrocartilage has a more rigid, less rubbery consistency than other cartilage types. Fibrocartilage forms the disks between vertebrae and may be found at other semimovable joints.

> **LANDMARK CHARACTERISTICS**
> Fibrocartilage may appear alongside hyaline cartilage and sometimes looks very much like it. However, the distinct fibrous appearance of fibrocartilage's matrix is the determining factor. See LABORATORY REFERENCE, Plate 20.

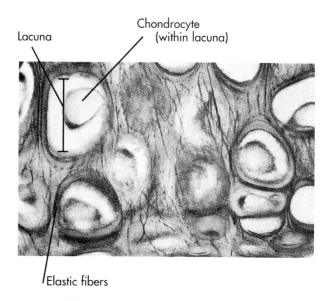

Figure 8-11 Elastic cartilage.

❏ 7 **Elastic cartilage**—As its name implies, elastic cartilage has a large proportion of elastin fibers in its matrix. Elastic cartilage is found in structures in which springiness is desirable in the support material. For example, the elasticity of the pinna (ear flap) is provided by elastic cartilage.

> **LANDMARK CHARACTERISTICS**
> In many preparations, elastin fibers stain very darkly. This makes their presence easy to detect and allows one to distinguish elastic cartilage without any problem. See LABORATORY REFERENCE, Plate 21.

❏ 8 **Compact bone**—Compact bone is formed by solid, cylindrical units called **osteons** packed tightly together. The osteon, or **Haversian system,** consists of multiple concentric layers of hard bone matrix, with cells sandwiched between each layer. Each layer is a **lamella** (plural *lamellae*). **Osteocytes** are literally trapped within lacunae between the lamellae. The osteocytes were once active **osteoblasts** but have trapped themselves in the solid matrix they formed. The lamellae are centered around the **Haversian canal's** blood vessels. The cells transport materials to and from the canal by way of tiny **canaliculi** ("small canals") that connect the osteocytes to each other and to the canal.

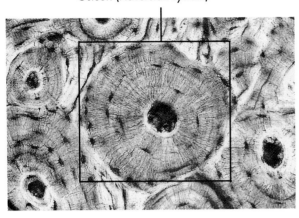

Figure 8-12 Compact bone.

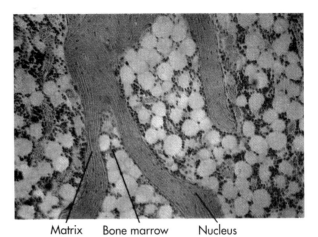

Figure 8-13 Cancellous bone and hematopoietic tissue.

LANDMARK CHARACTERISTICS

A cross section of compact bone has rings of lamellae surrounding several adjacent haversian, or *central,* canals. The lamellae resemble rings in an onion slice. The central canals are either clear or nearly black, the lamellae buff to orange, and the osteocytes brown or black. The canaliculi often appear as wavy hairlines radiating from the lacunae. See LABORATORY REFERENCE, Plate 22.

LANDMARK CHARACTERISTICS

Cancellous bone is distinguished by its rather disorganized array of trabecular beams of bone surrounded by myeloid tissue. The bone pieces may look like slivers of compact bone, with lamellae that often do not form complete circles. The myeloid tissue is a scattering of blood cells, which appear as tiny, dark circles. Myeloid (hematopoietic) tissue may also have a netlike formation of very thin collagen fibers called *reticular fibers.* In some preparations, the bone tissue is pink, and the myeloid cells are dark red. See LABORATORY REFERENCE, Plate 23.

❑ **9 Cancellous bone and hematopoietic tissue**—Cancellous bone is easily identified by its open, latticelike structure. Thin plates of bone matrix, with a scattering of osteocytes trapped within lacunae, form structural beams that have great strength despite the open spaces. These beams of hard bone are called **trabeculae.** Because cancellous bone has open spaces, it is sometimes called *spongy bone.* This name can be misleading because one might think spongy bone is as soft as a bath sponge. It is not soft at all because it has hard trabeculae. The spaces are filled with *hematopoietic* or *myeloid tissue,* a special type of blood tissue that produces new blood cells. Hematopoietic tissue is also called *red bone marrow.*

❑ **10 Blood**—Blood tissue is a fluid matrix connective tissue characterized by a variety of cell types. Tiny **red blood cells (RBCs),** tinier fragments called **platelets,** and larger **white blood cells (WBCs)** may all be seen in a blood-smear preparation. Of course, blood is found within the circulation vessels and has many functions. Blood transports and exchanges materials, serves in immune protection of the body, and helps regulate body temperature, among other functions. We will study blood in more detail in Lab Exercise 34.

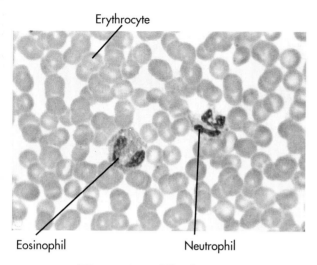

Figure 8-14 Blood tissue.

LANDMARK CHARACTERISTICS
A prepared blood smear is a drop of blood smeared on a slide and stained. The red blood cells are the more numerous, smaller cells stained pink to orange-red. RBCs have no nuclei. The white blood cells are much larger, with distinct, often distorted nuclei. Lab Exercise 34 has more complete details. See LABORATORY REFERENCE, Plate 24.

C. Gross tissue examples

Gross (large) examples of connective tissues often give one a more complete mental image of connective tissue types. If gross specimens are available, explore the texture, consistency, and other physical characteristics of the connective tissue listed.

SAFETY FIRST! Any animal tissue at room temperature can harbor growing colonies of microorganisms. As a safety precaution, *treat the samples as if they are contaminated* with harmful microbes. Use gloves or dissection tools when handling the samples and disinfect any surfaces that they touch.

Fill in the following table with the physical characteristics of different types of tissue.

Type	Physical characteristics
Dense fibrous (regular)	
Dense fibrous (irregular)	
Loose fibrous (areolar)	
Adipose	
Hyaline cartilage	
Fibrocartilage	
Elastic cartilage	
Compact bone	
Cancellous bone	

Connective Tissues

COLORING EXERCISE Using colored pens or pencils, shade in the figure and accompanying labels in contrasting colors of your choice as indicated by the red numerals.

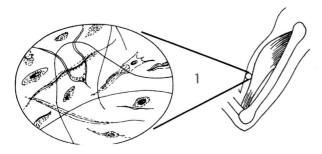
LOOSE, ORDINARY (AREOLAR)₁

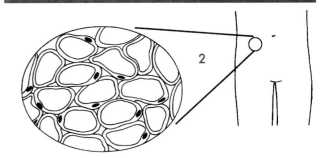
ADIPOSE₂

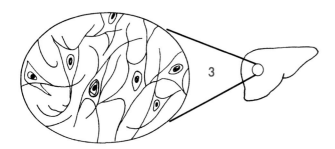
RETICULAR₃

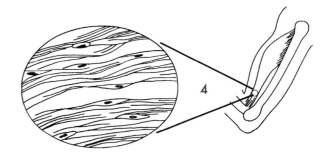
DENSE FIBROUS (REGULAR)₄

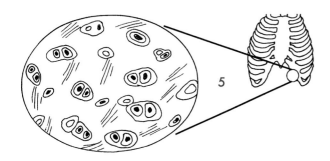
HYALINE CARTILAGE₅

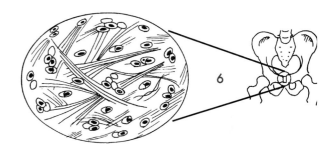
FIBROCARTILAGE₆

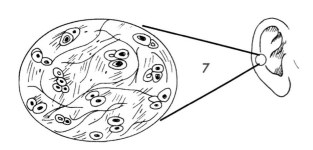
ELASTIC CARTILAGE₇

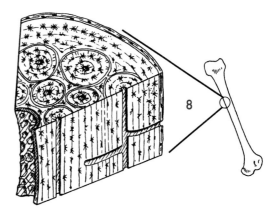
COMPACT BONE₈

Figure 8-15

NAME _____ DATE _____ SECTION _____

LAB REPORT 8

Connective Tissue

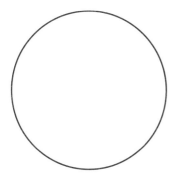

Specimen: *dense fibrous (regular) connective tissue*

Total Magnification: _____

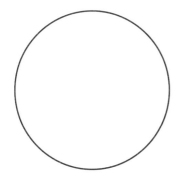

Specimen: *dense fibrous (irregular) connective tissue*

Total Magnification: _____

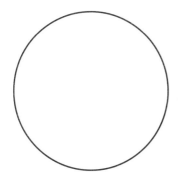

Specimen: *loose fibrous (areolar) tissue*

Total Magnification: _____

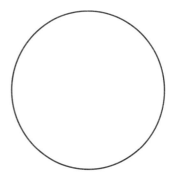

Specimen: *adipose tissue*

Total Magnification: _____

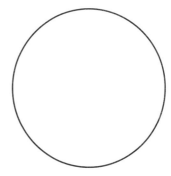

Specimen: *hyaline cartilage*

Total Magnification: _____

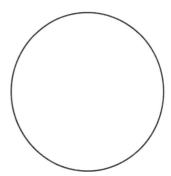

Specimen: *fibrocartilage*

Total Magnification: _____

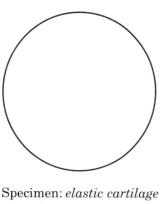

Specimen: *elastic cartilage*

Total Magnification: _____

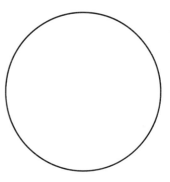

Specimen: *cancellous bone*

Total Magnification: _____

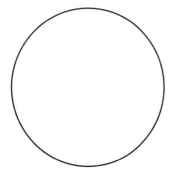

Specimen: *compact bone*

Total Magnification: _____

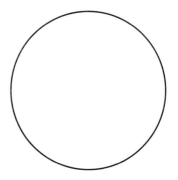

Specimen: *blood smear*

Total Magnification: _____

Figure 8-2

_____ 1
_____ 2
_____ 3
_____ 4
_____ 5

Figure 8-3

_____ 1
_____ 2
_____ 3
_____ 4
_____ 5
_____ 6

Matching

___ 1   ___ 6
___ 2   ___ 7
___ 3   ___ 8
___ 4   ___ 9
___ 5   ___ 10

Matching (may be used more than once or not at all)

a. dense fibrous connective tissue
b. adipose
c. hyaline cartilage
d. fibrocartilage
e. elastic cartilage
f. compact bone
g. cancellous bone

1. Cartilage type with a great deal of collagen in the matrix
2. Tissue type that is actually modified areolar tissue
3. Tissue type that stores lipid molecules
4. Tissue type composed of haversian systems
5. The most common type of cartilage
6. Strong tissue that forms tendons and ligaments
7. Tissue type associated with red bone marrow
8. Tissue type that forms strong membranes
9. Connective tissue that forms the disks between vertebrae
10. Tissue type that forms hard mineral trabeculae

LAB EXERCISE 9

Muscle and Nerve Tissue

Our exploration of the four basic human tissue types concludes with this study of muscle and nerve tissue. Detailed study of these tissue types is more appropriate for a later part of this course, but it is essential to achieve a basic understanding of these tissues now.

Before you begin

❏ Read the appropriate chapter in your textbook.

❏ Set your learning goals. When you finish this exercise, you should be able to
  - distinguish the three basic types of muscle tissue
  - identify the primary components of nerve tissue
  - recognize muscle and nerve tissue types in figures and specimens

❏ Prepare your materials:
  - microscope
  - prepared microslides:
    Skeletal muscle l.s.
    Cardiac muscle l.s.
    Smooth muscle l.s.
  - spinal cord smear

❏ Read the directions and safety tips for this exercise **carefully** before starting any procedure.

A. Muscle tissue

All three muscle tissue types contain cells with the ability to contract. Contraction in muscle tissue allows a muscle organ to pull with great force. Such pulling can move the skeleton or squeeze the contents of a hollow organ. In addition to this common function, all muscle tissues are composed of long, cylindrical cells called **muscle fibers.** Muscle fibers are living cells, unlike connective tissue fibers, which are nonliving protein fibers. Locations and functions of the major types of muscle tissue (as well as nervous tissue) are summarized in Table 9-1.

SAFETY FIRST! Observe the usual precautions when using the microscope and slides.

❏ 1 **Skeletal muscle** gets its name because it forms muscular organs that attach to the skeleton and move its parts. This tissue is also called *striated voluntary muscle* because it has striations (stripes) and can be controlled by the conscious mind. Skeletal muscle cells are large, fiberlike cells with fine cross stripes and many nuclei per cell. Skeletal muscle fibers and bundles of muscle fibers often have a coating of fibrous connective tissue.

LANDMARK CHARACTERISTICS
Skeletal muscle preparations include very large cells with the characteristic striped pattern. The striping is very fine, so good focusing technique is critical with this specimen. The muscle fibers are generally parallel to one another in a longitudinal section, with the stripes at right angles across each fiber. Unlike the other types of muscle, skeletal muscle fibers have many nuclei per cell. The nuclei are often up against the cell membrane. Some fibrous connective tissue may be seen between the muscle cells. See LABORATORY REFERENCE, Plate 25.

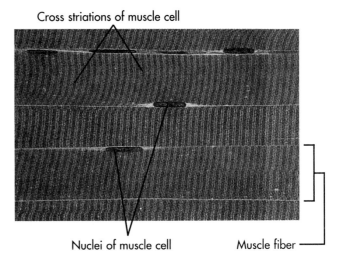

Figure 9-1 Skeletal muscle.

Table 9-1 Examples of major muscle and nerve tissues.

Tissue	Location	Function
MUSCLE		
Skeletal (striated voluntary)	Muscles that attach to bones Extrinsic eyeball muscles Upper third of esophagus	Movement of bones Eye movements First part of swallowing
Smooth (nonstriated, involuntary, or visceral)	In walls of tubular viscera of digestive, respiratory, and genitourinary tracts	Movement of substances along respective tracts
	In walls of blood vessels and large lymphatic vessels	Change diameter of blood vessels, thereby aiding in regulation of blood pressure
	In ducts of glands	Movement of substances along ducts
	Intrinsic eye muscles (iris and ciliary body)	Change diameter of pupils and shape of lens
	Arrector muscles of hairs	Erection of hairs (gooseflesh)
Cardiac (Striated involuntary)	Wall of heart	Contraction of heart
NERVOUS	Brain Spinal cord Nerves	Excitability Conduction

❑ **2 Cardiac muscle** is also known as *striated involuntary muscle*. This tissue also has striations, though less distinct than in skeletal muscle. Cardiac muscle is involuntary in the sense that subconscious mechanisms regulate its contraction. Cardiac muscle is found only in the walls of the heart. Because it must encircle and compress the heart chambers with great strength to pump blood, cardiac muscle requires some features not found in other muscle types. For example, the individual fibers are branched, allowing the fibers to mesh with other cells at different layers. Also, cardiac muscle fibers are fused end to end by **intercalated discs.** The branching and fusing gives a group of cells the ability to functionally imitate a giant cell encircling one or more chambers of the heart.

> **LANDMARK CHARACTERISTICS**
> Cardiac muscle striations are less distinct than those in skeletal muscle. Cardiac fibers have single nuclei, usually have branches, and do not taper at their ends. Many fibers attach end-to-end via intercalated disks, which appear as fine, dark (sometimes purple) lines at a right angle to a seemingly continuous fiber. See LABORATORY REFERENCE, Plate 26.

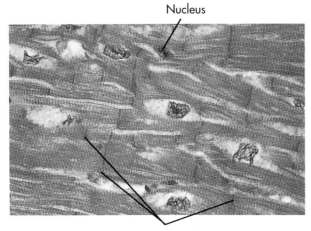

Figure 9-2 Cardiac muscle.

B. Nerve tissue

Nerve tissue composes organs of the nervous system: the brain, spinal cord, and nerves. Two basic types of cells are found in this tissue: **neurons** (impulse-conducting cells) and **glia** (support cells). Neurons are large cells with nucleated *bodies* and projections called *axons* and *dendrites*. There are many type of glia, or *neuroglia* as they are still sometimes called, but they are generally smaller than neurons, which they outnumber by several times. Glia surround and support neurons physically or biochemically.

❑ 3 **Smooth muscle** gets its name because it has no distinct striations. Like cardiac muscle, it also is an involuntary muscle type. Smooth muscle is found in the walls of hollow organs, such as digestive organs and blood vessels. This tissue type is composed of long, threadlike cells, each with a single nucleus. The cells are generally parallel with one another and with the edge of the wall in which they are imbedded.

LANDMARK CHARACTERISTICS
In a smear of spinal cord tissue, the neurons and glia are scattered randomly on the slide. The neurons are extremely large, each with a body having a single nucleus. Neuron projections crisscross throughout. Glia appear as tiny, dark dots. See LABORATORY REFERENCE, Plate 28.

LANDMARK CHARACTERISTICS
Unlike other muscle cells, smooth muscle fibers are unstriated and have single nuclei. In some preparations, the cells are pink and the nuclei purple or black. At first glance, you may confuse smooth muscle tissue with dense fibrous (regular) connective tissue. Smooth muscle is not wavy, like dense fibrous tissue sometimes is, and has a more even distribution of nuclei than dense fibrous tissue. See LABORATORY REFERENCE, Plate 27.

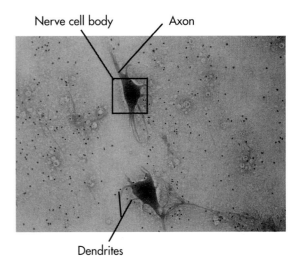

Figure 9-4 Nerve tissue: spinal cord smear.

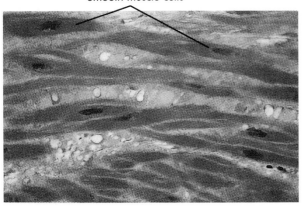

Figure 9-3 Smooth muscle.

Muscle and Nerve Tissues

COLORING EXERCISE Using colored pens or pencils, shade in the figure and accompanying labels in contrasting colors of your choice as indicated by the red numerals.

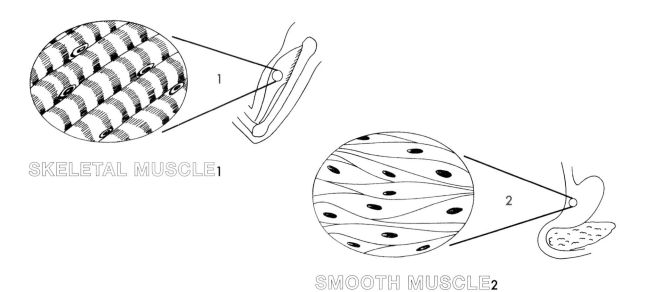

SKELETAL MUSCLE₁

SMOOTH MUSCLE₂

CARDIAC MUSCLE₃

NERVE TISSUE₄

Figure 9-5

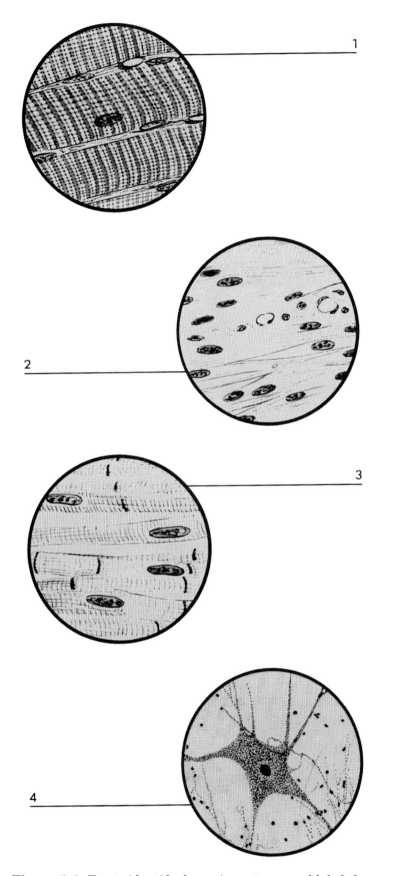

Figure 9-6 Try to identify these tissue types and label them.

NAME _____ DATE _____ SECTION _____

LAB REPORT 9

Muscle and Nerve Tissue

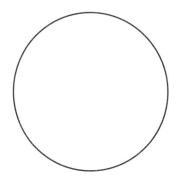

Specimen: *skeletal muscle*

Total Magnification: _____

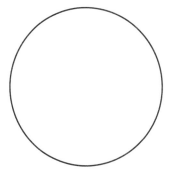

Specimen: *smooth muscle*

Total Magnification: _____

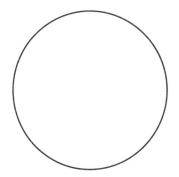

Specimen: *cardiac muscle*

Total Magnification: _____

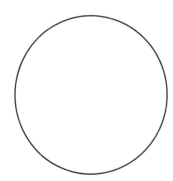

Specimen: *spinal cord smear*

Total Magnification: _____

Figure 9-6

_____ 1

_____ 2

_____ 3

_____ 4

Multiple choice

____ 1

____ 2

____ 3

____ 4

____ 5

____ 6

____ 7

____ 8

Multiple choice (only one response is correct)

1. A voluntary type of muscle tissue is
    a. skeletal
    b. cardiac
    c. smooth
    d. a and b are correct
    e. a and c are correct

2. An involuntary muscle tissue is
    a. skeletal
    b. cardiac
    c. smooth
    d. b and c are correct
    e. a and c are correct

3. Which of these muscle tissues is striated?
    a. skeletal
    b. cardiac
    c. smooth
    d. a and b are correct
    e. all are correct

4. Which of these is most likely to be found in the wall of the urinary tract?
    a. skeletal
    b. cardiac
    c. smooth

5. Cardiac muscle fibers are connected end to end by means of
    a. Velcro
    b. intercalated discs
    c. striations
    d. branching

6. Which of these cells is likely to have multiple nuclei?
    a. skeletal muscle
    b. cardiac muscle
    c. smooth muscle
    d. neuron
    e. a and c are correct
    f. a, c, and d are correct

7. Nerve tissue contains cells called
    a. fibers
    b. neurons
    c. glia
    d. striated cells
    e. b and c are correct

8. Nerve tissue forms the bulk of the
    a. brain
    b. heart
    c. digestive tract
    d. vertebrae
    e. a and d are correct

LAB EXERCISE 10

The Skin

The **skin** is the primary organ of the *integumentary system* and is the largest organ of the body. Forming the outer protective covering of the body, the skin is a continuous sheet of **cutaneous membrane.**

Before you begin

❑ Read the appropriate chapter in your textbook.

❑ Set your learning goals. When you finish this exercise, you should be able to
- describe the major structures of the skin and identify their functions
- identify important skin structures in a diagram, model, and prepared specimen
- compare and contrast features of thick skin and thin skin
- describe the gross structure of skin in a preserved mammalian specimen

❑ Prepare your materials:
- model or chart of a skin cross section
- microscope
- prepared microslides:
  Thin skin section
  Thick skin section
  Hair follicles c.s.
- preserved specimen: cat or fetal pig (unskinned)
- dissection tools and trays
- storage container (if specimen is to be reused)

❑ Read the directions and safety tips for this exercise **carefully** before starting any procedure.

A. Basic skin structure

First in a model, then in prepared microscopic specimens, identify the elements of skin structure described in the following steps.

> **SAFETY FIRST!** Don't forget the rules for safe use of the microscope.

❑ 1 The skin has two distinct layers. The superficial layer is a sheet of keratinized stratified squamous epithelium called the **epidermis.** The epidermis is itself divided into distinct histological regions, or **strata** (meaning "layers"):
- **Stratum basale** is the deepest stratum of the epidermis. It consists of a single sheet of columnar cells that continue to divide. As the daughter cells are formed, they are pushed upward, becoming part of the next stratum.
- **Stratum spinosum** is noted for its multilayer of distorted ("spined") cells. The cells become distorted as they are pushed up from the deeper stratum basale. Stratum basale and stratum spinosum together are often called *stratum germinativum.*
- **Stratum granulosum** is superficial to stratum spinosum. It contains flattened cells pushed up from the deeper strata. As the cells are pushed up through this stratum, they form the protein granules that give it the name *granulosum*. By the time the cells leave this stratum, they have died.
- **Stratum lucidum** (meaning "light layer") is a very thin layer present only in **thick skin.** Thick skin is found only in high-wear areas such as the palms and soles. The more flexible **thin skin** is found over most other areas of the body. This stratum's name comes about because it is translucent, allowing light to pass through it easily.
- **Stratum corneum** is the layer of dead, keratinized tissue already identified in Lab Exercise 7. Stratum corneum is extremely thick in thick skin, providing a great deal of protection. Stratum corneum protects deeper tissues from mechanical injury, inward or outward diffusion of water and other molecules, and invasion by microorganisms.

❑ 2 The layer of skin deep to the epidermis is a sheet of irregular fibrous connective tissue called the **dermis.** The dermis is usually much thicker than the epidermis. Like most connective tissues, the dermis has a scattering of blood vessels and nerves. The blood vessels supply both the dermis and epidermis. Because blood cools when it travels through the skin, the body varies the amount of blood sent to the skin to regulate loss of heat by the entire body. The dermis contains many sensory nerve endings. Sensations such as *heat, cold, touch,* and *pressure* are mediated by dermal nerve endings. There are two regions of the dermis:

■ The **reticular layer** of the dermis is a thick region of irregularly arranged protein fibers. Most of the fibers are collagenous, but a few are made of elastin.

■ The **papillary layer** is the bumpy superficial portion of the dermis attached to the epidermis. The bumps, called **papillae** (meaning "nipples"), form regular rows in thick skin but are rather irregularly arranged in thin skin. For this reason, thick skin can be observed to have distinct ridges, such as fingerprints. These ridges give the hands and feet better gripping ability.

❑ 3 Deep to the skin is a layer of **subcutaneous tissue,** sometimes called the **hypodermis** or **superficial fascia.** Although not a part of the skin, it is often studied along with skin. Subcutaneous tissue is loose, fibrous (areolar) connective tissue that connects the skin to underlying muscles and bone. Some of the areolar tissue has been modified to become adipose tissue. Adipose tissue's protective and insulating characteristics complement the protection and temperature regulation roles of the skin.

> **HINT** → LABORATORY REFERENCE, Plates 29-33 show light micrographs of skin structures. Plates 29 and 30 show thin skin, plate 31 thick skin, plate 32 skin with hair follicles and other accessory structures, and plate 33 thin skin with sweat glands.

B. Hair, nails, and glands

The skin has a variety of accessory structures, including **hair** and **nails.** Both hair and nails are modified forms of stratum corneum, or keratinized tissue. Hair is a cylinder of compact keratinized material, and a nail is a plate of compact keratinized material. Identify the structures described next in a model and in prepared specimens.

❑ 1 Each hair is formed within a separate **hair follicle.** The follicle is a sheathlike indentation of the epidermis. At the bottom of the follicle, a **hair papilla** covered with stratum germinativum produces the hair. The portion of each hair within the follicle is called the **hair root,** whereas the portion that has been pushed out of the follicle is called the **hair shaft.** The hair has a very dense cortex and a less dense medulla.

❑ 2 Attached to each follicle is an exocrine (ducted) **sebaceous gland.** This gland produces the fatty substance, **sebum,** that coats the hair and skin. Sebum prevents moisture loss and conditions the hair and skin so that they do not become brittle and easily broken.

❑ 3 The **arrector pili** muscle is a strap of smooth muscle tissue connecting the side of a follicle to the superficial surface of the dermis. When contracted, the muscle pulls the follicle so that it is nearly perpendicular to the skin's surface. This increases the air spaces among the hairs, improving its insulation quality. Contraction of the arrector pili also dimples the epidermis, raising a ridge at the edge of the follicle (a "goosepimple").

❑ 4 The toenail or fingernail is also formed by a modified portion of stratum germinativum. In the case of either hair or nail, this modified tissue is often called **matrix.** A portion of the **nail bed** (skin under the nail) is matrix that produces the nail plate. Part of the matrix may be visible through the nail as a pale crescent, or **lunula.** Nail formation begins under a fold of epidermis. The portion of the nail under the fold is the **root,** and the visible portion is the nail **body.** A **cuticle,** or **eponychium,** may extend from the fold onto the nail body.

❑ 5 **Sweat glands** are found in many areas of the skin. They are exocrine glands that produce a watery solution, *sweat,* that coats the skin. Sweat serves primarily to improve heat loss by the skin through evaporation. *Eccrine* sweat glands produce thin, watery sweat in many areas of the body. *Apocrine* sweat glands, found in the axillary and pubic regions, secrete a thicker sweat that is rich in complex organic molecules.

> **SAFETY FIRST!** Observe the usual precautions for dissection activities. Use gloves and take care to avoid injuries.

C. Dissection: cat integument

If you are going to use the cat as a dissection specimen throughout this course, it must first be skinned. This exercise not only directs you on how to perform this essential step, it also provides an

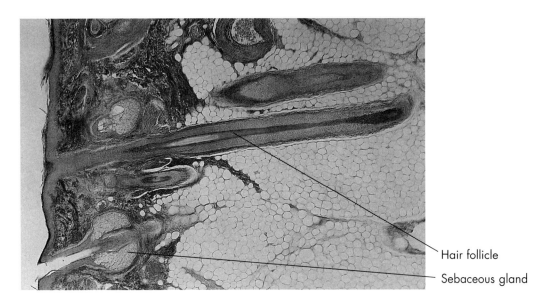

Figure 10-1 Hair follicle and sebaceous gland. 200×.

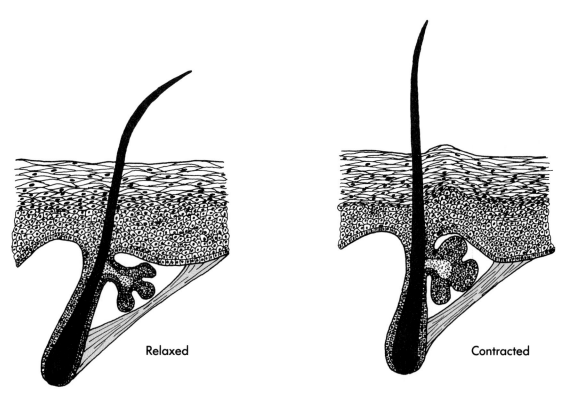

Figure 10-2 When the arrector pili contracts, it pulls the follicle and hair into a perpendicular position, improving the insulation quality of the hair. Notice how a "goosepimple" is raised around the follicle opening.

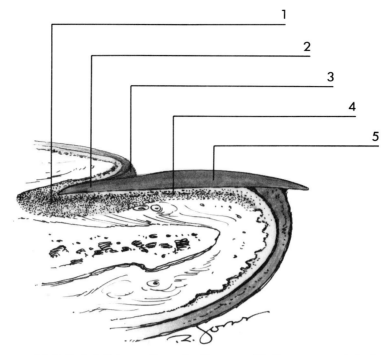

Figure 10-3 Label this section of a finger with the appropriate terms.

excellent opportunity to practice identifying surface anatomy—as well as to see "the big picture" of the integumentary system.

❏ 1 Examine the external aspect of your specimen:
  ■ Determine the anatomical orientation of the specimen. Which direction is anterior? posterior? Which direction is ventral? dorsal? Identify sagittal, transverse, and frontal planes in your specimen.
    ■ Identify these externally visible features:
      ■ **Pinna (auricle, or external ear)**
      ■ **External nares (nostrils)**
      ■ **Vibrissae (whiskers)**
      ■ **Integument (skin)**
      ■ **Forelimbs**
      ■ **Hindlimbs**
      ■ **Thoracic region**
      ■ **Abdominal region**
      ■ **Genitals (sex organs)**
      ■ **Anus**
      ■ **Tail**
  ■ Determine the sex of your specimen by examining the external genitals:
    ■ **Female**—Immediately anterior to the anus, on the ventral surface, is the **vulva** with an opening to the **vagina** and the **urethra.** The vulva is also called the *urogenital opening.*
    ■ **Male**—In the male, there is a pouch of skin immediately ventral to the anus. This pouch—the **scrotum**—contains the **testes,** or male primary sex organs. Anterior to the scrotum is the end of the **penis** with its **prepuce,** or skinfold covering. Locate the opening of the **urethra** in the penis.

❏ 2 Place the animal in the tray with its ventral surface facing you. Insert the tip of a scissors into the hole already present under the chin. This hole was used to inject the cat's vessels with latex. If you are using a cat that was not injected, then lift up a fold of skin and puncture it at the same site. Slide the bottom tip of the scissors into the **subcutaneous** area under the skin.

❏ 3 Begin cutting along the lines indicated in Figure 10-5. Notice that you must cut all the way around the neck, and the distal ends of each limb (just proximal to the feet, or paws). All the other cuts are to be made only on the ventral aspect of the cat. When you make the median cut along the posterior abdomen, toward the tail, *cut around* the genitals rather than through them. Be careful not to cut into the skeletal muscles under the skin.

❏ 4 With your forceps, pull the flaps of skin over the neck away from the cat's body. Notice the **areolar tissue** under the skin that is pulled apart as you remove the skin. Sometimes it helps if you scrape at the loose connective tissue under the skin with your scalpel as you peel the skin away. The skinning process is difficult unless you have patience and proceed slowly. Peel the other flaps of skin in a similar fashion until the cat is completely skinned.

The Skin

EPIDERMIS
STRATUM CORNEUM₁
STRATUM LUCIDUM₂
STRATUM GRANULOSUM₃
STRATUM SPINOSUM₄
STRATUM BASALE₅

DERMIS
PAPILLARY LAYER₆
RETICULAR LAYER₇
SWEAT GLAND₈
BLOOD VESSEL₉
NERVE₁₀

HAIR
ROOT₁₁
SHAFT₁₂
FOLLICLE₁₃
  PAPILLA₁₄
  MATRIX₁₅
ARRECTOR PILI₁₆
SEBACEOUS GLAND₁₇

HYPODERMIS₁₈

COLORING EXERCISE Using colored pens or pencils, shade in the figure and accompanying labels in contrasting colors of your choice as indicated by the red numerals.

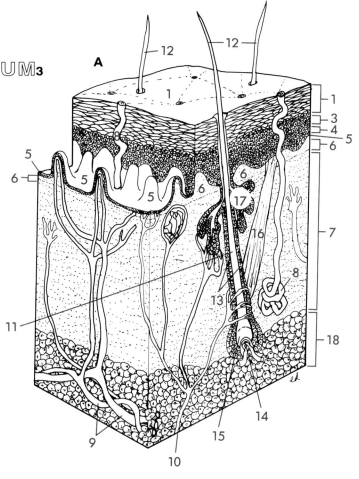
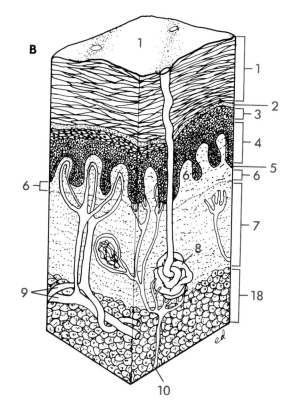

Figure 10-4

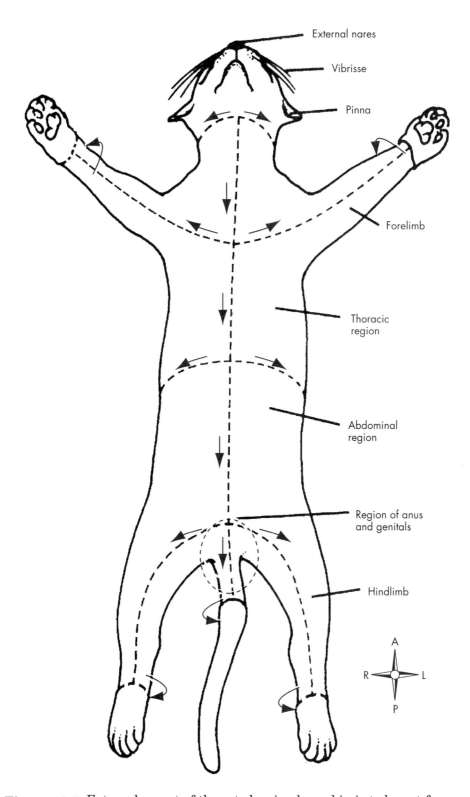

Figure 10-5 External aspect of the cat showing how skin is to be cut for removal.

❑ 5 Examine the skin, identifying these features:
- **Dermis**—The thick inner layer of the skin
- **Epidermis**—The thinner outer layer of the skin
- **Hypodermis**—The *subcutaneous tissue* under the skin proper, made of areolar and adipose tissue

❑ 6 Wrap the skin around the cat again and place the entire animal in its storage bag or other container provided.

> **HINT →** After each dissection activity later in this course, always wrap the skin around your specimen before storing it. This will help prevent your specimen from drying out—a common problem with preserved specimens.

D. Dissection: fetal pig integument

In this (and all succeeding) whole-animal dissection activities, the instructions for both cat and fetal pig specimens will be given. Both specimens are popular in anatomy and physiology courses.

❑ 1 Examine the external aspect of your specimen:
- Determine the anatomical orientation of the specimen. Which direction is anterior? posterior? Which direction is ventral? dorsal? Identify sagittal, transverse, and frontal planes in your specimen.
- Identify these externally visible features:
  - **Pinna (auricle, or external ear)**
  - **External nares (nostrils)**
  - **Integument (skin)**
  - **Umbilicus (umbilical cord)**
  - **Forelimbs**
  - **Hindlimbs**
  - **Thoracic region**
  - **Abdominal region**
  - **Nipples**
  - **Anus**
  - **Tail**
- Determine the sex of your specimen by examining the external genitals:
  - **Female**—Immediately anterior to the anus, on the ventral surface, is the **vulva** with an opening to the **vagina** and the **urethra.** The vulva is also called the *urogenital opening.*
  - **Male**—In the male fetal pig, there is a rather loose area of skin immediately posterior to the anus, perhaps even hiding the anus from view. Around the time of birth, each **testis** will descend from its position inside the body into the space under this skin. The skin will pouch out to form the outer wall of the **scrotum.** Just posterior to the umbilical cord is the distal end of the **penis** with its **prepuce,** or skinfold covering. Locate the opening of the **urethra** in the penis.

❑ 2 Place the animal in the tray with its ventral surface facing you. Pull up on the skin over the neck and puncture it with the tip of a scissors. Slide the bottom tip of the scissors into the **subcutaneous** area under the skin.

❑ 3 Begin cutting along the lines indicated in Figure 10-6. Notice that the posterior cuts are different for male and female specimens. Be careful not to cut into the skeletal muscles under the skin or through the base of the umbilical cord.

❑ 4 With your forceps, pull the two flaps of skin over the neck away from the animal's body. Notice the areolar tissue under the skin that is pulled apart as you remove the skin. Sometimes it helps if you scrape at the loose connective tissue under the skin with your scapel as you peel the skin away. The skinning process is difficult unless you have patience and proceed slowly. Pull the flaps of skin over the abdomen and over the **groin** area away in a similar fashion. If you have a male pig, leave the skin posterior to the umbilical cord in place for now. Cut the skin away from the body entirely.

❑ 5 Examine the skin, identifying these features:
- **Dermis**—Thick inner layer of the skin
- **Epidermis**—The thinner outer layer of the skin
- **Hypodermis**—The *subcutaneous tissue* under the skin proper, made of areolar and adipose tissue

> **HINT →** After each dissection activity later in this course, always wrap the skin around your specimen before storing it. This will help prevent your specimen from drying out—a common problem with preserved specimens.

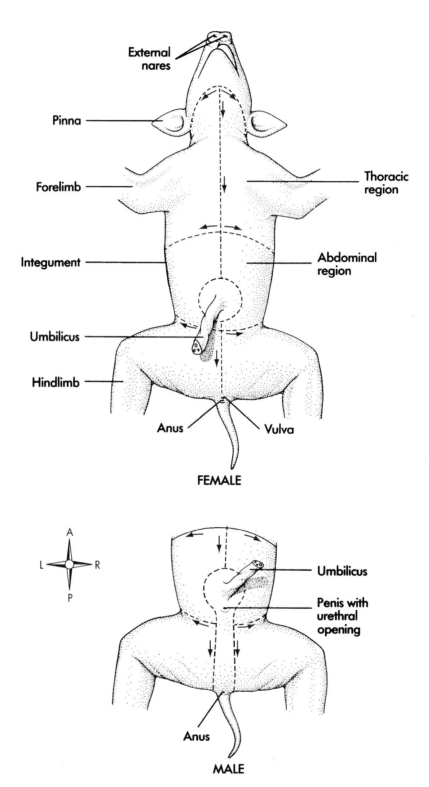

Figure 10-6 External aspect of the female (left) and male (right).

NAME _____ DATE _____ SECTION _____

LAB REPORT 10

The Skin

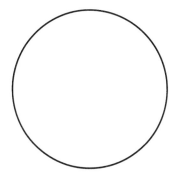

Specimen: *thin skin section*

Total Magnification: _____

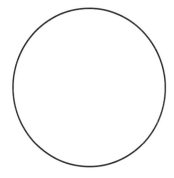

Specimen: *hair follicles c.s.*

Total Magnification: _____

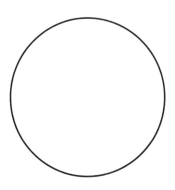

Specimen: *thick skin section*

Total Magnification: _____

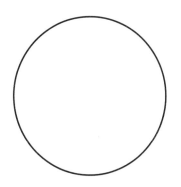

Specimen: _____

Total Magnification: _____

Figure 10-3

1. _____
2. _____
3. _____
4. _____
5. _____

Fill-in

1. _____
2. _____
3. _____
4. _____
5. _____
6. _____
7. _____
8. _____
9. _____
10. _____
11. _____
12. _____
13. _____
14. _____
15. _____
16. _____
17. _____
18. _____
19. _____
20. _____

Fill-in (complete each item with the correct term)

1. The ___?___ is the superficial layer (of two) in the skin.
2. The hypodermis is mainly areolar and ___?___ tissue.
3. Fingerprint ridges are formed as a result of the orderly arrangement of dermal ___?___.
4. Stratum corneum is composed of dead cell parts and a tough, waterproof protein called ___?___.
5. Stratum ___?___ is the epidermal stratum in which one could find many phases of mitosis.
6. Cells are pushed into stratum ___?___ from stratum basale.
7. When you sense cold air around you, nerve endings in the ___?___ are probably involved.
8. When you sense cold air around you, bumps may appear on the skin. This is caused by the contraction of the ___?___.
9. The subcutaneous tissue is also called the superficial fascia, or ___?___.
10. Physiologists believe that some sweat glands produce aromatic, organic molecules called *pheromones*. The type of sweat gland (of the two discussed in this exercise) most likely to produce pheromones is ___?___.
11. Strong shampoo tends to remove ___?___ from the hair and scalp, making it dry and easily damaged.
12. The eponychium, or ___?___, covers part of the nail plate.
13. When you have your hair cut, the portion of the hair trimmed is called the ___?___.
14. Blood vessels that supply the epidermis are found in the ___?___ region of the dermis.
15. Stratum basale and stratum spinosum together may be called stratum ___?___.
16. Keratinized ___?___ epithelium forms the epidermis.
17. The dermis is composed mainly of ___?___ fibers, with a few elastic fibers.
18. Stratum ___?___ is usually absent in thin skin.
19. ___?___ glands produce sebum.
20. The most widely distributed type of sweat gland is the ___?___ type of gland.

Table (fill in the spaces that are left blank)

Characteristic	Epidermis	Dermis
Tissue type		
Presence of blood vessels		
Relative thickness		
Permeability		
Relative strength		

Short answer (write a few complete sentences)

How does thin skin differ from thick skin?

LAB EXERCISE 11

Overview of the Skeleton

This exercise is the first of several that concerns the *skeletal system*. The skeletal system's major organs are the **bones** and **ligaments**. Ligaments are simply cords of regular dense fibrous tissue that bind the bones to one another. Bones are more complex in their structure, so we will spend some time investigating the nature of a typical bone. Before we move on to a detailed study of all the bones of the skeleton, we will survey the basic plan of the skeleton.

Investigation of the gross and microscopic structure of the typical bone and of the basic skeletal plan will be sound preparation for the exercises that follow.

Before you begin

❑ Read the appropriate chapter in your textbook.

❑ Set your learning goals. When you finish this exercise, you should be able to
- describe the organs of the skeletal system
- describe the gross and microscopic structure of bone tissue
- list the primary functions of the skeletal system
- outline the organization of the skeletal system

❑ Prepare your materials:
- microscope
- prepared microslides:
  Compact bone (ground bone) c.s.
  Cancellous bone
  Epiphyseal plate c.s.
- human skeleton (disarticulated)
- human skeleton (articulated)
- long bone (fresh, whole)
- whole long bone (fresh, l.s.)

❑ Read the directions and safety tips for this exercise **carefully** before starting any procedure.

> **HINT** → The concepts of this exercise will be easier to understand if you briefly review the microscopic organization of bone tissue presented in Exercise 8.

A. Bone types

The 206 bones in the standard human skeleton can be classified by their shapes. The best way to learn this classification scheme is by trying to classify the bones yourself, as outlined in the following steps.

❑ 1   Unpack the bones of a disarticulated (taken apart) human skeleton and spread the bones over your workplace.

❑ 2   Divide the group of bones into four piles, according to the categories given here. Do not use books or other aids to help you.
- **Long bones** are cylindrical bones that are longer than they are wide.
- **Short bones** are as long as they are wide, sometimes having an almost cuboidal shape.
- **Flat bones** arise when bone tissue invades and hardens fibrous membranes, so they are sheetlike in shape. They are usually curved, rather than absolutely flat.
- **Irregular bones** don't quite fit any of the other categories because of the complexity of their shapes.

❑ 3   Compare your results to the results of others in your lab section. Does everyone agree?

B. Gross structure of a bone

All bones have the same general structural pattern. Some bones have more "optional features" than other bones, some less. The long bone is often used as a general specimen for study because it has all the features that any bone can have.

> **SAFETY FIRST!** Because fresh animal tissues at room temperature can harbor dangerous bacterial colonies, specimens should only be handled when wearing disposable, nonporous gloves.

☐ 1 Obtain fresh long bone specimens from a large animal. One should be whole (uncut), and the other cut along its long axis (longitudinal section, or l.s.).

☐ 2 Examine the external aspect of a whole bone. Find the features described.
- **Ligament**—Although actually a separate organ, some bits of these fibrous straps that hold bones together may still be attached to your specimen.
- **Periosteum**—The periosteum is a sheet of irregular dense fibrous connective tissue continuous with the ligaments. It covers the shaft and part of the heads of a long bone. Try to scrape some of the periosteum away from the underlying bone. How strongly is it attached?
- **Articular cartilage**—The articular cartilage is a smooth cap of hyaline cartilage found where the bone *articulates* (forms a joint) with another bone. Joints, or connections between bones, are often movable. Which function of the skeletal system benefits by the presence of movable joints?
- **Diaphysis**—The diaphysis is the whole central shaft of the long bone. Only the external part of the shaft, made of solid bone tissue, is visible from the external aspect. For what skeletal functions is the hard shell of the diaphysis specialized?
- **Epiphysis**—The epiphyses are the "heads" of a long bone, one proximal to the diaphysis, one distal. Only the external portions are visible in a whole specimen.

☐ 3 Use the sectioned bone specimen to identify the structures listed.
- **Medullary cavity**—The medullary cavity, as its name implies, is a space within the center of the diaphysis. The walls surrounding the space are made of both cancellous and compact bone. In the adult, the cavity generally contains *yellow bone marrow,* which is a mass of fatty tissue. What is the purpose of yellow marrow?
- **Endosteum**—The endosteum is a thin epithelial membrane that lines the medullary cavity.
- **Cancellous bone**—The epiphyses like diaphyses, have a compact bone cortex, but its medulla is often different. The inside of each epiphysis has cancellous, or spongy, bone. The soft tissue in the spaces of the cancellous bone is often *red bone marrow,* which produces blood cells.

C. Microscopic structure of a bone

The microscopic structure of the long bone reflects the general nature of any type of bone. As you learned in Exercise 8, there are two basic types of bone tissue within a bone organ: **compact bone** and **cancellous bone.** In this exercise, you will build on what you learned about these bone tissue types. Then you will be able to integrate this information with what you already know about the gross structure of bone so that you can see "the big picture" of bone structure and function.

> **SAFETY FIRST!** Avoid electrical hazards while using the microscope. Be sure to exercise care in dealing with broken glass slides. Do not reach for slides or other objects while you are looking into the ocular, or you may knock over something (or someone).

☐ 1 **Compact bone**—Compact bone is found mainly in the hard, outer shell of a bone organ. Compact bone tissue is formed by solid, cylindrical units called **osteons** packed tightly together. The osteon, or **Haversian system,** consists of multiple concentric layers of hard bone matrix, with cells sandwiched between each layer. This bone matrix is made up of collagen fibers encrusted with crystals of a calcium-containing mineral called *apatite*. Each layer of bone matrix is a **lamella** (plural *lamellae*). **Osteocytes** are literally trapped within lacunae between the lamellae. The osteocytes were once active **osteoblasts** but have trapped themselves in the solid matrix they formed. Notice in Figure 11-1 that the **periosteum** that surrounds each bone is made up of an inner layer that contains active osteoblasts and an outer layer of dense fibrous connective tissue. The lamellae are centered around the **central (Haversian) canal**'s blood vessels. There are also transverse canals connecting the central canals of adjacent osteons. These transverse canals are called **transverse** or **Volkmann's canals.** They are also sometimes called **perforating canals.** The osteocytes trapped within lacunae transport materials to and from the canal by way of tiny **canaliculi** ("small canals") that connect the osteocytes to each other and to the canal. Observe a prepared slide of ground bone (compact bone) and try to identify as many features as possible.

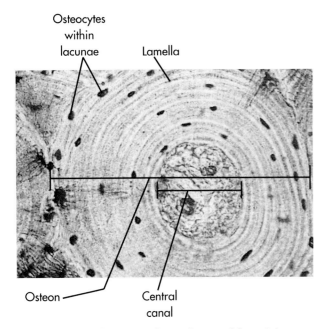

Figure 11-1 Compact bone (ground bone) in cross section. High power.

LANDMARK CHARACTERISTICS

A cross section of compact bone has rings of lamellae surrounding several adjacent Haversian, or *central,* canals. The lamellae resemble rings in an onion slice. The central canals are either clear or nearly black, the lamellae buff to orange, and the osteocytes brown or black. The canaliculi often appear as wavy hairlines radiating from the lacunae. See LABORATORY REFERENCE, Plate 22.

❏ 2 **Cancellous bone and hematopoietic tissue**—*Cancellous bone* is found in the inner portions of a bone organ. Cancellous bone is easily identified by its open, latticelike structure. Thin plates of bone matrix, with a scattering of osteocytes trapped within lacunae, form structural beams that have great strength despite the open spaces. These branching beams of hard bone are called **trabeculae.** Because cancellous bone has open spaces, it is sometimes called **spongy bone.** This name can be misleading because one might think spongy bone is as soft as a bath sponge; it is not soft at all because it has hard trabeculae. The spaces are filled with *hematopoietic* or *myeloid tissue,* a special type of blood tissue that produces new blood cells. Hematopoietic tissue is also called *red bone marrow.*

LANDMARK CHARACTERISTICS

Cancellous bone is distinguished by its rather disorganized array of trabecular beams of bone surrounded by myeloid tissue. The bone pieces may look like slivers of compact bone, with lamellae that often do not form complete circles. The myeloid tissue is a scattering of blood cells, which appear as tiny, dark circles. Myeloid (hematopoietic) tissue may also have a netlike formation of very thin collagen fibers called *reticular fibers*. In some preparations, the bone tissue is pink, and the myeloid cells are dark red. See LABORATORY REFERENCE, Plate 23.

❏ 3 **Epiphyseal plate**—Until a long bone has stopped growing in length, a layer of cartilage called the **epiphyseal plate** remains between each epiphysis and the diaphysis. During periods of growth, proliferation of epiphyseal cartilage cells brings about a thickening of this layer. **Ossification** (bone formation) of the additional cartilage nearest the diaphysis then follows; that is, osteoblasts make new bone matrix. As a result of this process, the bone becomes longer.

LANDMARK CHARACTERISTICS

The epiphyseal plate shown in Figure 11-2 is made up of four regions, each region in turn made of several layers of cells. The layer closest to the epiphysis (top of figure) are not changing or growing, so are said to be at rest (r). The letter p in the figure marks the *zone of proliferation,* which includes cells undergoing mitotic division. This is where the plate becomes thicker. The letter h in the figure marks the *zone of hypertrophy,* where older, enlarged cells degenerate before calcification occurs in the *zone of calcification* marked c in the figure. Below the zone of calcification, new cancellous bone can be seen. Typically, the epiphyseal plate is seen as a region of hyaline cartilage separated from a region of cancellous bone by a region of degenerating chondrocytes (cartilage cells) in lacuna that seem to be "stacked" in roughly parallel columns.

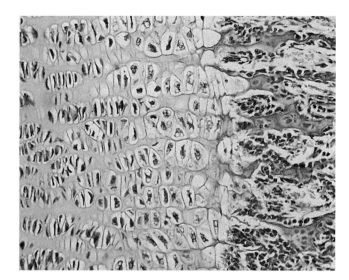

Figure 11-2 Detail of epiphyseal plate. Epiphyseal plate cartilage at right transforms into zones of proliferating chondrocytes with primary ossification occurring on their calcified remnants. Newly formed bone appears at left (50×).

D. The plan of the skeleton

The usual number given for bones in the human skeleton is 206. This is by no means the absolute normal number, however. Most people have more bones, but each person has different types, locations, and numbers of "extra" bones. Some people may be missing a bone or two. In this activity, you will examine both the standard 206 bones and the extra bones that may be present.

❏ 1 Obtain an articulated (connected) human skeleton.

❏ 2 The standard **axial skeleton** consists of 80 bones that form the central axis of the skeleton. These 80 bones include 28 skull bones, 1 unattached bone in the throat, 26 vertebrae, and 25 rib cage bones. Locate the bones of the axial skeleton in your specimen. Do not worry about learning the names of individual bones now. That will come later. For now, concentrate on "the big picture" of skeletal organization.

> **SAFETY FIRST!** Be cautious when handling the articulated skeleton. The bones or mounting hardware may become loose and fall from the support frame, injuring you or your lab mates.

❏ 3 Locate the bones of the **appendicular skeleton** in your specimen. The appendicular skeleton, comprising the 126 nonaxial bones, includes the bones of the appendages, or extremities (arms and legs). Sixty-four of these bones are in the *upper extremities* (shoulders and arms). Sixty-two bones are in the *lower extremities* (hips and legs).

❏ 4 Ask your lab instructor if there are any extra standard bones or any standard bones missing in your specimen. What difficulties could such differences have caused the individual during life?

❏ 5 Determine whether your specimen has any of the extra bone types typically found in skeletons:
- **Sesamoid bones** are so called because they resemble sesame seeds: tiny rounded specks. Sesamoid bones are often found within tendons of the hand and foot.
- **Wormian bones,** also called *sutural bones,* are flat bones that form in the sutures (joints) between the cranial bones of the skull.

E. Bone markings and features

As you have already observed on your specimen, bones do not generally have a smooth surface. There are many bumps, holes, and projections on the bones of the human skeleton. These *bone markings* are named with terms that describe their shape and location. As a preview to the next few exercises, review the terms used to name bone markings listed in Table 11-2.

Long Bone

GROSS STRUCTURE

EPIPHYSIS₁
EPIPHYSEAL PLATE₂
DIAPHYSIS₃
PERIOSTEUM₄
ARTICULAR CARTILAGE₅
MEDULLARY CAVITY₆
ENDOSTEUM₇
CANCELLOUS BONE₈
COMPACT BONE₉
BLOOD VESSEL₁₀

COLORING EXERCISE Using colored pens or pencils, shade in the figure and accompanying labels in contrasting colors of your choice as indicated by the red numerals.

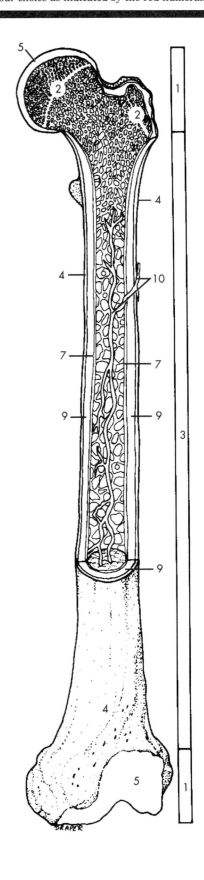

Figure 11-3

Table 11-1 Bones of Skeleton (206 total)*

Axial skeleton (80 bones total)		Appendicular skeleton (126 bones total)	
Part of body	**Name of bone**	**Part of body**	**Name of bone**
Skull (28 bones total)		**Upper extremities** (including shoulder girdle) (64 bones total)	Clavicle (2)
Cranium (8 bones)	Frontal (1)		
	Parietal (2)		
	Temporal (2)		Scapula (2)
	Occipital (1)		Humerus (2)
	Sphenoid (1)		Radius (2)
	Ethmoid (1)		Ulna (2)
Face (14 bones)	Nasal (2)		Carpals (16)
	Maxillary (2)		Metacarpals (10)
	Zygomatic (malar) (2)		Phalanges (28)
	Mandible (1)	**Lower extremities** (62 bones total)	Coxal bones (2)
	Lacrimal (2)		
	Palatine (2)		Femur (2)
	Inferior conchae (turbinates) (2)		Patella (2)
			Tibia (2)
	Vomer (1)		Fibula (2)
Ear bones (6 bones)	Malleus (hammer) (2)		Tarsals (14)
	Incus (anvil) (2)		Metatarsals (10)
	Stapes (stirrup) (2)		Phalanges (28)
Hyoid bone (1)			
Spinal column (26 bones total)	Cervical vertebrae (7)		
	Thoracic vertebrae (12)		
	Lumbar vertebrae (5)		
	Sacrum (1)		
	Coccyx (1)		
Sternum and ribs (25 bones total)	Sternum (1)		
	True ribs (14)		
	False ribs (10)		

*Excluding variable sesamoid and wormian bones.

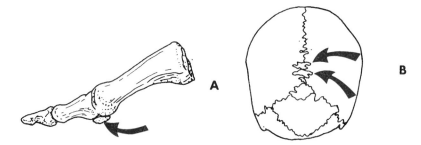

Figure 11-4 A, Sesamoid bone at the base of the thumb. **B,** Wormian bones along a suture joint.

Table 11-2

Marking	Meaning
Angle	A corner
Body	The main portion of a bone
Condyle	Rounded bump; usually fits into a fossa on another bone, forming a joint
Crest	Moderately raised ridge; generally a site for muscle attachment
Epicondyle	Bump near a condyle; often gives the appearance of a "bump on a bump;" for muscle attachment
Facet	Flat surface that forms a joint with another facet or flat bone
Fissure	Long, cracklike hole for blood vessels and nerves
Foramen	Round hole for vessels and nerves (pl. *foramina*)
Fossa	Depression; often receives an articulating bone (pl. *fossae*)
Head	Distinct epiphysis on a long bone, separated from the shaft by a narrowed portion (or neck)
Line	Similar to a crest but not raised as much (is often rather faint)
Margin	Edge of a flat bone or flat portion of an irregular bone
Meatus	Tubelike opening or channel (pl. *meati*)
Neck	A narrowed portion, usually at the base of a head
Notch	A V-like depression in the margin or edge of a flat area
Process	A raised area or projection
Ramus	Curved portion of a bone, like a ram's horn (pl. *rami*)
Sinus	Cavity within a bone
Spine	Similar to a crest but raised more; a sharp, pointed process; for muscle attachment
Sulcus	Groove or elongated depression (pl. *sulci*)
Trochanter	Large bump for muscle attachment (larger than tubercle or tuberosity)
Tubercle	Smaller version of a tuberosity
Tuberosity	Oblong, raised bump, usually for muscle attachment

Overview of the Skeleton

AXIAL SKELETON₁
APPENDICULAR SKELETON₂

COLORING EXERCISE Using colored pens or pencils, shade in the figure and accompanying labels in contrasting colors of your choice as indicated by the red numerals, then label the bones indicated by numbers 3-19.

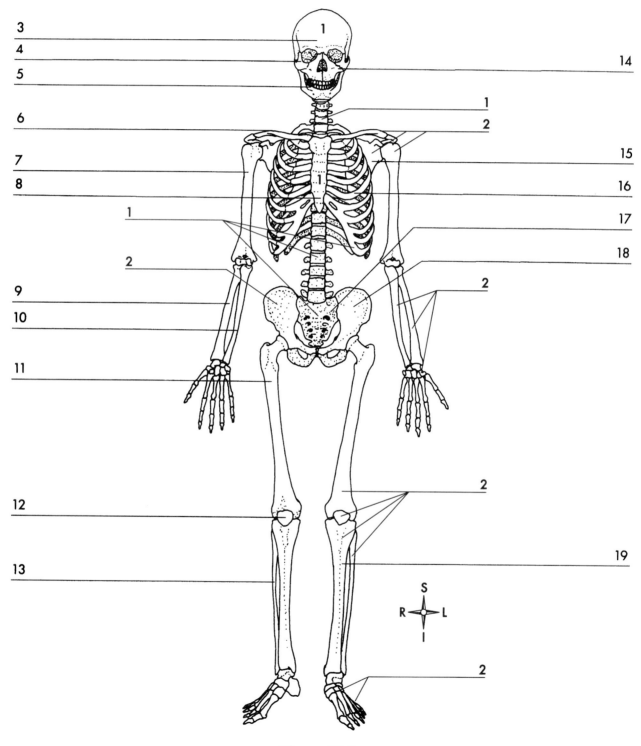

Figure 11-5

LAB REPORT 11

Overview of the Skeleton

Multiple choice

1. _____
2. _____
3. _____
4. _____
5. _____
6. _____

Figure 11-5

3. _____
4. _____
5. _____
6. _____
7. _____
8. _____
9. _____
10. _____
11. _____
12. _____
13. _____
14. _____
15. _____
16. _____
17. _____
18. _____
19. _____

Multiple choice (only one response is correct)

1. The inner lining of the medullary cavity is
    a. made of compact bone
    b. called the *endosteum*
    c. called the *periosteum*
    d. a and c are correct

2. Which of these tissues is present in a typical long bone?
    a. blood tissue
    b. cancellous bone
    c. compact bone
    d. dense fibrous tissue
    e. hyaline cartilage
    f. all of the above

3. In the coloring figure of the long bone, the epiphyseal plate is shown. What is the reason for its presence?
    a. it is scar tissue from a previous fracture
    b. it is an area of growth between the epiphysis and diaphysis during bone development
    c. it is callous tissue from overuse of the bone
    d. it is the site of a current fracture

4. The human skeleton functions to
    a. produce blood tissue
    b. store fat and minerals
    c. protect vital organs
    d. allow movement of the body
    e. provide a supporting framework
    f. all of the above

5. Your physician has just informed you that you have 40 bones in your skull. This means
    a. you have the standard number of skull bones
    b. you have some sesamoid bones in your skull
    c. you have some sutural bones in your skull
    d. you are missing some skull bones

6. The same physician tells you that all of your knee ligaments have been severed. This means that
    a. your leg bones are not being held together very well
    b. your leg muscles have become separated from the bone
    c. your patella (kneecap) is fractured
    d. your femur (thigh bone) is fractured

Fill-in

1. _____
2. _____
3. _____
4. _____
5. _____
6. _____
7. _____
8. _____
9. _____
10. _____
11. _____
12. _____
13. _____
14. _____
15. _____

Axial bones

1. _____
2. _____
3. _____
4. _____
5. _____
6. _____
7. _____

Appendicular bones

1. _____
2. _____
3. _____
4. _____
5. _____
6. _____
7. _____
8. _____

Fill-in (complete each item with the correct term)

1. Tiny round specks of bone found in a tendon are often called __?__ bones.
2. The fibrous covering of a long bone is called the __?__.
3. The shaft portion of a long bone is termed the __?__ of the bone.
4. The head region on the end of a long bone shaft is the __?__.
5. The __?__ on the outside of the long bone is made of hyaline cartilage.
6. Yellow bone marrow is made of __?__ tissue.
7. A bone that is as long as it is wide is classified as a __?__ bone.
8. When classified according to shape, the pelvic bone is considered to be __?__.
9. A human skeleton that is taken apart is called a(n) __?__ skeleton.
10. There are __?__ bones in a *standard* human skeleton.
11. The bones of the upper and lower extremities compose the __?__ skeleton.
12. Red bone marrow is associated with __?__ bone tissue.
13. The __?__ cartilage articulates with another bone or bone process.
14. When classified according to shape, the femur is a(n) __?__ bone.
15. The __?__ is the central space of the long bone.

Skeletal plan

Proceed through this list of bones, beginning at the top, and sort them according to how they fit in the skeletal plan (axial or appendicular). You may need to refer to your textbook or later exercises in this manual.

femur

fifth metatarsal

humerus

hyoid

mandible

occipital

patella

pelvic (coxal)

radius

rib

sacrum

scapula

second thoracic vertebra

sternum

tibia

LAB EXERCISE 12

The Skull

This exercise is the first of three exercises that challenge you to learn the bones and important markings of the human skeleton.

The **skull** is the superior portion of the *axial skeleton*. For ease of study, the 28 bones of the skull are divided into three categories. The **cranial bones** form a roughly spherical case for the brain called the *cranium*. The **facial bones** include most of the remaining skull bones, except the six **auditory ossicles** of the middle ear. In this exercise, all 28 skull bones are presented for study.

Before you begin

❑ Read the appropriate chapter in your textbook.

❑ Set your learning goals. When you finish this exercise, you should be able to
- distinguish between cranial bones and facial bones of the skull
- identify all the bones of the skull and their important markings in a specimen or figure
- describe the important features of a fetal skull
- name the major joints of the skull and identify them in a specimen or figure

❑ Prepare your materials:
- atlas or chart of the skull (or your textbook)
- articulated human skull (removable top)
- articulated fetal skull
- demonstration pointer

❑ Read the directions and safety tips for this exercise **carefully** before starting any procedure.

> HINT → Refer to the LABORATORY REFERENCE, Plates 34 to 37, for color photographs of the skull's bones and markings. Figures in your textbook will also be useful.

A. Bones of the cranium

Eight of the 28 skull bones form the cranium of the skull. Locate, in a specimen and in a figure, each bone and marking listed, noting the distinguishing characteristics of each.

As with all bones and markings studied in this course, try to understand the structural and functional relationships with surrounding structures.

> **SAFETY FIRST!** Use a disposable, plastic pipette (dropper), a pipe cleaner, or another soft *demonstration pointer* to study the features of the skeleton. If you use a pen or pencil, you may mark the skull or damage its delicate parts.

❑ 1 The single **frontal** bone forms the anterior third of the cranial dome. Locate these features on your specimen:
- **Supraorbital foramen**—Hole on the superior edge of the orbit
- **Frontal sinus**—Air space on the midline, between and superior to the orbits

❑ 2 The left and right **parietal** bones form the middle segment of the cranial dome, joined with each other along the midline by the **sagittal suture.** They form a **coronal suture** with the frontal bone and a **lambdoidal suture** with the occipital bone.

❑ 3 The **sphenoid** bone is a butterfly-shaped bone that forms part of the anterior floor and sides of the cranium. The **body** is the hollow, cubelike central portion. Locate some of the sphenoid bone's other features:
- **Greater wings**—Lateral, winglike extensions of the body that form part of the lateral wall of the orbit (eye socket) wall
- **Lesser wings**—Narrow, winglike extension of the upper part of the body that form the posterior part of the orbit's roof

- **Optic foramen**—Round hole (for the optic nerve) through the back wall of the orbit
- **Superior orbital fissure**—Cracklike opening lateral to the optic foramen
- **Inferior orbital fissure**—Cracklike opening in the inferior back wall of the orbit
- **Foramen rotundum**—Round opening seen from the inside of the skull, anterior to the foramen ovale
- **Foramen ovale**—Oval opening seen from the inside of the skull, posterior to the foramen rotundum
- **Sella turcica**—Literally "turkish saddle," a double projection on the superior aspect of the sphenoid
- **Sphenoid sinus**—Air space inferior to the sella turcica, within the body of the sphenoid

❏ 4 The paired **temporal** bones are at the sides of the cranium, extending inward to form part of the cranial floor. Find these temporal structures:
- **Zygomatic process**—Narrow, bridgelike projection of the temporal bone—articulates with the temporal process of the zygomatic bone
- **Mastoid process**—Large bump posterior and inferior to the external auditory meatus
- **Styloid process**—Thin process anterior to the mastoid process
- **External auditory meatus**—Tubelike opening of the ear canal
- **Petrous portion**—Can only be seen from the inside of the skull, it looks like a "rocky cliff" anterior to the occipital bone
- **Squamosal suture**—Joint along the top, curved edge of the temporal bone
- **Mandibular fossa**—Depression that receives the condyle of the mandible
- **Temporomandibular joint (TMJ)**—Joint formed by the mandibular fossa and the mandibular condyle
- **Internal auditory meatus**—Tubelike opening in the petrous portion
- **Jugular foramen**—Round hole at the junction of the petrous portion and the occipital bone
- **Carotid foramen**—Canal through the petrous portion, slightly medial to the jugular foramen

❏ 5 The **occipital** bone forms the posterior portion of the cranial dome, curving inferiorly to the base of the cranium. Locate these important features:
- **Foramen magnum**—Very large hole for the spinal cord
- **Occipital condyle**—Flattened bump lateral to the foramen magnum

❏ 6 The **ethmoid** bone forms the middle portion of the anterior cranial floor, extending inferiorly between the eye *orbits* to also form the roof of the nasal cavity. Find these important ethmoid structures:
- **Ethmoid sinuses (air cells)**—Air spaces within the main portion of the bone
- **Perpendicular plate**—Thin plate projection downward along the midsagittal plane, it forms the upper nasal septum
- **Crista galli**—Literally "cock's comb," it is a crest on the superior midline of the ethmoid
- **Superior nasal concha (turbinate)**—Superior of three curved projections from the nasal cavity's lateral wall
- **Middle nasal concha (turbinate)**—Middle of three curved projections from the nasal cavity's lateral wall
- **Cribriform plate**—Thin, perforated plate forming the ethmoid's superior surface
- **Olfactory foramina**—Tiny holes in the cribriform plate (for olfactory nerves from the upper nasal cavity)

B. Bones of the face

Fourteen bones of the skull are classified as facial bones. Find each facial bone and the markings listed in a specimen and figure.

❏ 1 The two **maxillae,** or maxillary bones, are also called the upper jaw bones. They support the face from the eyes down to the mouth, across the front of the cheek. Locate these maxillary features:
- **Palatine process**—Platelike posterior projection of the upper jaw, forms the anterior part of the hard palate
- **Incisive foramen**—Anterior hole in the palatine process, near the body's midline
- **Alveolar process**—Ridge in which the teeth of the upper jaw are anchored
- **Infraorbital foramen**—Hole on the anterior "cheek," below the orbit
- **Maxillary sinus**—Air space within the maxilla's "cheek" portion

❏ 2 Each of the two **lacrimal bones** are on the medial margin of an eye orbit, between the ethmoid bone and an upward projection of the maxilla. Each has a sulcus where a tear duct is located (in life).

❏ 3 The left and right **nasal** bone are joined at the midline, forming the superior margin of the nasal opening.

❏ 4 The two **palatine** bones join together at the midline to form the posterior third of the *hard palate* (roof of the mouth). Locate these features:

- **Transverse palatine suture**—Joint between the palatine bone and the maxilla's palatine process
- **Greater palatine foramen**—Larger, anterior of two holes in each palatine bone

❏ 5 The **inferior nasal conchae,** or turbinate bones, are a pair of thin, curved bones. They project medially from the lateral walls of the nasal cavity, curving toward the nasal floor.

❏ 6 The **mandible,** or lower jaw bone, is an oddly shaped bone forming the lower part of the face. Locate these elements of the mandible:
- **Mandibular arch (notch)**—Large, U-shaped curve of the superior edge of the ramus
- **Mandibular condyle**—Rounded, posterior of two projections on the superior part of the ramus
- **Coronoid process**—Anterior of two superior projections of the ramus
- **Alveolar process**—Ridge along which the teeth are anchored
- **Ramus**—The posterior arms of the mandible that angle upward
- **Angle**—The corner formed where the ramus begins its upward angle
- **Body**—The entire anterior portion of the mandible, to which the rami are attached
- **Mandibular foramen**—Hole on the inside surface of the ramus, near the angle
- **Mental foramen**—Hole on the outside surface of the body, just lateral to the midline

❏ 7 Each of two **zygomatic** bones form the upper lateral corner of a cheek, from the lower eye orbit around to the temporal bone. Locate:
- **Temporal process**—Wedge-shaped process that joins the zygomatic process of the temporal bone
- **Zygomatic arch**—Complete bridgelike structure formed by the zygomatic process of the temporal bone and the temporal process of the zygomatic bone

❏ 8 The single **vomer** forms the lower portion of the *nasal septum* that divides the nasal cavity.

C. Auditory ossicles

Each middle ear, within the petrous portion of the temporal bone, contains three auditory ossicles. They are listed here, but we will postpone studying them until Lab Exercise 28. (They are not visible in your specimen.)

❏ 1 The **malleus** is also called the *hammer.*

❏ 2 The **incus** is also called the *anvil.*

❏ 3 The **stapes** is known as the *stirrup.*

D. The fetal skull

The fetal skull features partly ossified skull bones and a large proportion of fibrous and cartilaginous tissue. Because the flat bones of the cranium have not met to form sutures, there are six fibrous areas called **fontanels.** Identify each in a fetal skull specimen or figure.

❏ 1 The two **anterolateral,** or sphenoid, fontanels are at the future junction of the sphenoid, temporal, parietal, and frontal bones.

❏ 2 The two **posterolateral,** or mastoid, fontanels are at the junction of the temporal, parietal, and occipital bones.

❏ 3 The **anterior,** or coronal, fontanel can be found where the right and left parietal bones are to meet the frontal bone.

❏ 4 The **posterior** fontanel forms where the parietal bones are to meet the occipital bone.

The Skull

FRONTAL 1
PARIETAL 2
OCCIPITAL 3
TEMPORAL 4
SPHENOID 5
ETHMOID 6
MAXILLA 7

LACRIMAL 8
NASAL 9
PALATINE 10
INFERIOR NASAL CONCHA 11
MANDIBLE 12
ZYGOMATIC 13
VOMER 14

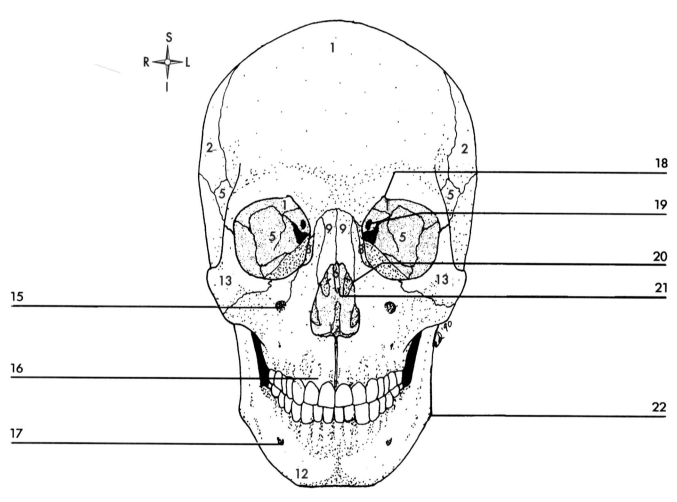

Figure 12-1

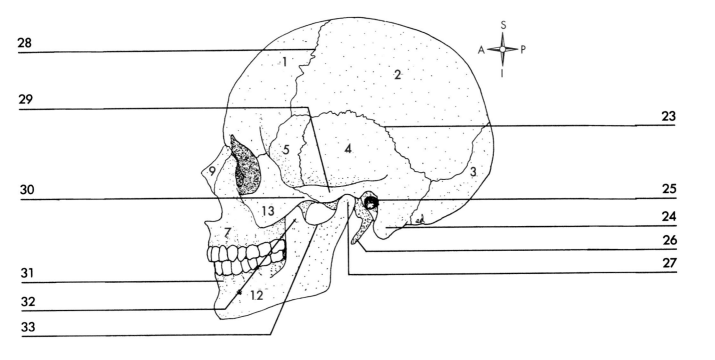
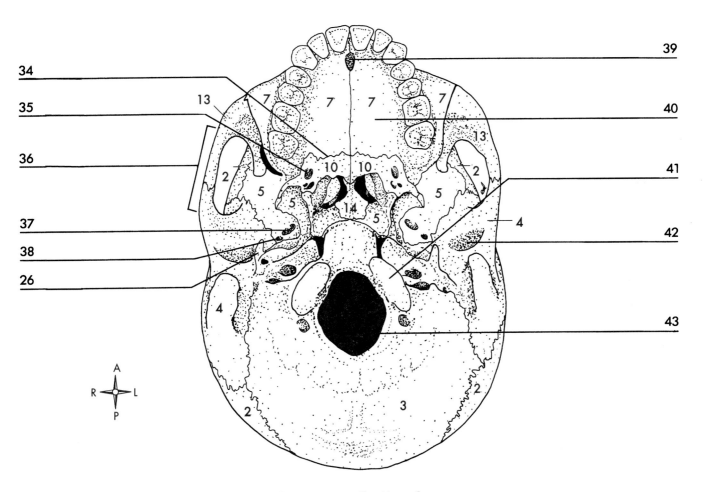

Figure 12-1, Continued.

113

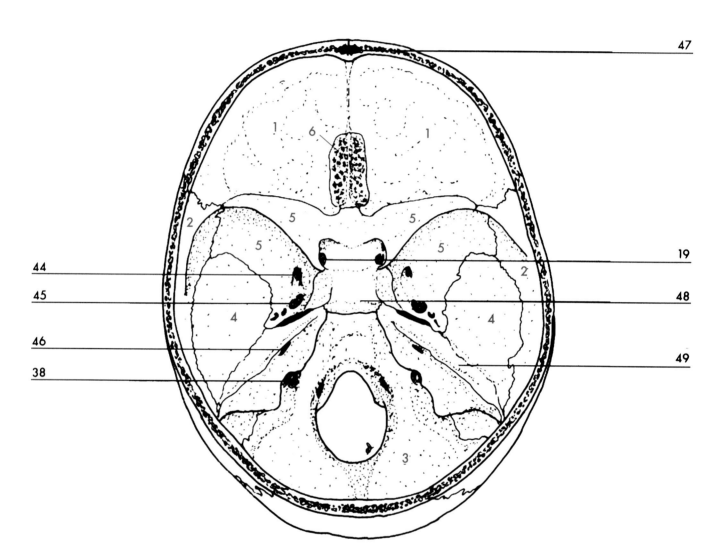

Figure 12-1, Continued.

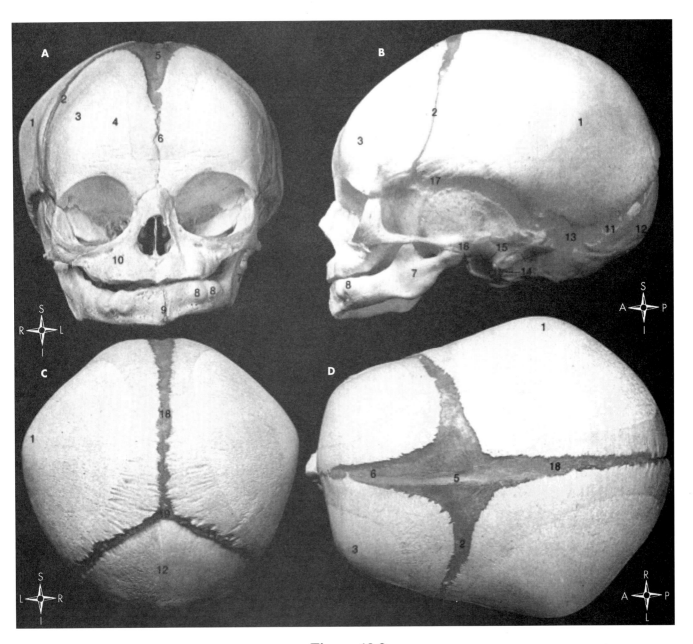

Figure 12-2

115

NAME_____ DATE_____ SECTION_____

LAB REPORT 12

The Skull

Figures 12-1 and 12-2

(Note: Numbers 1 through 14 are used for coloring labels.)

_____ 15
_____ 16
_____ 17
_____ 18
_____ 19
_____ 20
_____ 21
_____ 22
_____ 23
_____ 24
_____ 25
_____ 26
_____ 27
_____ 28
_____ 29
_____ 30
_____ 31
_____ 32
_____ 33
_____ 34
_____ 35
_____ 36
_____ 37
_____ 38
_____ 39
_____ 40
_____ 41
_____ 42
_____ 43
_____ 44
_____ 45
_____ 46
_____ 47
_____ 48
_____ 49

List the bones of the cranium

_____ 1
_____ 2
_____ 3
_____ 4
_____ 5
_____ 6

Name the facial bones

_____ 1
_____ 2
_____ 3
_____ 4
_____ 5
_____ 6
_____ 7
_____ 8

Name the auditory ossicles

_____ 1
_____ 2
_____ 3

Name the major fontanels

_____ 1
_____ 2
_____ 3
_____ 4

Sutures (determine the name of the suture joint described)

1. Joins the parietal bones together
2. Joins the superior margin of the temporal bone to the frontal, parietal, and occipital bones
3. Joins the palatine bone to the maxilla
4. Joins the frontal bone to the two parietal bones

Markings (give the name of the bone marking or feature described in each item)

1. Large process of the temporal bone, just posterior to the external auditory meatus; it contains sinuses
2. A smaller, needle-shaped process just medial to the process described in the previous statement
3. A hole in the sphenoid bone that allows the optic nerve to exit the eye orbit
4. Same name for ridgelike processes on both the maxilla and mandible in which the teeth are embedded
5. A crest or projection on the superior surface of the ethmoid bone
6. A curved plate of bone projecting from the lateral wall of the nasal cavity, just above the inferior nasal conchae (part of the ethmoid bone)
7. One of two holes on the anterior portion of the mandibular body
8. A hole in the maxilla just below the orbit of the eye
9. The upper portion of the nasal septum is formed by this part of the ethmoid
10. This structure is formed by both the zygomatic process of the temporal bone and the temporal process of the zygomatic bone

Anatomical relationships (use a directional term to complete each item correctly)

1. The frontal bone is __?__ to the occipital bone.
2. The lacrimal bones are on the __?__ margin of the orbit.
3. The mandible is mostly __?__ to the maxilla.
4. The occipital condyles are __?__ to the foramen magnum.
5. The palatine bones are __?__ to the maxilla.
6. The superior conchae are __?__ to the nasal septum.
7. The mandibular condyle is __?__ to the coronoid process.
8. The parietal bones are __?__ to the frontal bone.
9. The coronal fontanel is __?__ to the posterior fontanel.
10. The incisive foramen is __?__ to the transverse palatine suture.

LAB EXERCISE 13

The Vertebral Column and Thoracic Cage

As you recall, the axial skeleton is composed of the skull, the vertebral column, the thoracic cage, and one unattached bone. All these structures form the central core, or *axis,* of the skeleton. In the previous exercise, we explored the skull in some detail. This exercise presents the remainder of the axial skeleton.

The **vertebral column** is a set of 26 bones stacked one on another to form a slightly curved, flexible support rod. The 25 bones of the rib cage, or **thoracic cage,** protect the lungs and heart within the thorax. The **hyoid** bone is not attached to any other bone but is very close to the skull and vertebral column and lies along the central axis of the body.

Before you begin

❑ Read the appropriate chapter in your textbook.

❑ Set your learning goals. When you finish this exercise, you should be able to
- name the component bones of the vertebral column and thoracic cage
- identify the bones and markings of the vertebral column and thorax on a specimen and in figures

❑ Prepare your materials:
- human skeleton (disarticulated)
- human skeleton (articulated)
- human vertebrae set (optional)
- demonstration pointer

❑ Read the directions and safety tips for this exercise **carefully** before starting any procedure.

> **SAFETY FIRST!** Be careful when handling the skeletal specimens. Loose hardware can injure both you and the specimen. Do not forget to use only approved demonstration pointers.

A. The vertebral column

The 26 bones of the vertebral column are divided among the bones listed here. Find each bone and its important features on both an articulated skeleton and among the separate bones of a disarticulated skeleton or vertebrae set.

❑ 1 The seven **cervical** vertebrae at the superior end of the vertebral column are designated individually by number. The most superior is C1 (cervical number one), the next is C2, and so on. Of C1 through C7, only the first two have commonly used alternate names:
- **Atlas**—C1, or the atlas, is a ringlike vertebra that supports the skull by forming a joint with the occipital condyles. The atlas has an **anterior arch** and **posterior arch** fused to left and right **lateral masses** to form a circle. Flat articulating **facets** are found on both the superior and inferior aspects.
- **Axis**—The axis (C2) is remarkable for its **dens** or *odontoid* (toothlike) process. The dens points superiorly through the atlas to act as a pivot for the rotation of C1 and the skull.

In an articulated skeleton or figure, notice the **cervical curve** of the vertebral column produced by these seven vertebrae.

❑ 2 The twelve **thoracic** vertebrae (T1 through T12) are inferior to the cervical vertebrae. As a group, they curve in the opposite direction of the cervical curve to form the **thoracic curve.**

❑ 3 Five large **lumbar** vertebrae (L1 through L5) form the curve of the lower back, or **lumbar curve.**

❑ 4 All vertebrae from C3 to L5 have certain common features. Find each of these features on examples of all three vertebral types.
- **Vertebral foramen**—Large hole (for the spinal cord) in the center
- **Spinous process**—Spine projection posteriorly (*bifid,* forked, spinous process in cervical vertebrae)

- **Body**—Thick disc of bone forming the anterior arch
- **Lamina**—Thin platelike section forming each leg of the posterior arch
- **Pedicle**—Lateral section joining the lamina to the body
- **Intervertebral foramen**—Hole formed between laminae of successive vertebrae (for spinal nerves)
- **Transverse process**—Spine that projects laterally
- **Transverse foramen**—Hole in a transverse process
- **Superior articular facets**—Flat surfaces on the superior aspect
- **Inferior articular facets**—Flat surfaces on the inferior aspect

❏ 5 The **sacrum** is a bone that develops as a set of five vertebrae that fuse to form one large bone inferior to L5. A slight **sacral** curve can be seen from the lateral perspective. Find these sacral features:
- **Dorsal foramina**—Double row of holes between fused segments that form the sacrum (dorsal aspect)
- **Pelvic foramina**—Similar to dorsal foramina, but on the ventral aspect
- **Sacral canal**—Canal formed by the vertebral foramina of the fused sacral segments
- **Sacral hiatus**—Open space at the caudal end of the sacral canal (where the last two sacral segments have incomplete spinous processes)
- **Median sacral crest**—Bumpy ridge formed by fusion of sacral spinous processes
- **Sacral promontory**—Jutting ventral lip of the superior sacral segment

❏ 6 The most inferior bone of the vertebral column is the **coccyx,** or tailbone. This bone, like the sacrum, is actually a fusion of several vertebrae.

❏ 7 The **hyoid** bone is not really part of the vertebral column but will be considered here. This bone is a U-shaped bone in the throat. It serves as an attachment for tongue muscles and connective tissue associated with the larynx (voice box). In most articulated skeletal specimens, it is suspended by wire or plastic in a position anterior to the cervical curve.

B. The thoracic cage

The 25 bones of the rib cage, or **thoracic cage,** form a partially flexible, protective shield for the heart, lungs, and other thoracic organs. The thoracic cage also helps protect some organs of the upper abdomen, such as the liver and spleen. Locate the major structures of the thoracic cage.

❏ 1 Each of the 12 pairs of **ribs** articulates with the vertebral column. The ribs curve around anteriorly. The superior seven pairs of ribs attach directly to the *sternum* by way of **costal cartilages.** These 14 ribs are also called **true ribs.** The inferior five pairs, or **false ribs,** do not directly connect to the sternum. The upper three pairs of false ribs indirectly connect to the sternum, via the costal cartilages of true ribs. The last two pairs of false ribs do not connect in any way, so they are also called **floating ribs.**

❏ 2 The **sternum,** or *breast bone,* receives the costal cartilages of the true ribs on the anterior aspect of the thoracic cage. In life, it forms an anterior protective wall over the heart and associated structures. Find these structures of the sternum:
- **Body (gladiolus)**—Main (middle) portion of the sternum
- **Manubrium**—Superior segment
- **Xiphoid process**—Inferior, pointed segment

The Thoracic Cage

RIBS
TRUE RIB₁
FALSE RIB₂
COSTAL CARTILAGE₃

STERNUM
MANUBRIUM₄
BODY₅
XIPHOID PROCESS₆

THORACIC VERTEBRA₇

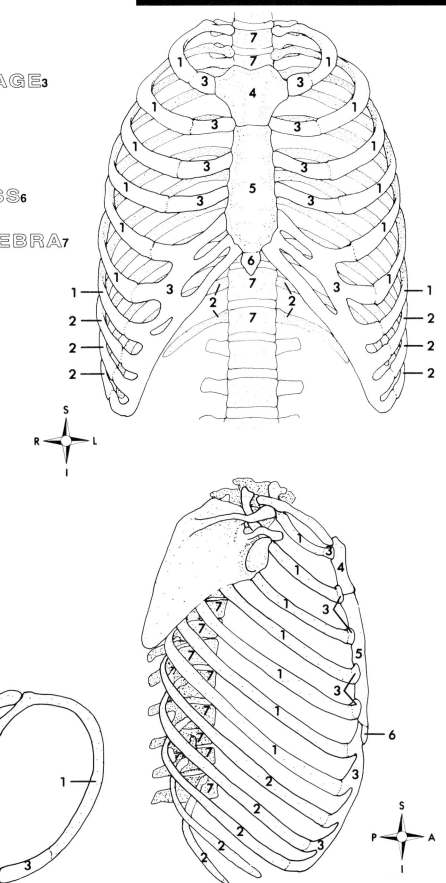

Figure 13-1

Typical Vertebra

VERTEBRAL ARCH₁
LAMINA₂
PEDICLE₃
SPINOUS PROCESS₄
TRANSVERSE PROCESS₅
INFERIOR ARTICULAR PROCESS₆
INFERIOR ARTICULAR FACET₇
SUPERIOR ARTICULAR PROCESS₈
SUPERIOR ARTICULAR FACET₉
VERTEBRAL FORAMEN₁₀
BODY OF VERTEBRA₁₁

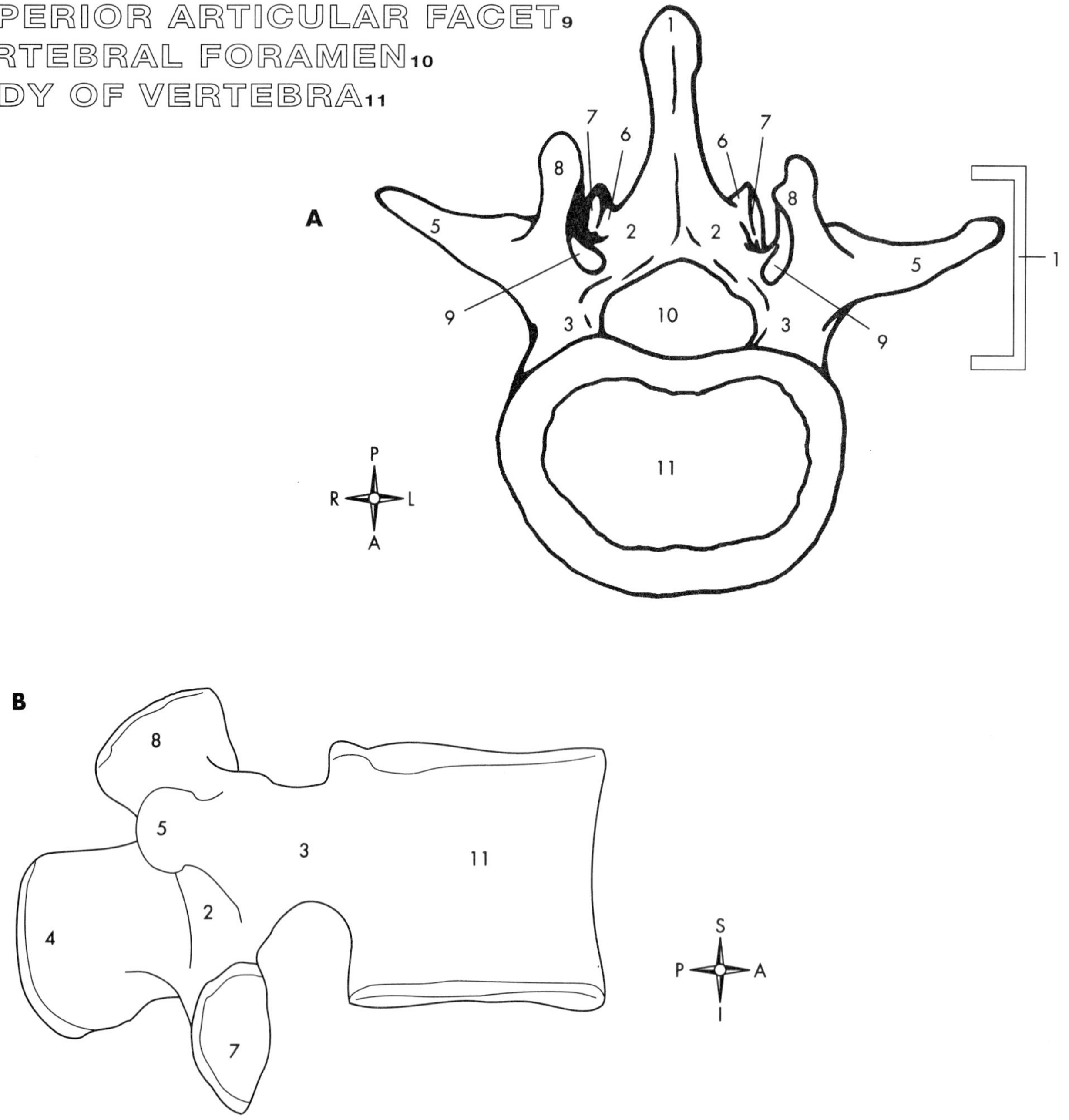

COLORING EXERCISE Using colored pens or pencils, shade in the figure and accompanying labels in contrasting colors of your choice as indicated by the red numerals.

Figure 13-2

The Vertebral Column

CERVICAL VERTEBRA₁
 CERVICAL CURVE₂
THORACIC VERTEBRA₃
 THORACIC CURVE₄
LUMBAR VERTEBRA₅
 LUMBAR CURVE₆
INTERVERTEBRAL DISCS₇
SACRUM₈
 SACRAL CURVE₉
COCCYX₁₀

COLORING EXERCISE Using colored pens or pencils, shade in the figure and accompanying labels in contrasting colors of your choice as indicated by the red numerals. Write the names of bone markings on the black label lines here and in Lab Report 13.

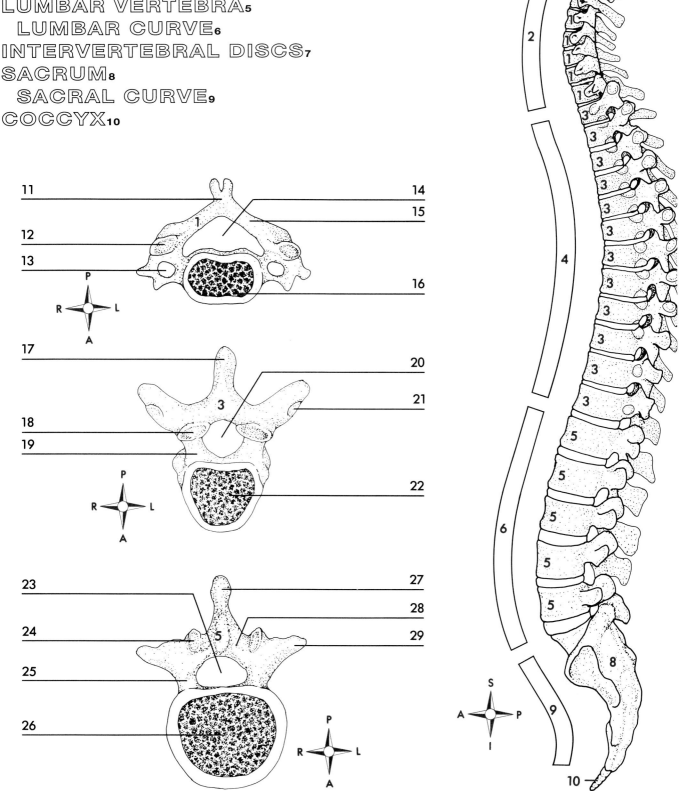

Figure 13-3

123

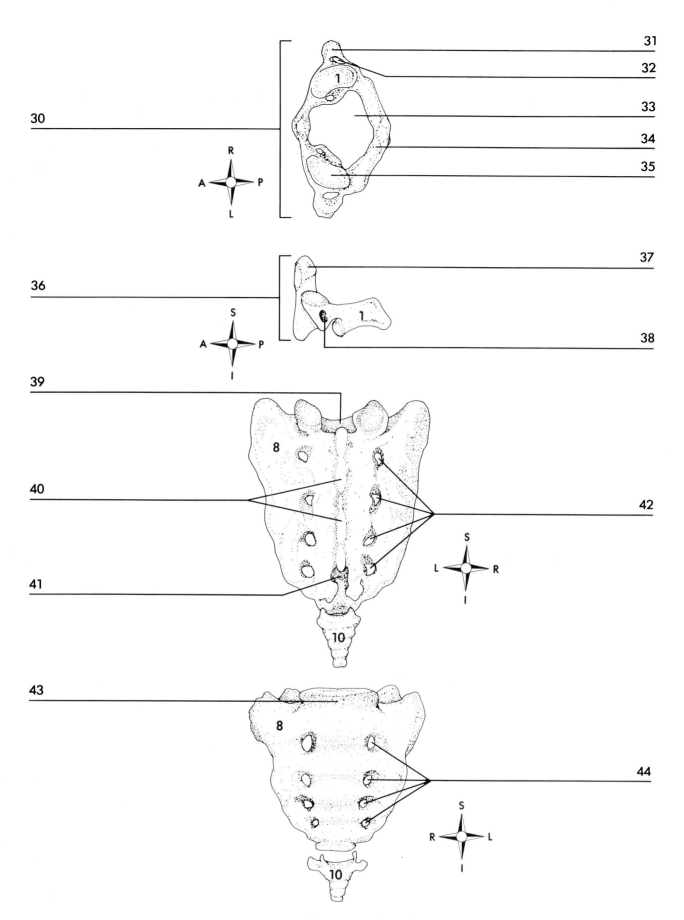

Figure 13-4

ANATOMICAL IMAGES

Perhaps the oldest method of producing images of internal structures of the human body is sketching dissected specimens. Over the past few decades, biotechnology has produced a number of advanced techniques of producing images of human structures. One such technique, first produced by Wilhelm Roentgen around the turn of the century, is **radiography.** This technique, which is also called *x-ray photography,* uses radiation waves that pass through the specimen and onto photographic film. Because the waves pass more easily through soft tissue than through dense tissue such as bone, shadows of some structures are visible when the film is developed. Figure 13-5, *A* shows a radiograph, or x-ray photograph, of the lumbar region. Can you identify the structures indicated?

A more recent variation of this technique is computed tomography (CT). A CT scanner emits a beam of x-rays in a circular path around the subject. After passing through the body, the beams are processed by a computer that is able to reconstruct a three-dimensional image of internal structures. The operator can view different sections of the body on a video monitor. Figure 13-5, *B* shows a CT image of a horizontal section of the upper abdomen. A vertebra is clearly shown; can you identify its parts?

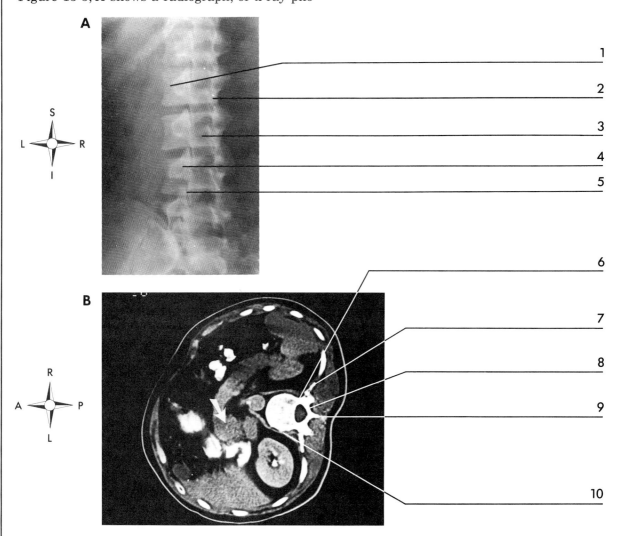

Figure 13-5 A, Standard radiograph of the lumbar region (notice the "tilted" perspective). **B,** CT image of the abdomen.

NAME_____ DATE_____ SECTION_____

LAB REPORT 13

The Vertebral Column and Thoracic Cage

Figures 13-3 and 13-4

_____ 11
_____ 12
_____ 13
_____ 14
_____ 15
_____ 16
_____ 17
_____ 18
_____ 19
_____ 20
_____ 21
_____ 22
_____ 23
_____ 24
_____ 25
_____ 26
_____ 27
_____ 28
_____ 29
_____ 30
_____ 31
_____ 32
_____ 33
_____ 34
_____ 35
_____ 36
_____ 37

Table (fill in the numbers as indicated)

Number	Structures
	Bones in the vertebral column
	Unfused vertebrae
	Cervical vertebrae
	Thoracic vertebrae
	Lumbar vertebrae
	Spinal curves
	Bones in the thoracic cage
	Pairs of ribs
	Pairs of true ribs
	Pairs of false ribs
	Pairs of floating ribs

Figures 13-3 and 13-4 (continued)

_____ 38
_____ 39
_____ 40
_____ 41
_____ 42
_____ 43
_____ 44

127

Figure 13-5

_____ 1
_____ 2
_____ 3
_____ 4
_____ 5
_____ 6
_____ 7
_____ 8
_____ 9
_____ 10

Fill-in

_____ 1
_____ 2
_____ 3
_____ 4
_____ 5
_____ 6
_____ 7
_____ 8
_____ 9
_____ 10
_____ 11
_____ 12
_____ 13
_____ 14
_____ 15
_____ 16
_____ 17
_____ 18
_____ 19
_____ 20

Fill-in (complete each statement with the appropriate term)

1. The first cervical vertebra is called C1, or __?__.
2. The second cervical vertebra is called C2, or __?__.
3. The inward curve of the lower back is called the __?__ curve.
4. All ribs articulate with __?__ .
5. Both radiographs and CT scans use __?__ to form images of anatomical structures.
6. Kyphosis is a condition in which the vertebral column is distorted, giving a person a humpback. This condition involves an exaggerated __?__ curve.
7. Both the coccyx and __?__ are vertebral bones that are actually fused vertebrae.
8. The thoracic vertebrae are __?__ to the lumbar vertebrae.
9. The superior portion of the sternum is called the __?__.
10. The only bone that does not articulate with another bone is the __?__.
11. The __?__ is a vertical tunnel through the sacrum for the passage of nerves descending from the spinal cord.
12. The medial bumps along the midline of your back are caused by the __?__ of each vertebra.
13. The __?__ ribs do not articulate, even indirectly, with the sternum.
14. The __?__ ribs connect directly to the sternum.
15. Costal cartilages are composed of __?__ cartilage tissue.
16. __?__ disks (appearing in Figure 13-2) form a type of joint between the bodies of vertebrae.
17. The __?__ is a horseshoe-shaped bone.
18. The tailbone is more properly known as the __?__.
19. Spinal nerves leave the protection of the vertebral column by way of lateral holes or gaps between the vertebrae called __?__ foramina.
20. The medial ridge along the posterior surface of the sacrum is called the __?__.

LAB EXERCISE 14

The Upper Extremities

Remember that the human skeleton is divided into the axial skeleton and the upper extremities (appendicular skeleton). You are now ready to begin learning the upper extremities.

The appendicular skeleton consists of all 126 bones that form the upper and lower extremities. The *shoulder girdle* and arms have 64 bones altogether, whereas the *pelvic girdle* and legs have a total of 62 bones. This exercise presents all the bones and many important bone features of the upper extremities.

Before you begin

❑ Read the appropriate chapter in your textbook.

❑ Set your learning goals. When you finish this exercise, you should be able to
- name the bones of the upper extremity skeleton
- identify upper extremity bones and markings on a specimen and in figures

❑ Prepare your materials:
- human skeleton (disarticulated)
- human skeleton (articulated)
- demonstration pointer

❑ Read the directions and safety tips for this exercise **carefully** before starting any procedure.

> **SAFETY FIRST!** Beware of loose parts in the articulated skeleton. Remember to use only approved demonstration pointers.

A. The shoulder girdle

Find each bone and marking listed in a specimen and in figures.

❑ 1 The **scapula,** or shoulder blade, forms part of the shoulder girdle that supports the arms. It is an irregular bone on the superior, posterior aspect of the rib cage that extends laterally to form a joint with the arm. Find these features of the scapula:
- **Axillary border**—Edge of the triangular scapula that faces the armpit region
- **Vertebral border**—Scapular edge that faces the vertebral column
- **Superior border**—Superior edge of the scapula
- **Acromion process**—Large process projecting laterally from the scapular spine to form part of the shoulder joint
- **Coracoid process**—Process anterior to, but smaller than, the acromion process
- **Glenoid cavity**—Depression at the upper, lateral corner of the scapular triangle; it forms part of the shoulder socket
- **Spine**—Large, dorsal crest extending from the vertebral border to the acromion process

❑ 2 The **clavicle,** or collar bone, is the other bone of the shoulder girdle. The clavicle is a long bone whose long axis lies along a horizontal axis on the anterior, superior aspect of the thoracic cage.

B. The arm

Identify the bones and markings listed.

❑ 1 The **humerus** is the large, long bone of the upper arm. Find these parts of the humerus:
- **Head**—Enlarged proximal portion
- **Anatomical neck**—Diagonal line just inferior to the articular surface of the head
- **Surgical neck**—The point at which the bone narrows significantly distal to the head
- **Greater tubercle**—Large bump near the lateral aspect of the head
- **Lesser tubercle**—Bump near the midline of the head's anterior aspect
- **Intertubercular sulcus**—Groove between the greater and lesser tubercles
- **Deltoid tuberosity**—Rough muscle attachment site near the middle of the lateral surface of the shaft
- **Coronoid fossa**—Depression on the anterior, distal surface (between the epicondyles) that articulates with the ulna's coronoid process

- **Olecranon fossa**—Depression on the posterior, distal surface that articulates with the ulna's olecranon
- **Medial epicondyle**—Blunt projection of bone on the medial aspect of the distal end
- **Lateral epicondyle**—Blunt projection of bone on the lateral aspect of the distal end
- **Trochlea**—Spool-shaped articulating surface on the distal end that articulates with ulna's trochlear notch
- **Capitulum**—Ball-like articulating surface on the distal end, lateral to the trochlea, that articulates with the head of the radius

☐ 2 The **radius** is one of two long bones of the lower arm. It is lateral to the other lower arm bone (when in the anatomical position), and articulates with the capitulum of the humerus. Find these features:
- **Head**—Proximal disclike enlargement
- **Radial tuberosity**—Oval bump just distal to the head, on the medial surface (facing the ulna)
- **Styloid process**—Pointed bump on the lateral aspect of the distal end

☐ 3 The **ulna,** the other long bone of the lower arm; articulates with the trochlea of the humerus. Find these markings of the ulna:
- **Olecranon**—Large proximal process that forms the blunt point on the posterior aspect of the flexed elbow joint
- **Semilunar (trochlear) notch**—Curved anterior indentation on the olecranon that articulates with the trochlea of the humerus
- **Coronoid process**—Sharp, anterior lip of the trochlear notch that articulates with the coronoid fossa of the humerus
- **Head**—Distal enlargement
- **Styloid process**—Pointed bump on the medial aspect of the head

C. The wrist and hand

Identify the bones and markings listed.

☐ 1 The **carpus,** or wrist, is composed of eight small bones:
- **Pisiform**
- **Triquetrum (triangular)**
- **Lunate**
- **Scaphoid (navicular)**
- **Hamate**
- **Capitate**
- **Trapezoid**
- **Trapezium**

☐ 2 The hand consists of five similar **metacarpal** bones, numbered one through five (starting from the thumb side). Articulating with these are the **phalanges,** or finger bones. The phalanges are named *proximal, middle,* and *distal* and numbered one through five, according to their position. The first middle phalanx (of the thumb) is not present.

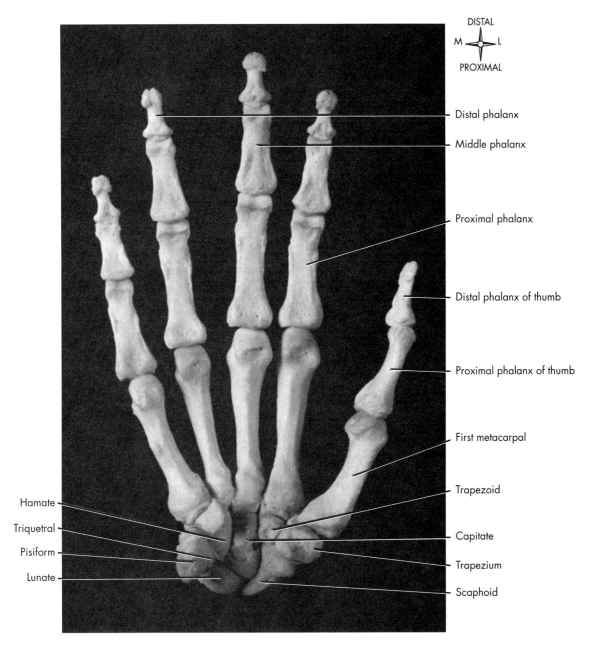

Figure 14-1

The Upper Extremity

SCAPULA 1
CLAVICLE 2
HUMERUS 3
RADIUS 4
ULNA 5
PISIFORM 6
TRIQUETRUM 7
LUNATE 8
SCAPHOID 9

HAMATE 10
CAPITATE 11
TRAPEZOID 12
TRAPEZIUM 13
METACARPAL 14
PROXIMAL PHALANX 15
MIDDLE PHALANX 16
DISTAL PHALANX 17

COLORING EXERCISE Using colored pens or pencils, shade in the figure and accompanying labels in contrasting colors of your choice as indicated by the red numerals.

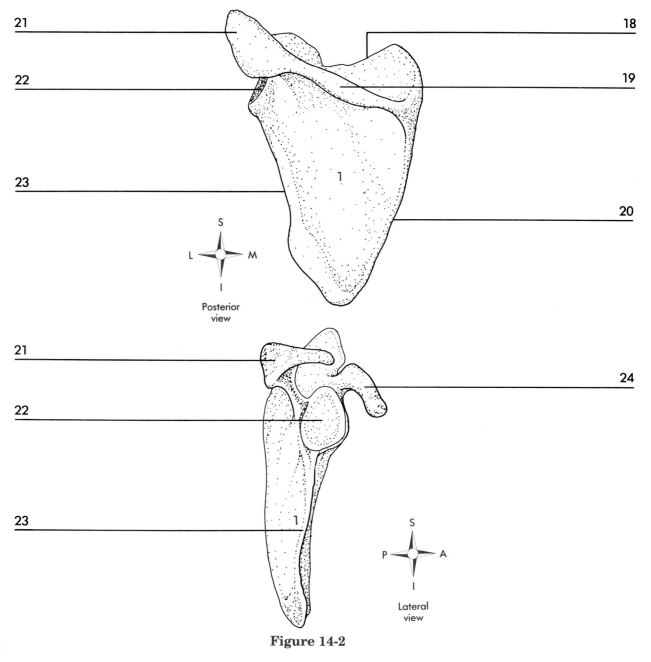

Figure 14-2

132

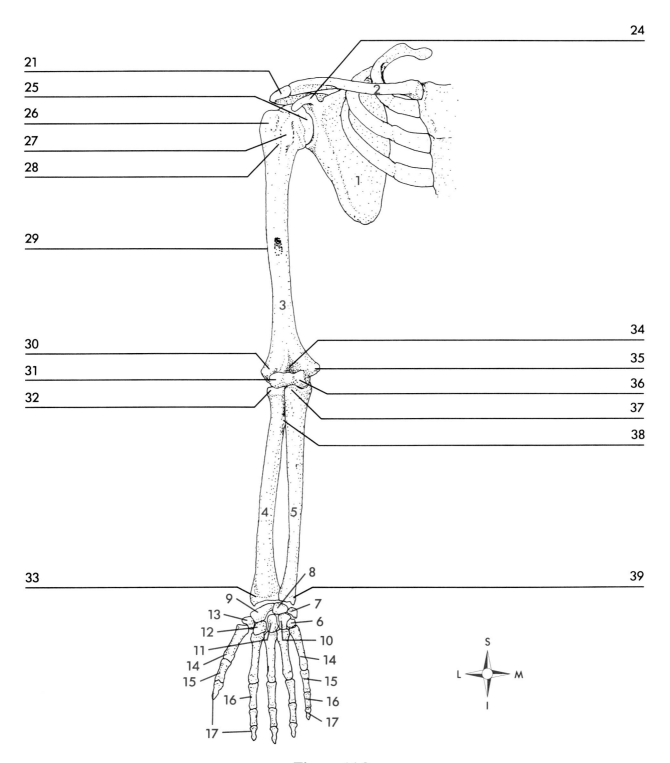

Figure 14-3

133

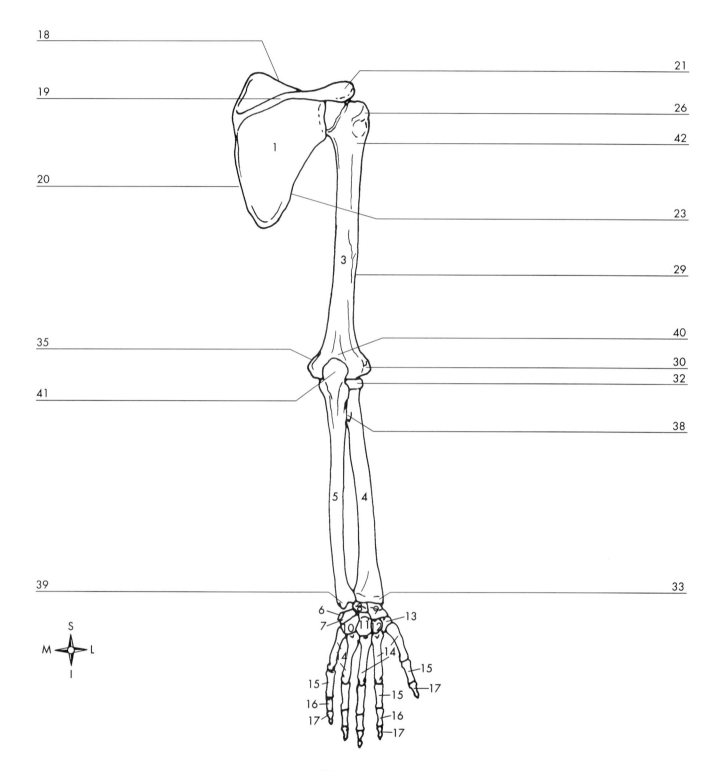

Figure 14-4

NAME _____ DATE _____ SECTION _____

LAB REPORT 14

The Upper Extremities

Figures 14-2, 14-3, and 14-4

_____ 18
_____ 19
_____ 20
_____ 21
_____ 22
_____ 23
_____ 24
_____ 25
_____ 26
_____ 27
_____ 28
_____ 29
_____ 30
_____ 31
_____ 32
_____ 33
_____ 34
_____ 35
_____ 36
_____ 37
_____ 38
_____ 39
_____ 40
_____ 41
_____ 42

Multiple choice

____ 1 ____ 4
____ 2 ____ 5
____ 3 ____ 6

Multiple choice (only one response is correct in each item)

1. Jeb accidentally smashed his hand in a car door, breaking one of his bones. If the injury is near the middle of his palm, which type of bone must be involved?
   a. proximal phalanx
   b. metacarpal
   c. metatarsal
   d. distal phalanx
   e. carpal

2. The humeral process that articulates with the ulna is the
   a. trochlea
   b. capitulum
   c. lateral epicondyle
   d. medial epicondyle
   e. a and b are correct

3. The largest long bone of the upper extremities is
   a. humerus
   b. tibia
   c. femur
   d. ulna
   e. clavicle

4. The forearm contains the
   a. ulna
   b. tibia
   c. radius
   d. humerus
   e. a and b are correct
   f. a and c are correct

5. The large, proximal process of the ulna that forms the posterior "point" of the flexed elbow is the
   a. head
   b. styloid process
   c. trochlea
   d. olecranon
   e. radial tuberosity

6. While skiing in the Alps last winter, Laura fell and fractured her collar bone. To which bone(s) does this collar bone attach?
   a. scapula
   b. sternum
   c. floating rib
   d. a and b are correct
   e. a, b, and c are correct

Table (fill in the numbers as indicated)

Number	Structures
	Bones in the appendicular skeleton
	Bones in the upper extremities
	Wrist bones (total)
	Hand/finger bones (total)

LAB EXERCISE 15

The Lower Extremities

Recall that the appendicular skeleton consists of all 126 bones that form the upper and lower extremities. The *shoulder girdle* and arms have 64 bones altogether, whereas the *pelvic girdle* and legs have a total of 62 bones. This exercise presents all the bones and many important bone features of the lower extremities.

Before you begin

❑ Read the appropriate chapter in your textbook.

❑ Set your learning goals. When you finish this exercise, you should be able to
- name the bones of the lower extremity
- identify lower extremity bones and markings on a specimen and in figures

❑ Prepare your materials:
- human skeleton (disarticulated)
- human skeleton (articulated)
- demonstration pointer

❑ Read the directions and safety tips for this exercise **carefully** before starting any procedure.

> **SAFETY FIRST!** Beware of loose parts in the articulated skeleton. Remember to use only approved demonstration pointers.

A. The pelvic girdle

Find the bones and markings listed.

❑ 1 A left and right **coxal bone,** or pelvic bone, form the pelvic girdle. The *coxae* articulate with the sacrum and support the legs. The coxae are formed by the fusion of three bones early in development: the **ilium, ischium,** and **pubis.** These bones are still identifiable as pelvic regions:
- **Ilium**—Lateral, superior region of the bone; it includes the bowl-like wall of the upper pelvis
- **Ischium**—Inferior, posterior, lateral coxal region
- **Pubis**—Inferior medial region, comprising an upper and lower *ramus;* forms the pubic arch at the body's anterior midline

❑ 2 Find these other features of the coxal bone:
- **Acetabulum**—Cuplike socket formed where the ilium, ischium, and pubis articulate with the head of the femur
- **Iliac crest**—Upper, curving "lip" of the "bowl" formed by the iliac portion
- **Iliac spines**—The "corners" of the iliac portion:
  - **Anterior superior spine**—Blunt anterior point at the end of the iliac crest
  - **Anterior inferior spine**—Smaller blunt point a short distance below the anterior superior spine
  - **Posterior superior spine**—Point at the posterior end of the iliac crest
  - **Posterior inferior spine**—Point a short distance below the posterior superior spine
- **Ischial spine**—Pointed process on the inferior edge of the ischial region that protrudes into the pelvic outlet
- **Obturator foramen**—Large hole formed by the curve of the pubis (medially) and the ischium
- **Pubic arch**—Arch formed by both the left and right pubis's lower angles, or rami
- **Pelvic brim (inlet)**—Boundary of the opening into the **true pelvis;** formed by the upper margins (**pubic crests**) of the superior pubic rami, the *ileopectineal lines* that run along the inside face of the ileum, and the *sacral promontory*
- **True pelvis**—Also called the *pelvis minor,* it is the space below the pelvic brim in which the pelvic organs usually lie
- **False pelvis**—Also called the *pelvis major,* it is the broad, shallow space above the pelvic brim and below the iliac crests (actually part of the abdominal cavity)

❑ 3 Each coxal bone forms a joint directly with two other bones: the sacrum (posteriorly) and the opposite coxal bone (anteriorly).

- **Symphysis pubis**—Joint between the left and right pubis; also called the *pubic symphysis*
- **Sacroiliac joint**—Joint between the coxal's ilium and the lateral surface of the sacrum

B. The leg

Find the bones and markings listed.

❑ 1 The **femur** is the long bone of the upper leg, or thigh bone. Find these features of the femur:
- **Head**—Large, spherical enlargement at the proximal end
- **Neck**—Narrow region just distal to the head
- **Greater trochanter**—Large bump that forms a lateral, proximal "corner"
- **Lesser trochanter**—Bump on the medial, proximal aspect (just distal to the neck)
- **Intertrochanteric line**—Line that runs between the greater and lesser trochanter
- **Linea aspera**—Ridge that runs lengthwise along posterior surface of shaft
- **Medial condyle**—Medial, distal bump that articulates with the tibia
- **Lateral condyle**—Lateral, distal bump that articulates with the tibia
- **Medial epicondyle**—Blunt projection from the side of the medial condyle
- **Lateral epicondyle**—Blunt projection from the side of the lateral condyle

❑ 2 The **patella** is a large sesamoid bone forming the anterior bone of the knee joint.

❑ 3 The **tibia** is one of two long bones of the lower leg. The tibia is the larger, medial lower leg bone. Find these tibial features:
- **Lateral condyle**—Lateral, proximal articular surface
- **Medial condyle**—Medial, proximal articular surface
- **Crest**—Sharp ridge running longitudinally along the anterior surface
- **Tibial tuberosity**—Rough bump on the anterior aspect, just distal to the condyles
- **Medial malleolus**—Pointed bump on the medial aspect of the distal end

❑ 4 The **fibula** is the narrower, lateral bone of the lower leg, having these parts:
- **Head**—Proximal enlargement
- **Lateral malleolus**—Pointed bump on the lateral aspect of the distal end

C. The ankle and foot

Find the bones and markings listed.

❑ 1 The seven **tarsal** bones form the ankle:
- **Talus**
- **Calcaneus**
- **Cuboid**
- **Navicular**
- **Medial cuneiform**
- **Intermediate cuneiform**
- **Lateral cuneiform**

❑ 2 The foot comprises five **metatarsal bones,** similar to the metacarpals of the hand. Also like the hand, the foot has 28 **phalanges.**

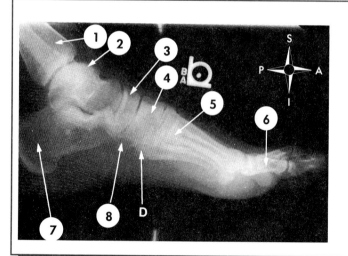

RADIOGRAPH OF THE FOOT

The radiographic, or x-ray, image of the foot seen here is a medial view of the left foot. Try to identify the labeled structures:

_____ 1   _____ 5
_____ 2   _____ 6
_____ 3   _____ 7
_____ 4   _____ 8

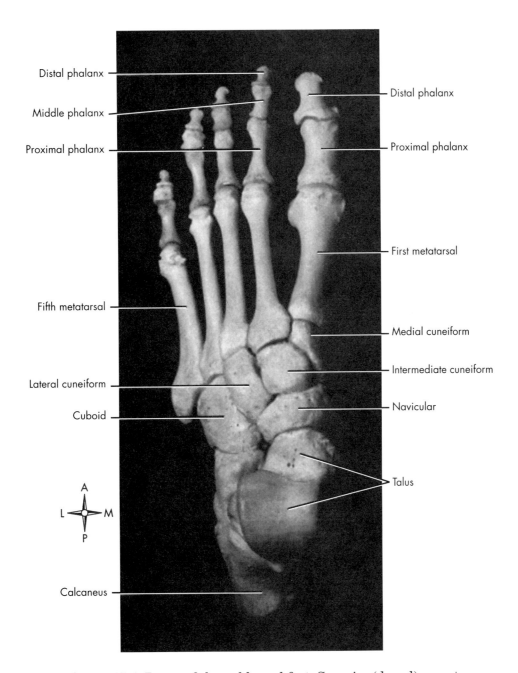

Figure 15-1 Bones of the ankle and foot. Superior (dorsal) aspect.

The Lower Extremity

COLORING EXERCISE Using colored pens or pencils, shade in the figure and accompanying labels in contrasting colors of your choice as indicated by the red numerals.

COXAL 1
   ILIUM 2
   ISCHIUM 3
   PUBIS 4
FEMUR 5
PATELLA 6
TIBIA 7
FIBULA 8
TALUS 9
CALCANEUS 10

CUBOID 11
NAVICULAR 12
MEDIAL CUNEIFORM 13
INTERMEDIATE CUNEIFORM 14
LATERAL CUNEIFORM 15
METATARSAL 16
PROXIMAL PHALANX 17
MIDDLE PHALANX 18
DISTAL PHALANX 19

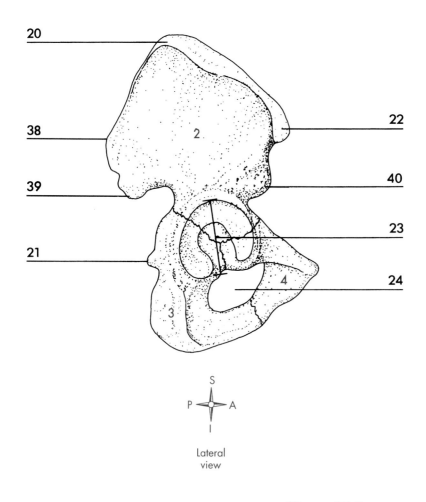
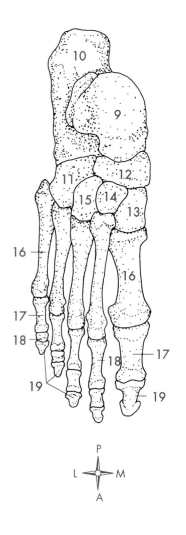

Figure 15-2

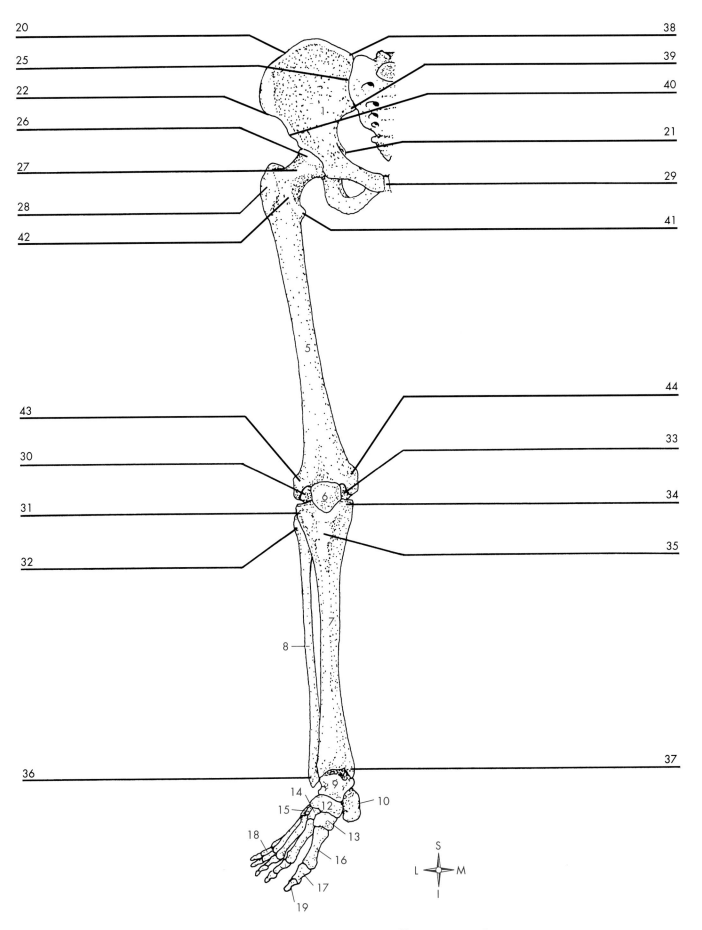

Figure 15-3 A, Anterior view of lower extremity.

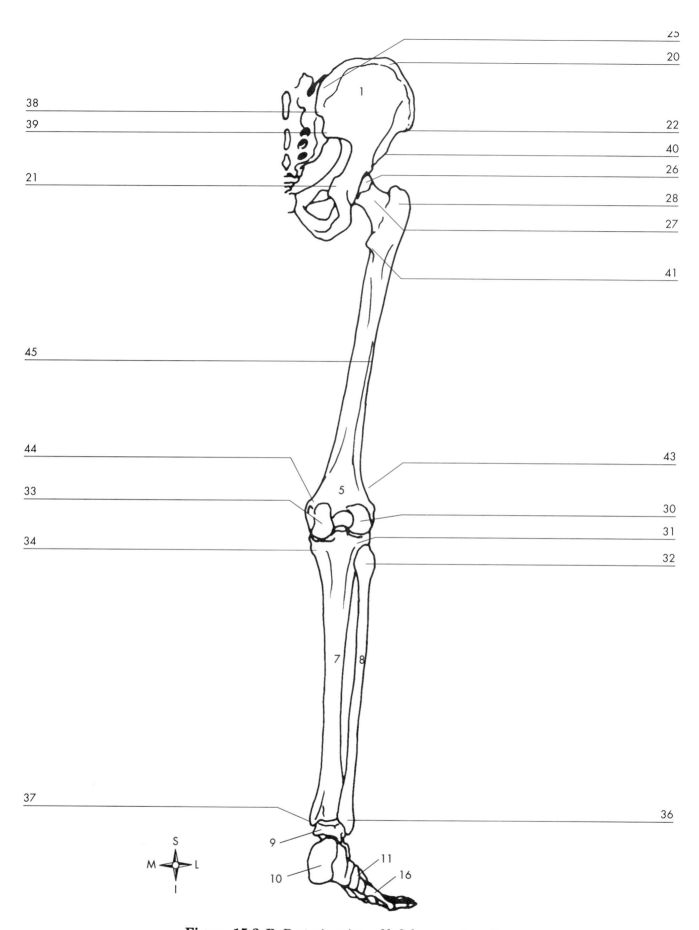

Figure 15-3 B, Posterior view of left lower extremity.

NAME _____ DATE _____ SECTION _____

LAB REPORT 15

The Lower Extremities

Multiple choice

_____ 1
_____ 2
_____ 3
_____ 4
_____ 5
_____ 6

Multiple choice (only one response is correct in each item)

1. The lower leg comprises the
    a. tibia
    b. radius
    c. ulna
    d. fibula
    e. b and c are correct
    f. a and d are correct

2. Lynn has lost quite a bit of weight recently and now notices that a bump is visible in each side of her lower abdomen. Palpating it, she feels that her pelvic bone is protruding. Which part?
    a. ischial spine
    b. symphysis pubis
    c. anterior superior spine
    d. iliac crest
    e. acetabulum

3. The sacrum and coxal bone are joined at the
    a. sacroiliac joint
    b. hip joint
    c. sacrocoxal joint
    d. ischial tuberosity
    e. symphysis pubis

4. The coxal bone's cuplike socket for the head of the femur is called the
    a. trochlear notch
    b. ischial notch
    c. acetabulum
    d. obturator foramen
    e. femoral fossa

5. Which one of these is *not* a tarsal bone?
    a. calcaneous
    b. navicular
    c. cuneiform
    d. hamate
    e. talus

6. The long ridge that runs along the front of the tibia is called the
    a. tibial tuberosity
    b. intertrochanteric line
    c. iliac crest
    d. tibial crest
    e. fibular ridge

Figures 15-2 and 15-3

_____ 20
_____ 21
_____ 22
_____ 23
_____ 24
_____ 25
_____ 26
_____ 27
_____ 28
_____ 29
_____ 30
_____ 31
_____ 32
_____ 33
_____ 34
_____ 35
_____ 36
_____ 37
_____ 38
_____ 39
_____ 40
_____ 41
_____ 42
_____ 43
_____ 44
_____ 45

Radiograph (foot)

_____ 1
_____ 2
_____ 3
_____ 4
_____ 5
_____ 6
_____ 7
_____ 8

Table (fill in the numbers as indicated)

Number	Structures
	Bones in the appendicular skeleton
	Bones in the upper extremities
	Bones in the lower extremities
	Wrist bones (total)
	Ankle bones (total)
	Hand/finger bones (total)
	Foot/toe bones (total)

LAB EXERCISE 16

Joints

Joints, or articulations, are points in the skeleton where two or more bones join together. Because different joints are built differently and therefore behave differently, anatomists have devised a scheme for classifying articulations. This exercise presents a common system of joint classification and challenges you to find examples of joint classes and the different types of movements that they allow. This exercise also offers you the opportunity to examine examples of specific joints of the body.

Before you begin

❏ Read the appropriate chapter in your textbook.

❏ Set your learning goals. When you finish this exercise, you should be able to
- distinguish the three classes of joints
- describe examples of each joint class
- identify different types of motion in joints
- describe the structure of the different joints

❏ Prepare your materials:
- human skeleton (articulated)
- synovial joints (model or fresh animal specimen l.s., knee or shoulder)

❏ Read the directions and safety tips for this exercise **carefully** before starting any procedure.

A. Classifying joints

Find an example of each type of joint described in the following steps. Try to find examples other than those given here.

> **SAFETY FIRST!** Use caution when handling the articulated skeleton.

❏ 1 **Fibrous joints** are found where fibrous connective tissue tightly binds the articulating bones. Identify these types of fibrous joints:
- **Suture**—This is a joint between two flat bones, as between the left and right parietal bones (sagittal suture)
- **Syndesmosis**—Bands of fibrous tissue bind bones, as between the distal ends of the radius and ulna
- **Gomphosis**—A fibrous membrane connects each tooth to its socket in a jaw's alveolar process

❏ 2 **Cartilaginous joints** are formed when a mass of cartilage joins bones.
- **Synchondrosis**—Hyaline cartilage connects bones; for example, the costal cartilage connection between a rib and the sternum
- **Symphysis**—Fibrocartilage forms the joint, as in the symphysis pubis, joining left and right coxae

> **HINT** → Fibrous and cartilaginous joints can be classified according to function, rather than structure. Thus, functional categories of **immovable** (synarthrotic) and **slightly movable** (amphiarthrotic) are sometimes used. Most fibrous joints are immovable, so they are called *synarthroses*. Likewise, many cartilaginous joints are slightly movable, so they are called *amphiarthroses*.

❏ 3 **Synovial joints** are always **freely movable** (*diarthrotic*) joints. A flexible *joint capsule,* composed of ligaments and other connective structures and lined with a lubricating *synovial membrane,* allows a wide range of movement. Categories of synovial joints are based on the way in which the articulating bones fit together:
- **Gliding joint**—Two flat surfaces slide past each other, as between two carpal bones
- **Hinge joint**—As with a door hinge, two bones are joined so that they can move in one plane only, as in the elbow
- **Ellipsoid joint**—An oval condyle fits into an oval fossa, allowing movement in two planes, as between the metatarsals and phalanges of the foot
- **Pivot joint**—One bone pivots on the axis of another, allowing rotation, as with the atlas and axis

Joint Classification

FIBROUS JOINT₁
CARTILAGINOUS JOINT₂
SYNOVIAL JOINT
    ARTICULATING BONE₃
    ARTICULAR CARTILAGE₄
    SYNOVIAL MEMBRANE₅
    SYNOVIAL CAVITY₆
    SYNOVIAL CAPSULE₇

SYNOVIAL JOINT TYPES
GLIDING₈
HINGE₉
ELLIPSOID₁₀
PIVOT₁₁
SADDLE₁₂
BALL AND SOCKET₁₃

COLORING EXERCISE Using colored pens or pencils, shade in the figure and accompanying labels in contrasting colors of your choice as indicated by the red numerals.

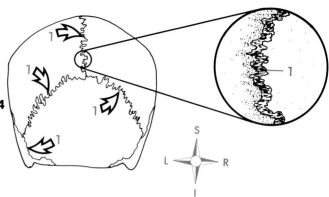

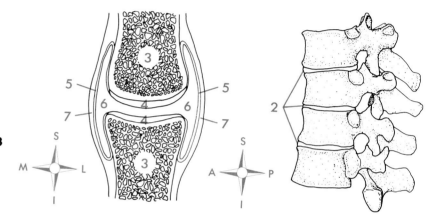

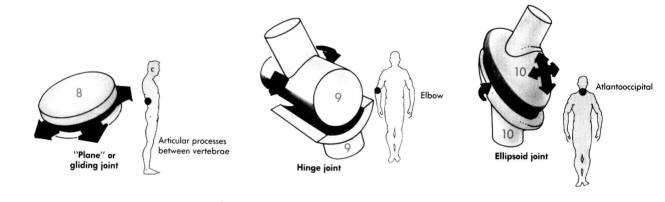

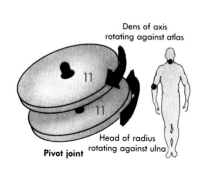

"Plane" or gliding joint — Articular processes between vertebrae

Hinge joint — Elbow

Ellipsoid joint — Atlantooccipital

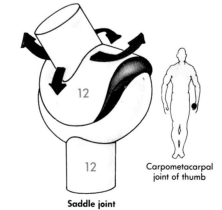

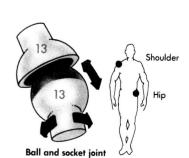

Pivot joint — Dens of axis rotating against atlas; Head of radius rotating against ulna

Saddle joint — Carpometacarpal joint of thumb

Ball and socket joint — Shoulder, Hip

Figure 16-1

- **Saddle joint**—Two saddle-shaped processes fit together to allow movement in two planes, as between the thumb's proximal phalanx and the trapezium of the wrist
- **Ball-and-socket joint**—A ball-shaped process fits into a rounded fossa, allowing almost unrestricted movement, as between the femur and the acetabulum

B. Examples of synovial joints

In this activity, you are challenged to explore the detailed anatomy of the human knee joint and the human shoulder joint. Find the structures listed in models or charts.

> **HINT** → Use the diagrams given in the figures here, as well as those found in your textbook, for help in locating the major structures listed in this activity.

❑ 1 In a model or chart of the human knee, identify these features:
- **Articular cartilage**
- **Lateral meniscus**
- **Synovial membrane**
- **Patella**
- **Condyles of tibia**
- **Condyles of femur**
- **Prepatellar bursa**
- **Suprapatellar bursa**
- **Posterior cruciate ligament**
- **Anterior cruciate ligament**
- **Tibial collateral ligament**
- **Fibular collateral ligament**
- **Patellar ligament**
- **Head of fibula**

How many of these features can you palpate in your own knee?

❑ 2 In a model or chart of the human shoulder joint identify these features:
- **Acromion process of scapula**
- **Coracoid process of scapula**
- **Articular cartilage of humerus**
- **Glenoid cavity**
- **Articular cartilage of glenoid cavity**
- **Medial glenohumeral ligament**
- **Superior glenohumeral ligament**
- **Acromiocoracoid ligament**
- **Subdeltoid bursa**
- **Synovial membrane**
- **Synovial cavity**

C. Dissection of a joint

In a preserved or fresh animal specimen (longitudinal section) of a synovial joint, such as the knee or shoulder, identify the features described. Sketch your observations on a piece of paper and attach it to Lab Report 16.

> **SAFETY FIRST!** Because fresh animal tissue at room temperature may harbor bacterial colonies, handle your specimen with gloved hands.

❑ 1 Identify the *articulating bones*. Which joint is this?

❑ 2 Find the *articular cartilage* on the articulating surfaces of the joint. How would you describe their texture? What is the function of articular cartilage?

❑ 3 Locate the *ligaments* and other connective tissues of the *joint capsule* holding the joint together. Note how strong and how flexible they are.

❑ 4 The *synovial membrane* lines the joint capsule. What is its consistency? Is any *synovial fluid* still in the *synovial cavity* formed by this membrane? What is this fluid's function?

D. Joint movement

Although synovial, or *diarthrotic* joints, can allow a wide range of motion, they are limited by the structure of the joint and surrounding body parts. Demonstrate the types of joint movement listed by performing them yourself.

> **SAFETY FIRST!** Be careful to avoid accidentally hitting someone as you do this activity. Do not attempt this activity if you have a physical condition that may be worsened by joint movements.

❑ 1 Some skeletal movements involve movement of body parts relative to a coronal (frontal) plane:
- **Flexion**—Decreasing the angle of a joint
- **Extension**—Increasing the angle of a joint

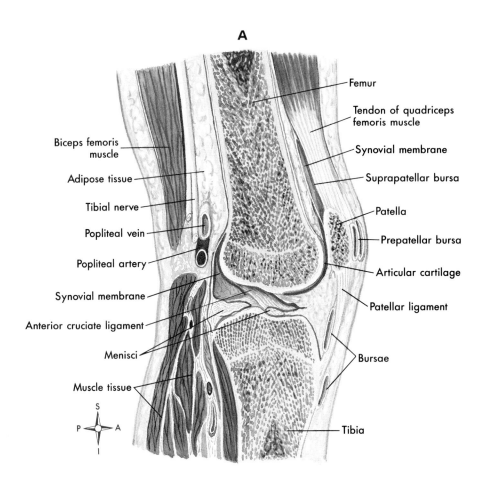
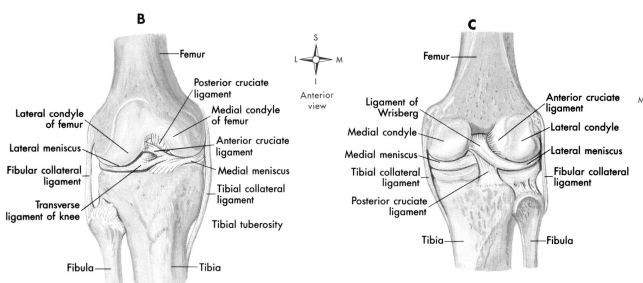

Figure 16-2 The human knee joint. **A,** Sagittal section of the knee. **B,** Anterior view, with superficial structures removed. **C,** Posterior view, with superficial structures removed.

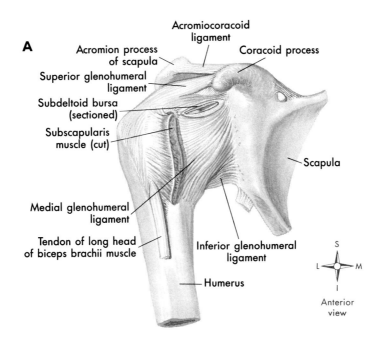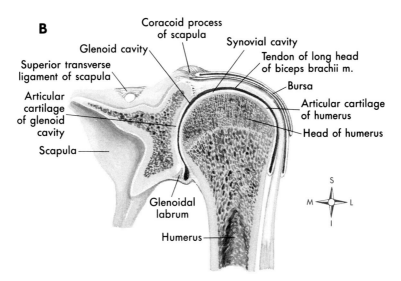

Figure 16-3 The shoulder joint. **A,** Anterior view. **B,** Viewed from behind through shoulder joint.

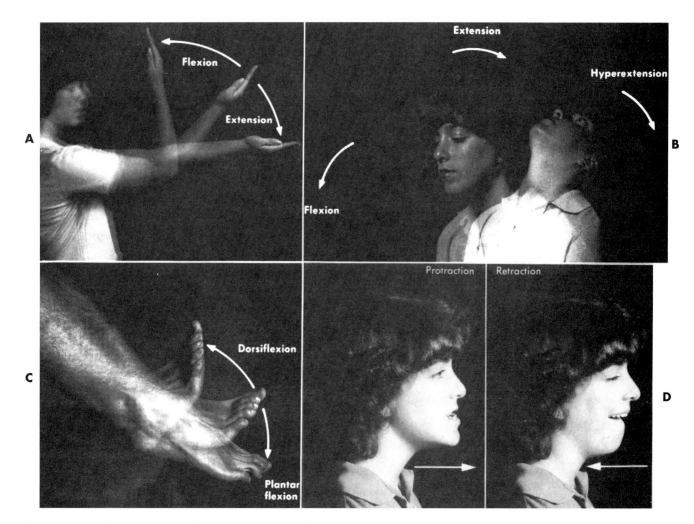

Figure 16-4 Movements relative to a coronal (frontal) plane. **A,** Flexion and extension of the elbow. **B,** Flexion, extension, and hyperextension of the neck. **C,** Dorsiflexion and plantar flexion. **D,** Protraction and retraction of the jaw.

- **Hyperextension**—Moving a joint beyond its normal range or beyond the anatomical position
- **Dorsiflexion**—Bending the ankle so that the toes point upward
- **Plantar flexion**—Bending the ankle so that the toes point downward
- **Protraction**—Moving a part anteriorly, along a horizontal plane
- **Retraction**—Moving a part posteriorly, along a horizontal plane

HINT → The angle of movement in joints is measured in the manner shown:

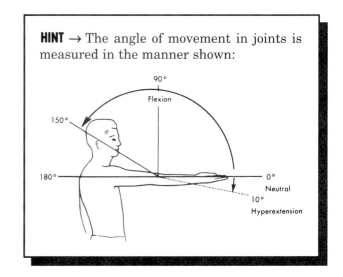

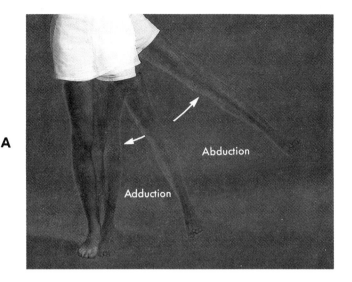 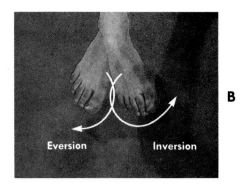

Figure 16-5 Movements relative to a sagittal plane. **A,** Adduction and abduction of the thigh. **B,** Eversion and inversion of the foot.

❏ 2 Some motions are done relative to a sagittal plane:
  ■ **Abduction**—Moving an appendage's distal end away from the midsagittal plane
  ■ **Adduction**—Moving an appendage's distal end toward the midsagittal plane
  ■ **Inversion**—Moving the foot from the anatomical position (sole downward) to a position in which the sole is facing the midsagittal plane
  ■ **Eversion**—Moving the foot from the anatomical position to a position in which the sole faces away from the midsagittal plane

❏ 3 Try these circular movements:
  ■ **Circumduction**—Moving the distal end of an appendage in a circle, making a cone-shaped sweep
  ■ **Rotation**—Moving a bone on its axis, as if on a pivot or axle
  ■ **Pronation**—Rotating the forearm from the anatomical position (palm forward) to reverse it (palm facing the posterior)
  ■ **Supination**—Rotating the forearm from the pronated position back to the anatomical position

❏ 4 Try these special movements:
  ■ **Elevation**—Raising a part
  ■ **Depression**—Lowering a part

> **HINT** → These joint movements are commonly classified in other ways as well. For example, what scheme of classification is used in the textbook?

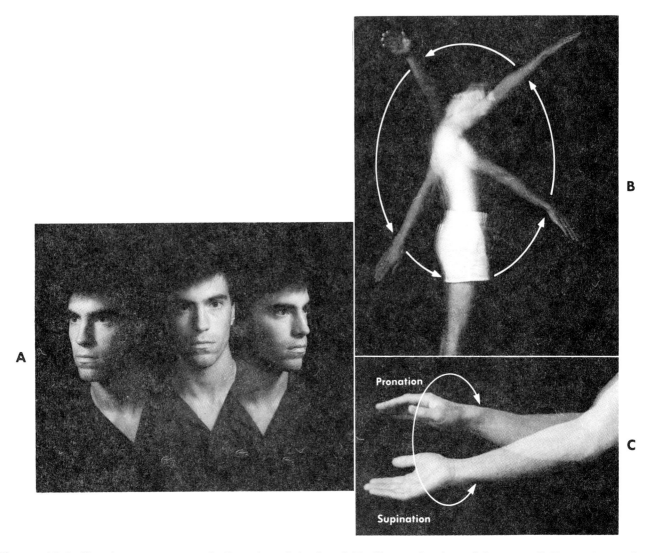

Figure 16-6 Circular movements. **A,** Rotation of the head. **B,** Circumduction of the arm. **C,** Pronation and supination of the hand.

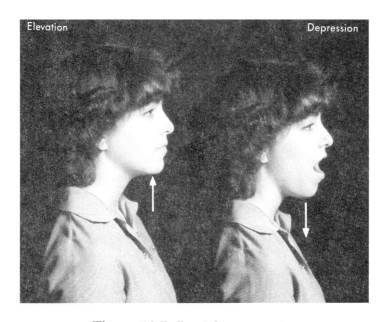

Figure 16-7 Special movements.

ARTHROGRAPHY

Radiographic examination of the soft tissues that form joints is called **arthrography.** In one method of arthrography, a **contrast medium** that absorbs x-rays is injected into the joint. The liquid medium coats the cartilage, ligaments, and other soft structures. When a regular radiography is taken, the coated soft tissues clearly appear. The image is called an **arthrogram.**

In the knee arthrograms shown in Figure 16-8, several features that are invisible in a regular radiograph can be seen:

- **Bursa**—A "pillow" made of synovial membrane and filled with synovial fluid
- **Meniscus**—A piece of cartilage shaped like a disc with a curved surface
- **Articular cartilage**—A cartilage coating on the articulating surfaces of bone

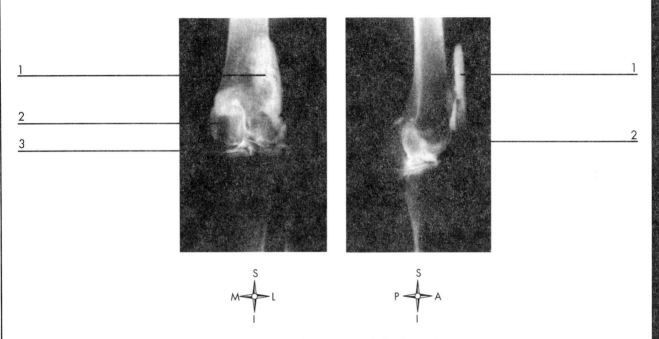

Figure 16-8 Arthrograms of the knee joint.

Can you identify these three joint structures in the arthrograms above?

1. What advantages does using contrast medium give when doing radiography?

2. Computed tomography (CT) has been more recently used to study joints. What advantages does the CT method have over the method described here?

LAB REPORT 16

Joints

Figure 16-8

_____ 1
_____ 2
_____ 3

Matching I

____ 1
____ 2
____ 3
____ 4
____ 5
____ 6
____ 7
____ 8
____ 9
____ 10

Matching II

____ 1
____ 2
____ 3
____ 4
____ 5
____ 6
____ 7
____ 8
____ 9
____ 10

Matching I (may be used more than once)

a. Fibrous joint
b. Synovial joint
c. Cartilaginous joint

1. Gliding joint
2. Synchondrosis
3. Freely movable
4. Suture
5. Gomphosis
6. Saddle joint
7. Hinge joint
8. Symphysis
9. Mainly hyaline cartilage or fibrocartilage
10. Ellipsoid joint

Matching II (match these examples to their types)

a. Suture
b. Syndesmosis
c. Gomphosis
d. Synchondrosis
e. Symphysis
f. Hinge
g. Gliding
h. Pivot
i. Saddle
j. Ball and socket

1. Between bodies of vertebrae
2. Between the distal ends of the tibia and fibula
3. At the base of the thumb's proximal phalanx
4. Between the articular facets of the vertebrae's processes
5. Between the true ribs and sternum
6. Between the talus and lower leg
7. Between C1 and C2, at the dens
8. The shoulder
9. The transverse palatine suture
10. Between the tooth and jaw

Identify

1 _____
2 _____
3 _____
4 _____
5 _____
6 _____
7 _____
8 _____
9 _____
10 _____

Identify (In the blanks that follow, write the type of motion illustrated in the matching figure to the right. For example, in #1 is *abduction* or *adduction* shown?)

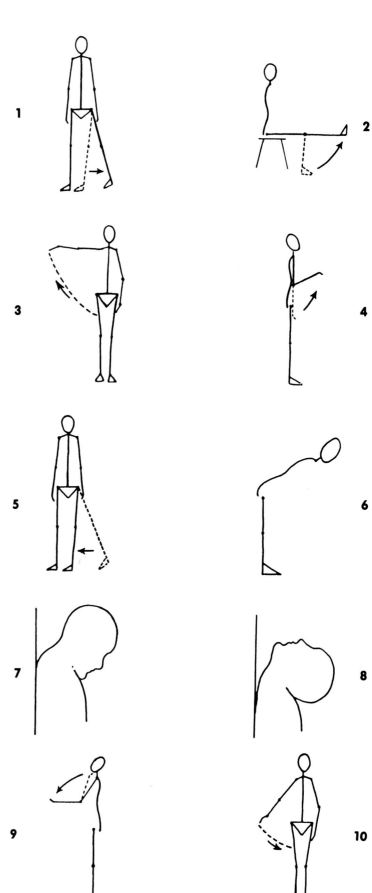

LAB EXERCISE 17

Organization of the Muscular System

The muscular system comprises all the skeletal muscles, here sometimes called muscle organs to distinguish them from muscle tissue and muscle cells. Skeletal muscles are attached to the skeleton in ways that move skeletal parts when the muscles contract. The chief function of this system, then, is skeletal movement. Secondarily, this system helps in temperature regulation by providing a source of metabolic heat.

This exercise presents some preliminary information about the muscular system. Although we have seen some of the histology of skeletal muscle tissue, we have not yet explored the muscle cell or the muscle organ. In this exercise we will do both. We will also take a brief look at how muscle organs are named so that the next exercise, which involves muscle identification, will be easier.

Before you begin

❑ Read the appropriate chapters in your textbook.

❑ Set your learning goals. When you finish this exercise, you should be able to
- describe the structure of the skeletal muscle cell
- identify and describe the principal structures of a skeletal muscle organ
- define the terms *origin* and *insertion*
- interpret the meaning of muscle names

❑ Prepare your materials:
- model or chart of a skeletal muscle cell
- model or chart of a skeletal muscle organ
- animal muscle specimen (fresh meat section)
- dissection tools

❑ Read the directions and safety tips for this exercise **carefully** before starting any procedure.

A. The skeletal muscle cell

Skeletal muscle cells provide the basic functional capability of the muscle organ because they can contract with great force. Such contraction requires a rather unusual structural pattern in muscle cells if it is to be efficient. Study these features of the skeletal muscle cell in a model or figure:

❑ 1 The **sarcolemma** is the cell membrane of the long, cylindrical **muscle fiber,** or cell. The sarcolemma has a *resting potential,* or electrical charge. This charge temporarily reverses during an *action potential,* or impulse, when the muscle fiber is stimulated.

❑ 2 The sarcolemma dips inward at several points to form internal transverse tubules called **T tubules.** An impulse traveling along the sarcolemma can thus also travel inside the cell.

❑ 3 Inside the cell are a number of membranous networks similar to ER. They are all part of the **sarcoplasmic reticulum (SR).** The SR receives the impulse from the T tubules, which are nearby, and release calcium ions into the **sarcoplasm** in response.

❑ 4 The calcium ions released from the SR diffuse through the sarcoplasm among parallel bundles of protein **myofilaments.** Each bundle is called a **myofibril** and is composed of an orderly arrangement of **thin filaments** and **thick filaments.**

❑ 5 The myofilaments are arranged in a repeating pattern called a **sarcomere.** When calcium ions react with some of the myofilament molecules, they slide past one another, shortening each sarcomere. Because all the sarcomeres in the cell's myofibrils shorten, the cell contracts.

❑ 6 As you recall, a muscle cell has multiple **nuclei.** The nuclei are against the inside of the sarcolemma.

B. Structure of a muscle organ

Each muscle organ has a unique shape and size, but the principal components of each are the same. Identify each of these in a model or chart. Some structures may be visible in a fresh muscle specimen.

> **SAFETY FIRST!** Use gloved hands when handling fresh meat at room temperature. You may wish to use dissection tools. If so, be careful to avoid cuts and puncture wounds.

❑ 1 The entire organ is covered with fibrous connective tissue that forms a sheath called the **epimysium.**

❑ 2 The fibrous tissue of the epimysium extends inward to form sheaths around bundles, or **fasciculi,** of muscle cells. This inner fibrous sheath is called the **perimysium.**

❑ 3 The perimysium's fibrous tissue continues inside each fascicle to wrap around each individual muscle cell. The connective tissue sheath around individual muscle fibers is called the **endomysium.** Just deep to the endomysium is the sarcolemma.

❑ 4 All the fibrous sheaths are continuous with one another and contain blood vessels that supply the muscle fibers. At the ends of the muscle organ, the muscle fibers stop while the connective sheaths continue. This forms an extension of dense fibrous connective tissue that attaches to the periosteum of a bone (or to another muscle). If the fibrous connection is in the shape of a strap or band, it is called a **tendon.** If it is a broad, flat sheet, it is called an **aponeurosis.**

❑ 5 Generally a muscle attaches to the skeleton at both of two ends. The end that attaches to the more stationary bone is called the **origin.** The other end, or **insertion,** attaches to the bone that moves as the muscle contracts.

C. Naming skeletal muscles

Muscle **nomenclature** is the system of naming skeletal muscle organs. The names of muscles, as with many organs, are in Latin rather than in English. By using only a handful of descriptive Latin words, in combinations of two or three, anatomists have named all the muscles of the human body. Review these terms used to name muscles because they will come in handy when you learn specific muscle names in the next exercise.

❑ 1 Muscles can be named for their overall shape. Review these terms related to muscle shape:

deltoid	shaped like *delta* (Δ)
orbicularis	circular
platy	flattened; platelike
quadratus	square
rhomboideus	diamond-shaped
trapezius	trapezoidal
trangularis	triangular

❑ 2 Some muscles are named for their points of attachment (*origin* and *insertion*). For example, the *sternocleidomastoid* muscle has attachments on the sternum, clavicle, and mastoid process of the temporal bone.

❑ 3 Muscles can be named according to relative size:

brevis	short
longus	long
magnus	large
maximus	large
medius	moderately sized
minimus	small

❑ 4 The direction of fibers visible in a muscle can be a basis for its name, using these terms:

oblique	diagonal to the body's midline
rectus	parallel to the midline
sphincter	circling an opening
transversus	at a right angle to the midline

❑ 5 Some muscle names are derived from the action(s) produced:

abductor	abducts a part
adductor	adducts a part
depressor	depresses a part
extensor	extends a part
flexor	flexes a part
levator	elevates a part
rotator	rotates a part

❑ 6 Some muscles are named for the region in which they are found. Some of these terms should be familiar to you:

brachialis	arm
frontalis	frontal (bone)
femoris	femur
gluteus	posterior of hip/thigh
oculi	eye
radialis	radius
ulnaris	ulna

Muscle Cell

🖉COLORING EXERCISE Using colored pens or pencils, shade in the figure and accompanying labels in contrasting colors of your choice as indicated by the red numerals.

SARCOPLASMIC RETICULUM₁
NUCLEUS₂
SARCOLEMMA₃
BLOOD CAPILLARY₄
MYOFIBRIL₅
MITOCHONDRION₆
TRANSVERSE TUBULE₇

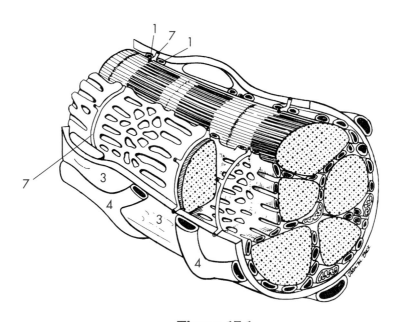

Figure 17-1

Skeletal Muscle

TENDON₁
EPIMYSIUM₂
PERIMYSIUM₃
ENDOMYSIUM₄
MYOFIBRIL₅
THIN FILAMENT₆
THICK FILAMENT₇
SARCOMERE₈
A BAND₉
I BAND₁₀

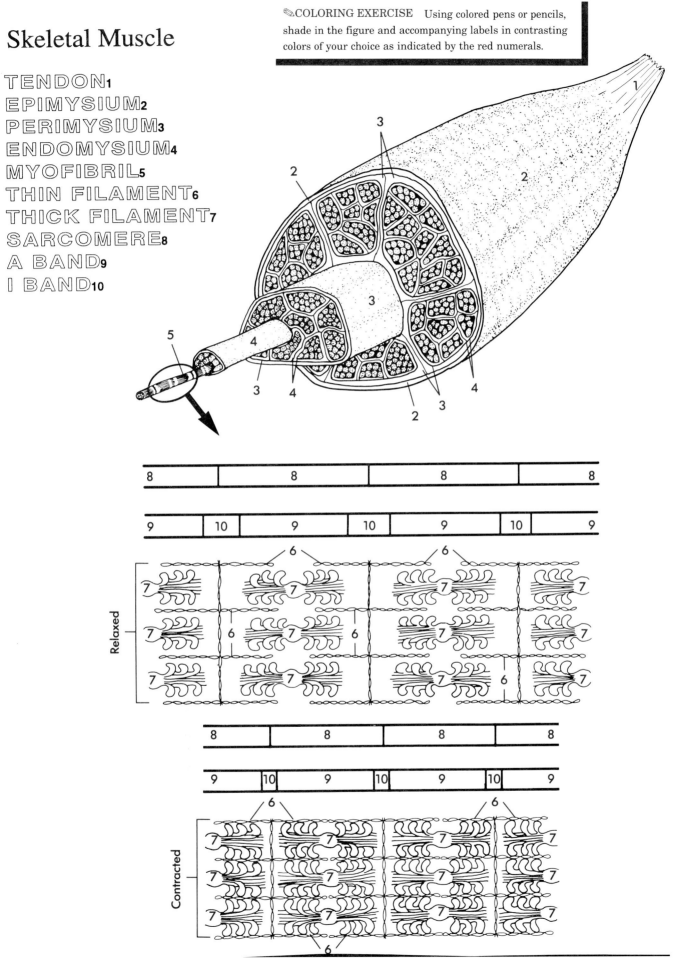

Figure 17-2

LAB REPORT 17

Organization of the Muscular System

Fill-in

1. _____
2. _____
3. _____
4. _____
5. _____
6. _____
7. _____
8. _____
9. _____
10. _____
11. _____
12. _____
13. _____
14. _____

Put in order

1. _____
2. _____
3. _____
4. _____

Fill-in (complete each item with the appropriate term)

1. The ___?___ is the cell membrane of a muscle fiber.
2. A ___?___ is a bundle of parallel myofilaments within a muscle fiber.
3. The ___?___ is similar to ER but collects and stores calcium ions in a resting muscle cell.
4. Two types of myofilaments in a skeletal muscle cell are ___?___ filaments and thick filaments.
5. A repeating pattern, or unit, of myofibrils within the myofilament is the ___?___.
6. A tube formed by the inward extension of the cell membrane is called a(n) ___?___.
7. An entire muscle is covered with a fibrous sheath called the ___?___.
8. A bone is anchored by its ___?___, whereas it pulls at its insertion.
9. Each fascicle, or fasciculus, is wrapped with a fibrous ___?___.
10. A broad, flat version of a tendon is more accurately called a(n) ___?___.
11. Each individual muscle cell has a fibrous ___?___ wrapped around its sarcolemma.
12. A muscle that flexes a joint may have the term ___?___ in its name.
13. A muscle around an opening may be named by the term ___?___.
14. A muscle associated with the femur may have ___?___ in its name.

Put in order (put these structures in anatomical order from superficial to deep)

sarcolemma
perimysium
endomysium
epimysium

Identify

1. _____
2. _____
3. _____
4. _____
5. _____
6. _____
7. _____
8. _____
9. _____
10. _____

Identify (determine the muscle naming term defined in each item)

1. Associated with the frontal bone
2. Adducts a part
3. Long
4. Encircles an opening or tube
5. Short
6. Flattened
7. Parallel to the midline of the body
8. Rotates a part
9. Associated with the ulna
10. Extends a part

LAB EXERCISE 18

Skeletal Muscle Identification

About 36% of the mass in female bodies and 42% of the mass in male bodies is composed of skeletal muscle. In this exercise, some of the larger skeletal muscles of the human body are presented for study.

Before you begin
❏ Read the appropriate chapter in your textbook.

❏ Set your learning goals. When you finish this exercise, you should be able to
- identify the following on a model and in figures:
  Muscles of the head and neck
  Muscles of the trunk
  Muscles of the upper extremity
  Muscles of the lower extremity
  Muscles of the pelvic floor
- name the origin and insertion of each major muscle of the body
- demonstrate the action of each muscle studied

❏ Prepare your materials:
- models and charts of human musculature
- demonstration pointers

❏ Read the directions and safety tips for this exercise **carefully** before starting any procedure.

> **HINT** → In this exercise, the muscles are presented by their location (e.g., head, trunk). Another common approach is to group muscles by the part they move (e.g., muscles that move the head, muscles that move the arm). Yet another manner of grouping muscles involves their action (e.g., flexors, extensors, adductors). You may want to keep these groupings in mind as you explore the muscular system.

A. Identifying muscles

Using models and charts of the human musculature, locate the skeletal muscles outlined in the succeeding pages. As you do, note the size and shape of each muscle.

> **HINT** → Individual muscles can be located on a model of human musculature by locating them in the diagrams on the following pages, then checking their origins, insertions, actions as listed on the tables given for this exercise. Another method—more challenging, but more informative—is to try to identify them by their attachment points and actions *before* identifying them in a diagram.

❏ 1 Identify these muscles of the *head and neck:*
- **Occipitofrontalis (epicranius)**
- **Orbicularis oculi**
- **Orbicularis oris**
- **Buccinator**
- **Zygomaticus (two muscles)**
- **Levator labii superioris,** *major* **and** *minor*
- **Depressor anguli oris**
- **Temporalis**
- **Masseter**
- **Pterygoids (two muscles,** *internal* **and** *external***)**
- **Sternocleidomastoid**
- **Trapezius**

❏ 2 Identify these muscles of the *trunk:*
- **Erector spinae (divides into three muscles)**
- **Deep back muscles**
- **External intercostals**
- **Internal intercostals**
- **Rectus abdominis**
- **External abdominal oblique**
- **Internal abdominal oblique**
- **Transversus abdominis**
- **Trapezius**
- **Levator scapulae**
- **Rhomboideus (two muscles),** *major* **and** *minor*
- **Serratus anterior**
- **Pectoralis minor**
- **Pectoralis major**
- **Teres major**
- **Latissimus dorsi**

Text continued on p. 176.

Muscles of the Head and Neck

🖉COLORING EXERCISE Using colored pens or pencils, shade in the figure and accompanying labels in contrasting colors of your choice as indicated by the red numerals.

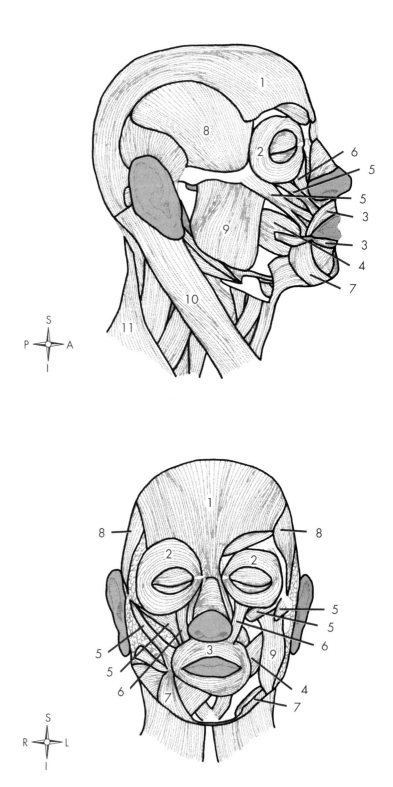

Figure 18-1

Muscle	Origin	Insertion	Action
OCCIPITOFRONTALIS₁	Occipital	Skin of eyebrow, nose	Elevates brows; moves scalp
ORBICULARIS OCULI₂	Maxilla, frontal	Encircles eye, near origin	Closes eye
ORBICULARIS ORIS₃	Maxilla, mandible	Lips	Closes lips
BUCCINATOR₄	Maxilla, mandible	Angle of mouth	Compresses cheeks
ZYGOMATICUS₅ (two)	Zygomatic bone	Angle of mouth, upper lip	Elevates angle of mouth, upper lip
LEVATOR LABII SUPERIORIS₆	Maxilla	Upper lip, nose	Elevates upper lip, nose
DEPRESSOR ANGULI ORIS₇	Mandible	Lower lip near angle	Depresses angle of mouth
TEMPORALIS₈	Temporal aspect of skull	Mandible	Closes jaw
MASSETER₉	Zygomatic arch	Mandible	Closes jaw
PTERYGOIDS(two)	Inferior aspect of skull	Mandible	Medial closes jaw; lateral opens jaw
STERNOCLEIDOMASTOID₁₀	Sternum, clavicle	Mastoid process (skull)	Rotates, extends head
TRAPEZIUS₁₁	Skull, upper vertebral column	Scapula	Extends head, neck

Use terms in OUTLINE type as coloring labels. Terms in **SOLID** type do not appear in the coloring plate.

Muscles of the Trunk

🖉 COLORING EXERCISE   Using colored pens or pencils, shade in the figure and accompanying labels in contrasting colors of your choice as indicated by the red numerals.

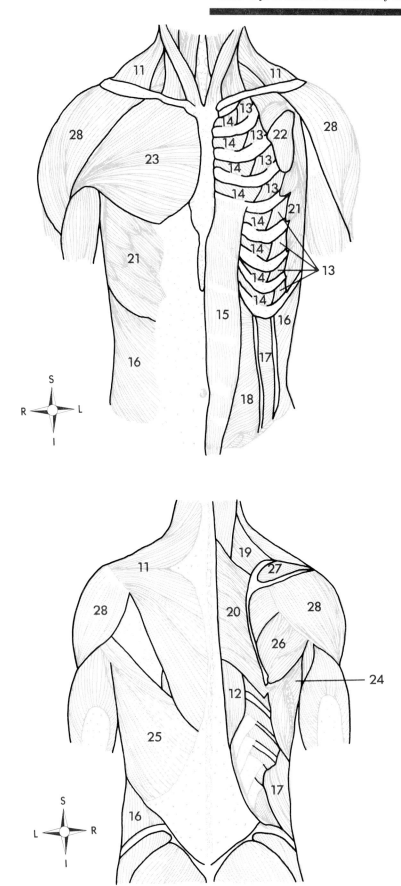

Figure 18-2

Muscle	Origin	Insertion	Action
ERECTOR SPINAE₁₂ (divides into three)	Vertebrae, pelvis	Superior vertebrae, ribs	Holds body upright
DEEP BACK MUSCLES	Vertebrae	Vertebrae	Flex or extend trunk
EXTERNAL INTERCOSTALS₁₃	Ribs	Edge of next rib (inferiorly)	Expand thorax
INTERNAL INTERCOSTALS₁₄	Ribs	Edge of next rib (superiorly)	Compress thorax
RECTUS ABDOMINIS₁₅	Pubis	Inferior thoracic cage	Flexes waist
EXTERNAL ABDOMINAL OBLIQUE₁₆	Inferior thoracic cage	Midline of abdomen	Compresses abdomen
INTERNAL ABDOMINAL OBLIQUE₁₇	Pelvis	Midline of abdomen	Compresses abdomen
TRANSVERSUS ABDOMINIS₁₈	Vertebrae, pelvis, ribs	Midline of abdomen	Compresses abdomen
TRAPEZIUS₁₁	Skull, upper vertebral column	Scapula	Extends head, neck; rotates scapula
LEVATOR SCAPULAE₁₉	Vertebrae	Scapula	Elevates scapula
RHOMBOIDEUS₂₀ (two)	Vertebrae	Scapula	Retract scapula
SERRATUS ANTERIOR₂₁	Ribs	Scapula	Protracts scapula
PECTORALIS MINOR₂₂	Ribs	Scapula	Depresses scapula
PECTORALIS MAJOR₂₃	Ribs, clavicle	Humerus	Adducts, flexes arm
TERES MAJOR₂₄	Scapula	Humerus	Extends, adducts, rotates arm
LATISSIMUS DORSI₂₅	Vertebrae	Humerus	Extends arm
INFRASPINATUS₂₆	Scapula	Humerus	Extends, rotates arm
SUPRASPINATUS₂₇	Scapula	Humerus	Abducts arm
SUBSCAPULARIS	Scapula	Humerus	Extends, rotates arm
TERES MINOR	Scapula	Humerus	Adducts, rotates arm
DELTOID₂₈	Scapula, clavicle	Humerus	Abducts arm

Use terms in OUTLINE type as coloring labels. Terms in **SOLID** type do not appear in the coloring plate.

Muscles of the Upper Extremity

COLORING EXERCISE Using colored pens or pencils, shade in the figure and accompanying labels in contrasting colors of your choice as indicated by the red numerals.

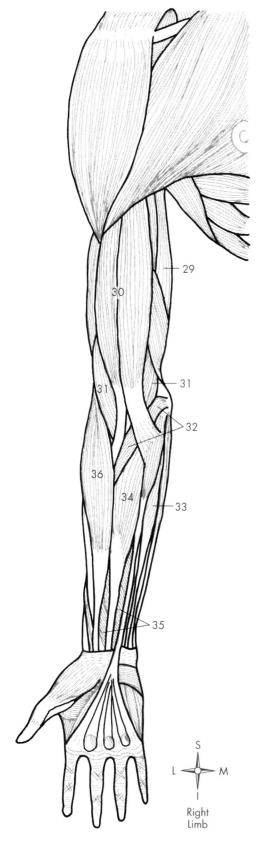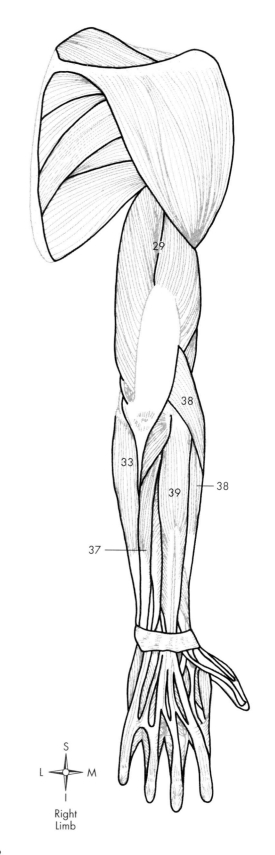

Figure 18-3

Muscle	Origin	Insertion	Action
TRICEPS BRACHII₂₉	Humerus, scapula	Ulna	Extend forearm
BICEPS BRACHII₃₀	Humerus, scapula	Radius	Flex, supinate forearm
BRACHIALIS₃₁	Humerus	Ulna	Flexes forearm
PRONATORS₃₂ (two)	Ulna, humerus	Radius	Pronate forearm
FLEXOR CARPI ULNARIS₃₃	Humerus (medial epicondyle)	Carpal bone	Flex, abduct wrist
FLEXOR CARPI RADIALIS₃₄	Humerus (medial epicondyle)	Metacarpal bones	Flex, abduct wrist
FLEXOR DIGITORUM₃₅ (two)	Humerus (medial epicondyle, ulna, radius)	Phalanges	Flex fingers
BRACHIORADIALIS₃₆	Humerus	Radius (distal)	Flex, pronate forearm
SUPINATOR	Ulna	Radius	Supinates forearm
EXTENSOR CARPI ULNARIS₃₇	Humerus (lateral epicondyle)	Metacarpal bones	Extends, abducts wrist
EXTENSOR CARPI RADIALIS₃₈ (two)	Humerus (lateral epicondyle)	Metacarpal bones	Extends, abducts wrist
EXTENSOR DIGITORUM₃₉	Humerus (lateral epicondyle)	Phalanges	Extends fingers

Use terms in OUTLINE type as coloring labels. Terms in **SOLID** type do not appear in the coloring plate.

Muscles of the Lower Extremity

COLORING EXERCISE Using colored pens or pencils, shade in the figure and accompanying labels in contrasting colors of your choice as indicated by the red numerals.

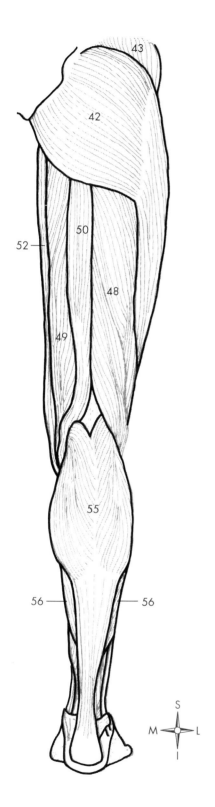
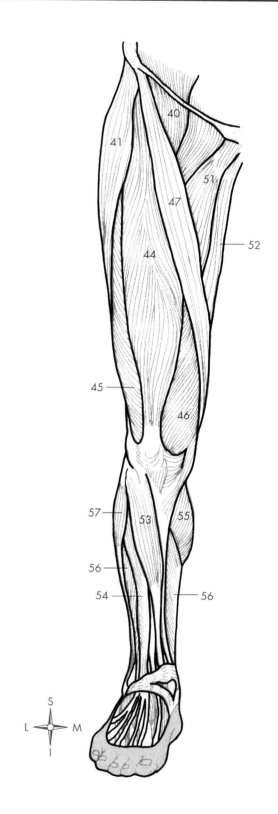

Figure 18-4

172

Muscle	Origin	Insertion	Action
ILIOPSOAS₄₀	Ilium, vertebrae	Femur	Flexes thigh
TENSOR FASCIAE LATAE₄₁	Hip	Tibia	Abducts thigh
GLUTEAL GROUP (a, b, c)			
a GLUTEUS MAXIMUS₄₂	Hip	Femur	Extends thigh
b GLUTEUS MEDIUS₄₃	Hip	Femur	Extends thigh
c GLUTEUS MINIMUS	Ilium	Femur	Abducts thigh
QUADRICEPS FEMORIS GROUP (a, b, c, d)			
a RECTUS FEMORIS₄₄	Ilium	Tibia	Extends leg; flexes thigh
b VASTUS LATERALIS₄₅	Femur	Tibia	Extends leg
c VASTUS MEDIALIS₄₆	Femur	Tibia	Extends leg
d VASTUS INTERMEDIUS	Femur	Tibia	Extends leg
SARTORIUS₄₇	Ilium	Tibia	Flexes thigh; flexes, rotates leg
HAMSTRING GROUP (a, b, c)			
a BICEPS FEMORIS₄₈	Ischium, femur	Fibula	Flexes leg; extends thigh
b SEMIMEMBRANOSUS₄₉	Ischium	Tibia	Flexes leg; extends thigh
c SEMITENDINOSUS₅₀	Ischium	Tibia	Flexes leg; extends thigh
ADDUCTOR GROUP (a, b)			
a ADDUCTOR LONGUS₅₁	Pubis	Femur	Adducts thigh
b GRACILIA₅₂	Pubis	Tibia	Adducts thigh
TIBIALIS ANTERIOR₅₃	Tibia	Metatarsal bones	Extends (dorsiflexes) foot
EXTENSOR DIGITORUM LONGUS₅₄	Tibia	Phalanges	Extends toes
GASTROCNEMIUS₅₅	Femur	Calcaneus	(Plantar) flexes foot
SOLEUS₅₆	Tibia, fibula	Calcaneus	(Plantar) flexes foot
PERONEUS₅₇ (three)	Tibia, fibula	Tarsal, metatarsal bones	Flex, evert foot

Use terms in OUTLINE type as coloring labels. Terms in **SOLID** type do not appear in the coloring plate.

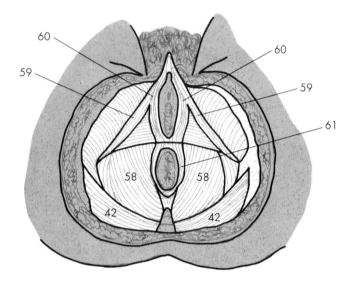

Female

SPECIAL CHALLENGE → By now you have noticed that several of the muscles illustrated in the coloring exercises do not have labels. Some of them are muscles that you were asked to identify but are not labeled for coloring because they are too far away from the outline labels. Go back to the figures and try to identify the unlabeled muscles. Use your textbook and anatomy reference books in your school library to help you.

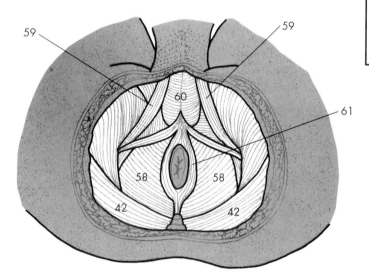

Male

Figure 18-5

Muscle	Origin	Insertion	Action
LEVATOR ANI[58]	Pubis, ischium	Sacrum, coccyx	Elevates anus
ISCHIOCAVERNOSUS[59]	Ischium	Clitoris or penis	Compresses base of clitoris or penis
BULBOSPONGIOSUS[60]	MALE: bulb of penis FEMALE: central tendon of perineum clitoris	MALE: central tendon of perineum FEMALE: base of clitoris	MALE: constricts urethra; erects penis FEMALE: erects
SPHINCTER EXTERNUS ANI[61]	Coccyx	Central tendon (median raphe)	Closes anal canal
GLUTEUS MAXIMUS[42]	Hip	Femur	Extends thigh

- **Infraspinatus**
- **Supraspinatus**
- **Subscapularis**
- **Teres minor**
- **Deltoid**

> **HINT** → Notice that many of the muscles listed for the trunk include shoulder (upper trunk) muscles that insert on the arm.

❑ 3 Identify these muscles of the *upper extremity:*
- **Triceps brachii**
- **Biceps brachii**
- **Brachialis**
- **Pronators** (two muscles, *teres* and *quadratus*)
- **Flexor carpi ulnaris**
- **Flexor carpi radialis**
- **Flexor digitorum (two muscles)**
- **Brachioradialis**
- **Supinator**
- **Extensor carpi ulnaris**
- **Extensor carpi radialis** (two muscles, *profundus* and *superficialis*)
- **Extensor digitorum**

❑ 4 Identify these muscles of the *lower extremity:*
- **Iliopsoas**
- **Tensor fasciae latae**
- **Gluteal group: gluteus maximus, gluteus medius, gluteus minimus**
- **Quadriceps femoris group: rectus femoris, vastus lateralis, vastus medialis, vastus intermedius**
- **Sartorius**
- **Hamstring group: biceps femoris, semimembranosus, semitendinosus**
- **Adductor group: adductor longus, gracilis**
- **Tibialis anterior**
- **Extensor digitorum longus**
- **Gastrocnemius**
- **Soleus**
- **Peroneus** (three muscles, *longus, brevis,* and *xertius*)

❑ 5 Identify these muscles of the *pelvic floor:*
- **Levator ani**
- **Ischiocavernosus**
- **Bulbospongiosus**
- **Sphincter externus ani**
- **Gluteus maximus**

B. Demonstrating muscle action

For each of the muscles found in Activity A of this exercise, demonstrate its action with your own muscle (if possible). As you contract the muscle, palpate it and note its size and location.

> **SAFETY FIRST!** Be careful to avoid injuring yourself and others as you perform each muscle action.

> **HINT** → For a more detailed description of the origin, insertion, action, and innervation of some of these muscles, please refer to your textbook.

LAB REPORT 18

Skeletal Muscle Identification

Table (each muscle name listed is used in the table once only)

- Adductor group
- Biceps brachii
- Deltoid
- Erector spinae
- Gastrocnemius
- Gluteus maximus
- Hamstring group
- Iliopsoas
- Latissimus dorsi
- Pectoralis major
- Pectoralis major/latissimus dorsi combination
- Quadriceps femoris group
- Rectus abdominus
- Tensor fasciae latae
- Tibialis anterior
- Triceps brachii

Part moved	Flexor	Extensor	Abductor	Adductor
Upper arm				
Lower arm				
Thigh				
Lower leg				
Foot				
Trunk				

Fill-in (complete each item with the correct term)

1. The occipitofrontalis, or epicranius, originates on the __?__ bone.
2. The masseter, temporalis, and pterygoids all insert on the __?__.
3. The sternocleidomastoid inserts on the __?__.
4. The temporalis closes the __?__.
5. The __?__ closes the eye.
6. The pectoralis __?__ inserts on the humerus.
7. The deltoid inserts on the __?__.
8. The __?__ intercostals help accomplish forced expiration.
9. The erector __?__ help maintain posture.
10. The pectoralis major is __?__ to the pectoralis minor.
11. The lastissimus dorsi is on the __?__ side of the trunk.
12. The __?__ intercostals contract when you take a deep breath.
13. The rippling effect seen on the lower midline of the abdomens of some athletes is caused by hypertrophy of the __?__ muscle.
14. The origin of the brachialis is __?__ to its insertion.
15. In grasping a baseball tightly in the hand, one would likely use the __?__ muscle.
16. Both the biceps brachii and the __?__ flex the forearm.
17. The extensor digitorum inserts on the __?__.
18. The adductor longus and the __?__ are both part of the *adductor group*.
19. The hamstring muscles all __?__ the leg.
20. The muscle that pulls the leg so that you can cross it over the other leg while sitting is called the __?__ muscle.
21. Both the gastrocnemius and the __?__ plantar flex the foot.
22. Muscles of the *quadriceps group* all __?__ the leg.
23. In the male, the __?__ pulls the penis erect but also can constrict the urethra.
24. The large muscle that extends the thigh is the __?__.
25. The levator ani raises the __?__.

LAB EXERCISE 19

Dissection: Skeletal Muscles

In the previous exercise, you identified the components of the human musculature on models and/or charts of the human muscular system. In this exercise, you are challenged to find some of the major muscles in a preserved mammalian specimen. While the musculature of these animals may differ somewhat from human musculature, a dissection exercise stimulates an appreciation for the highly variable, three-dimensional nature of muscular anatomy that no other type of learning activity can.

Activity A provides directions for studying the musculature of the cat. Activity B is an alternate activity, providing directions for studying the musculature of the fetal pig.

Before you begin

❑ Read the appropriate chapter in your textbook.

❑ Set your learning goals. When you finish this exercise, you should be able to
- dissect the musculature of a preserved cat or fetal pig
- identify the following in a dissected mammalian specimen:
  - muscles of the head and neck
  - muscles of the trunk
  - muscles of the upper extremity
  - muscles of the lower extremity
  - muscles of the pelvic floor

❑ Prepare your materials
- preserved (plain or injected) cat or fetal pig
- dissection tools and trays
- mounted skeleton of the cat or fetal pig
- storage container (if specimen is to be reused)

❑ Read the directions and safety tips for this exercise **carefully** before starting any procedure.

SAFETY FIRST! Observe the usual precautions when working with a preserved specimen. Heed the safety advice accompanying preservatives used with your specimen. Use protective gloves while handling your specimen. Avoid injury with dissection tools. Dispose of or store your specimen as instructed.

A. Cat musculature

If your cat was not skinned in Exercise 10, it must be done before beginning this activity. Refer to Exercise 10, for directions regarding the removal of the skin.

HINT → The muscles are held in place by various connective tissues. The best way to dissect muscles is to first scrape or cut away any large chunks of fat and fascia that cover the muscles of interest, *taking care not to cut into the muscle itself.* Next carefully slip a blunt probe under the muscle and slide it back and forth to break the loose fibers holding the muscle in place. It is best not to cut any of the muscles unless you absolutely must to see an important underlying muscle. When you do cut a muscle, cut it in a way that will permit you to later fold it back into its normal position. Remember, you will probably be studying these muscles over and over again as you prepare for your practical examination on cat musculature. A bag full of separate muscle organs will not help you identify them by location in another specimen.

❑ 1 Identify the muscles of the *head and neck* listed. Before you begin, you must place the cat on its back and remove the skin and underlying connective tissue up to the chin (see Figure 19-2).
- **Mylohyoid**
- **Masseter**
- **Digastric**
- **Sternohyoid**
- **Sternomastoid**

HINT → Refer to LABORATORY REFERENCE, Plates 39 to 44 for color photographs of dissected cat musculature.

ANATOMICAL ATLAS OF THE CAT
(MUSCULAR SYSTEM)

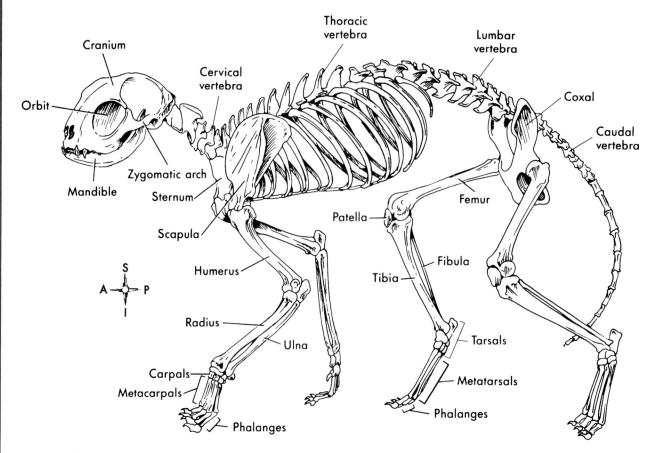

Figure 19-1 Cat skeleton. This figure is provided to facilitate location of muscle origins and insertions listed in accompanying tables.

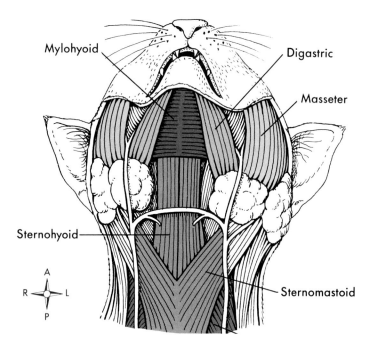

Figure 19-2 Muscles of head and neck (cat).

Table 19-1 Muscles of the head and neck (cat).

Muscle	Origin	Insertion	Action
Mylohyoid	Mandible (body)	Hyoid	Depresses mandible; elevates floor of mouth
Masseter	Zygomatic	Mandible	Elevates, protracts mandible
Digastric	Occipital	Mandible	Depresses, retracts mandible
Sternohyoid	Sternum	Hyoid	Depresses hyoid
Sternomastoid	Sternum	Temporal (mastoid)	Flexes, rotates head

❏ 2 With the cat still on its back, identify the superficial muscles of the *trunk and shoulder* (ventral aspect) listed.
- **Clavobrachialis (clavodeltoid)**
- **Pectoantebrachialis**
- **Pectoralis major**
- **Pectoralis minor**
- **Xiphihumeralis**
- **External oblique**

❏ 3 Make careful incisions near the midline, as shown in Figure 19-3, to expose the following deeper muscles of the *trunk* (ventral aspect):
- **Internal oblique**
- **Transversus abdominis**
- **Rectus abdominis**

❏ 4 Turn the cat onto its ventral side and identify the following superficial muscles of the *trunk and shoulder* (dorsal aspect):
- **Clavotrapezius**
- **Levator scapulae ventralis**
- **Acromiotrapezius**
- **Spinotrapezius**
- **Clavobrachialis**
- **Acromiodeltoid**
- **Spinodeltoid**
- **Latissimus dorsi**

❏ 5 With your scissors, cut the three trapezius muscles as shown in Figure 19-4, to expose the deeper muscles of the *trunk and shoulder*. Identify these muscles:
- **Supraspinatus**
- **Infraspinatus**
- **Teres major**
- **Rhomboideus major**
- **Rhomboideus minor**
- **Rhomboideus capitis**
- **Splenius**

❏ 6 Turn the cat so that it is again lying on its back and identify these muscles of the *forelimb* from the medial aspect:
- **Triceps brachii**
- **Epitrochlearis**
- **Biceps brachii**

❏ 7 Place the cat so that it is lying on its ventral surface and identify these muscles of the *forelimb* from the lateral aspect:
- **Brachialis**
- **Triceps brachii**

HINT → You will have to cut some of the superficial muscles of the upper limb to see the deeper muscles. See Figure 19-5 for directions.

❏ 8 Place the cat on its back and identify these superficial and deep muscles of the *hindlimb* from the medial aspect:
- **Sartorius**
- **Gracilis**
- **Rectus femoris**
- **Vastus medialis**
- **Adductor longus**
- **Semimembranosus**
- **Semitendinosus**
- **Tensor fasciae latae**
- **Flexor digitorum longus**
- **Gastrocnemius**
- **Soleus**
- **Tibialis anterior**

Text continued on p. 193.

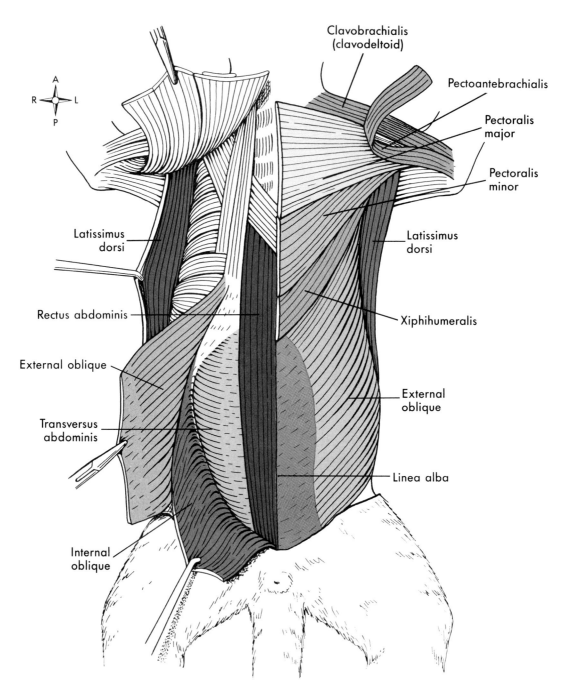

Figure 19-3 Muscles of trunk and shoulder, ventral aspect (cat). Superficial muscles on right side of body have been cut and folded back to expose deeper muscles.

Table 19-2 Muscles of the trunk and shoulder (cat).

Muscle	Origin	Insertion	Action
Clavobrachialis	Clavicle	Ulna (proximal end)	Extends humerus, flexes elbow, rotates head
Pectoantebrachialis	Sternum (anterior)	Ulnar fascia	Adducts humerus
Pectoralis major	Sternum, clavicle	Humerus	Adducts humerus
Pectoralis minor	Sternum	Humerus	Adducts humerus
Xiphihumeralis	Sternum (xiphoid)	Humerus	Adducts humerus
External oblique	Lumbodorsal fascia, posterior ribs	Linea alba	Compresses abdomen; flexes, rotates vertebral column
Internal oblique	Lumbodorsal fascia, pelvis	Linea alba	Compresses abdomen; flexes, rotates vertebral column
Transversus abdominis	Lumbar vertebra, costal cartilage	Linea alba	Compresses abdomen
Rectus abdominis	Coxal (pubis)	Sternum, costal cartilages	Compresses abdomen
Clavotrapezius	Back of skull, dorsal midline of neck	Clavicle	Moves head and clavicle toward each other
Levator scapulae ventralis	Occipital and vertebra C1 (atlas)	Scapula	Moves scapula anteriorly
Acromiotrapezius	Vertebrae C2 to T3 (spinous processes)	Scapula (spine)	Moves scapula toward midline of body
Spinotrapezius	Vertebrae C1 to C4 and T1 to T12 (spinous processes)	Scapula	Moves scapula toward midline of body and posteriorly
Acromiodeltoid	Scapula (acromion)	Humerus (proximal end)	Adducts, rotates humerus
Spinodeltoid	Scapula (spine)	Humerus	Elevates, rotates humerus
Latissimus dorsi	Lumbodorsal fascia and vertebrae T4 to L6	Humerus	Adducts, rotates, extends humerus
Supraspinatus	Scapula (above spine)	Humerus (greater tubercle)	Extends humerus; moves scapula toward head
Infraspinatus	Scapula (below spine)	Humerus (greater tubercle)	Rotates humerus

Continued

Table 19-2 (Continued.)

Muscle	Origin	Insertion	Action
Teres major	Scapula (lateral border)	Humerus (proximal end)	Rotates humerus
Rhomboideus major	Vertebrae (thoracic)	Scapuula (inferior angle)	Pulls scapula toward vertebral column
Rhomboideus minor	Vertebrae (posterior cervical and thoracic)	Scapula (vertebral border)	Pulls scapula toward vertebral column
Rhomboideus capitis	Occipital	Scapula (vertebral border)	Pulls scapula forward and medially
Splenius	Vertebrae (T1 and T2) and fascia of neck	Occipital	Extends and turns head laterally

Table 19-3 Muscles of the forelimb (cat).

Muscle	Origin	Insertion	Action
Triceps brachii	Long head: scapula (edge of glenoid cavity) Medial head: humerus (shaft) Lateral head: humerus (deltoid tuberosity)	Ulna (olecranon)	Extends distal forelimb
Epitrochlearis	Fascia of latissimus dorsi	Ulna (olecranon)	Extends distal forelimb
Biceps brachii	Scapula (edge of glenoid cavity)	Radius (radial tuberosity)	Flexes distal forelimb
Brachialis	Humerus (lateral surface)	Ulna (proximal end)	Flexes distal forelimb

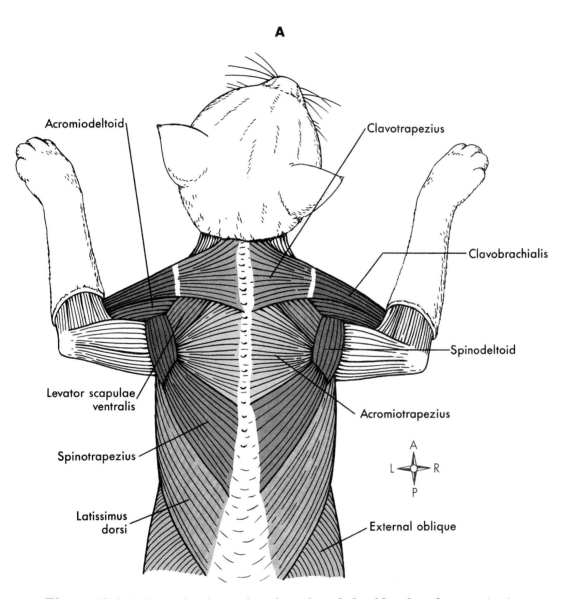

Figure 19-4 A, Superficial muscles of trunk and shoulder, dorsal aspect (cat).

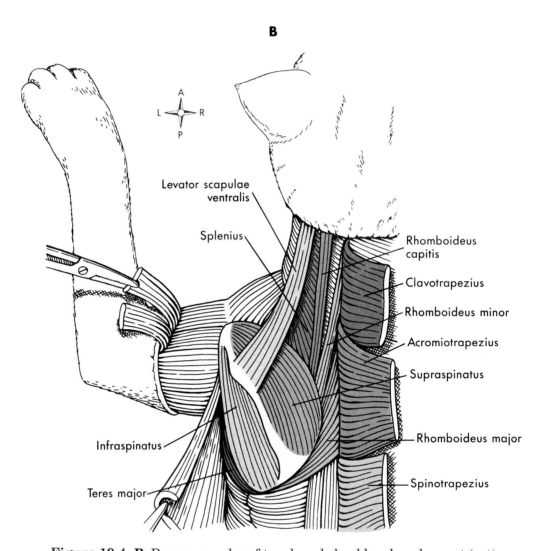

Figure 19-4 B, Deeper muscles of trunk and shoulder, dorsal aspect (cat).

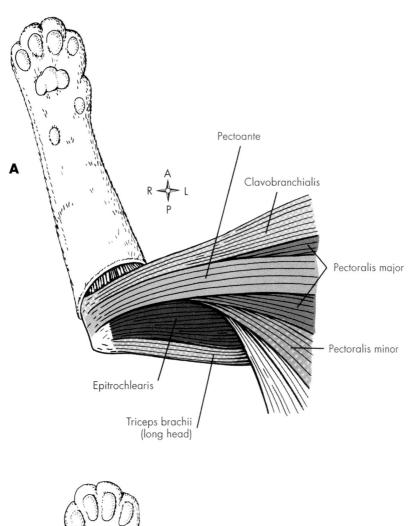
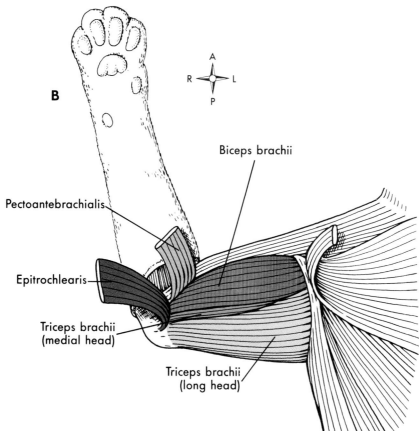

Figure 19-5 Muscles of forelimb (cat). **A,** Superficial muscles, medial aspect. **B,** Deep muscles, medial aspect.

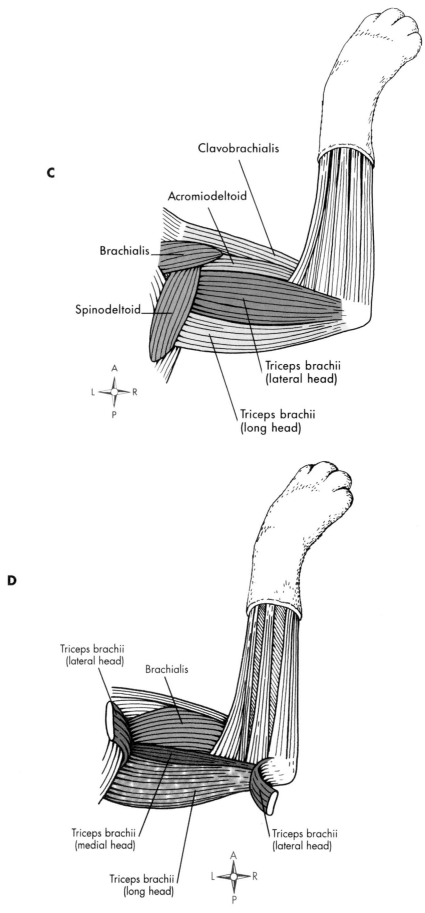

Figure 19-5 Muscles of forelimb (cat). **C,** Superficial muscles, lateral aspect. **D,** Deep muscles, lateral aspect.

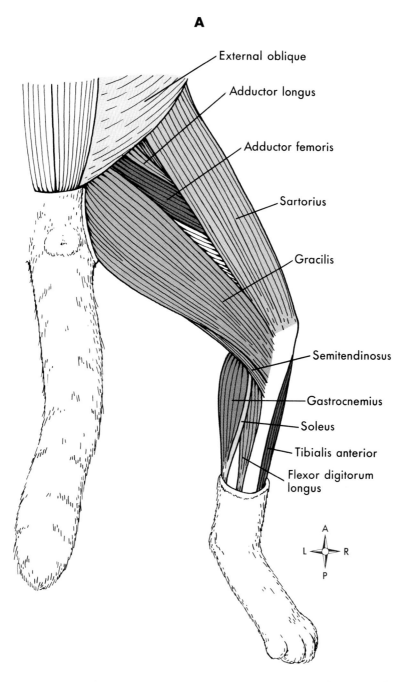

Figure 19-6 Muscles of hindlimb (cat). **A,** Superficial muscles, medial aspect.

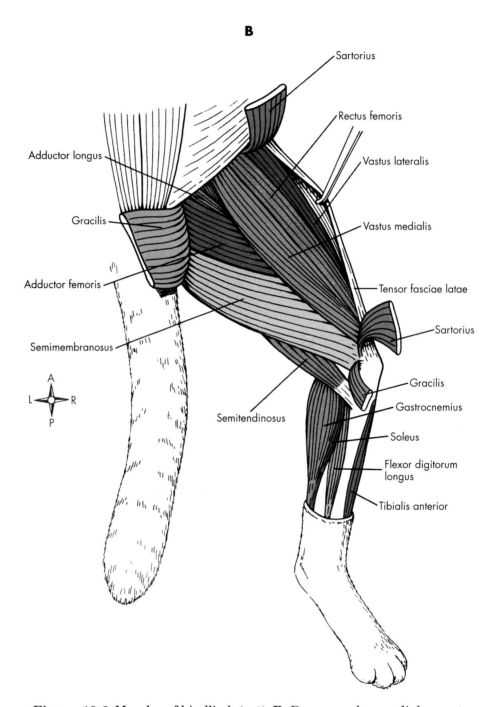

Figure 19-6 Muscles of hindlimb (cat). **B,** Deep muscles, medial aspect.

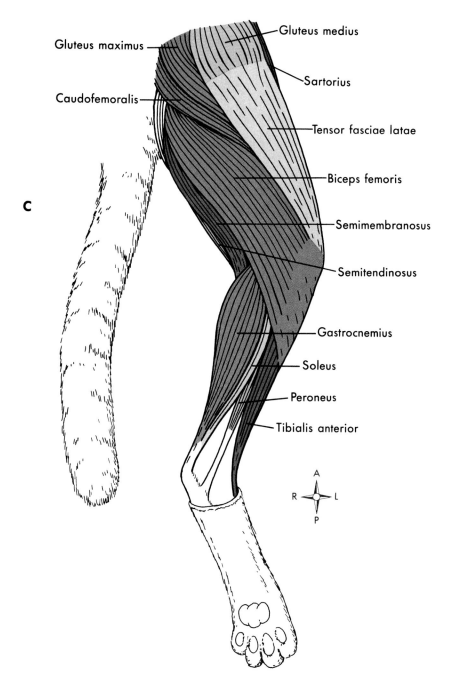

Figure 19-6 Muscles of hindlimb (cat). **C,** Superficial muscles, lateral aspect.

Table 19-4 Muscles of the hindlimb (cat).

Muscles	Origin	Insertion	Action
Sartorius	Coxal (iliac crest)	Tibia, patella	Adducts, rotates thigh and extends distal hindlimb
Gracilis	Coxal (pubis)	Tibia (medial surface)	Adducts hindlimb
Rectus femoris	Coxal (near acetabulum)	Patella	Extends hindlimb
Vastus medialis	Femur	Patella	Extends hindlimb
Adductor longus	Coxal (pubis)	Femur (proximal end)	Adducts hindlimb
Semimembranosus	Coxal (ischium)	Femur (distal end)	Extends hindlimb
Semitendinosus	Coxal (ischium)	Tibia	Flexes distal hindlimb
Tensor fascia latae	Coxal (ilium)	Fascia of thigh	Extends hindlimb
Flexor digitorum longus	Tibia, fibula	Distal phalanges	Flexes digits
Gastrocnemius	Femur (epicondyles)	Calcaneus	Extends foot
Soleus	Fibula (proximal end)	Calcaneus	Extends foot
Tibialis anterior	Tibia, fibula (proximal ends)	Metatarsal I	Flexes foot
Gluteus medius	Coxal (ilium), vertebrae (sacral and caudal)	Femur (proximal end)	Abducts hindlimb
Gluteus maximus	Coxal (ilium), vertebrae (sacral and caudal)	Femur (proximal end)	Abducts hindlimb
Caudofemoralis	Vertebrae (caudal)	Patella	Abducts hindlimb; extends distal hindlimb
Biceps femoris	Coxal (ischium)	Tibia, patella	Abducts hindlimb; extends distal hindlimb
Peroneus	Fibula	Metatarsals	Flexes foot

❏ 9 Place the cat on its ventral surface and identify these muscles of the *hindlimb* from the lateral aspect:
- **Gluteus medius**
- **Gluteus maximus**
- **Caudofemoralis**
- **Tensor fasciae latae**
- **Biceps femoris**
- **Semimembranosus**
- **Semitendinosus**
- **Gastrocnemius**
- **Soleus**
- **Peroneus**
- **Tibialis anterior**

B. Fetal pig musculature

This activity is a brief alternative to the study of cat musculature offered in Activity A. If your fetal pig was not skinned in Exercise 10, it must be done before beginning this activity. Refer to Exercise 10 for directions regarding the removal of the skin.

> **HINT** → The muscles are held in place by various connective tissues. Because the muscles of a fetal pig have not been used extensively by the animal, they often appear underdeveloped and thus difficult to distinguish. The best way to dissect muscles is to first scrape or cut away any fascia that cover the muscles of interest, *taking care not to cut into the muscle itself.* Next carefully slip a blunt probe under the muscle and slide it back and forth to break the loose fibers holding the muscle in place. It is best not to cut any of the muscles unless you absolutely must to see an important underlying muscle. When you do cut a muscle, cut it in a way that will permit you to later fold it back into its normal position. Remember, you will probably be studying these muscles over and over again as you prepare for your practical examination on fetal pig musculature. A bag full of separate muscle organs will not help you identify them by location in another specimen.

❏ 1 Identify the muscles of the *head and neck* listed. Before you begin, you must place the fetal pig on its back and remove the skin and underlying connective tissue up to the chin (see Figure 19-8, *B*).
- **Mylohyoid**
- **Masseter**
- **Digastric**
- **Sternohyoid**
- **Sternomastoid**

❏ 2 With the fetal pig still on its back, identify the superficial muscles of the *trunk and shoulder* (ventral aspect) listed.
- **Clavobrachialis**
- **Superficial pectoral**
- **Serratus anterior**
- **External oblique**
- **Rectus abdominis**

❏ 3 Make a careful cut across the superficial pectoral muscle near the midline, as shown in Figure 19-8, *B*, to expose the following deeper muscles of the *shoulder* (ventral aspect):
- **Anterior deep pectoral**
- **Posterior deep pectoral**
- **Teres major**

❏ 4 Make a cut along the midline of the abdomen to expose these deep muscles of the *trunk* (ventral aspect):
- **Internal oblique**
- **Transversus abdominis**

❏ 5 Turn the fetal pig onto its ventral side and identify the following superficial and deep muscles of the *trunk and shoulder:*
- **Clavotrapezius**
- **Supraspinatus**
- **Acromiotrapezius**
- **Spinotrapezius**
- **Deltoid**
- **Latissimus dorsi**

> **HINT** → Often, the deeper muscles can be located easily by cutting through some of the superficial muscles. Be careful not to remove the superficial muscles, however.

❏ 6 Turn the fetal pig so that it is again lying on its back and identify these muscles of the *forelimb:*
- **Triceps brachii**
- **Biceps brachii**
- **Brachialis**

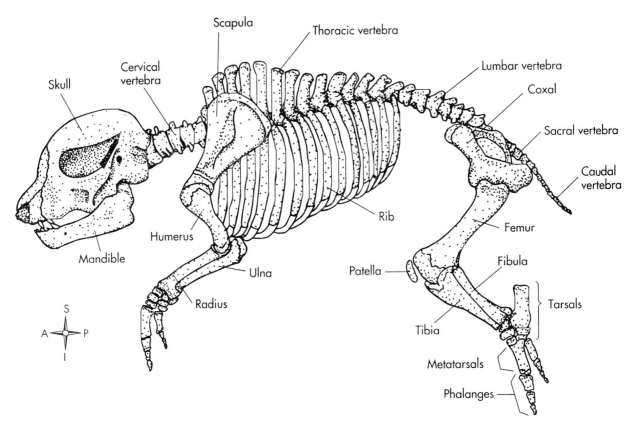

Figure 19-7

Table 19-5 Muscles of the head and neck (fetal pig).

Muscle	Origin	Insertion	Action
Mylohyoid	Mandible (body)	Hyoid	Depresses mandible; elevates floor of mouth
Masseter	Zygomatic	Mandible	Elevates, protracts mandible
Digastric	Occipital	Mandible	Depresses, retracts mandible
Sternohyoid	Sternum	Hyoid	Depresses hyoid
Sternomastoid	Sternum	Temporal (mastoid)	Flexes, rotates head

❑ 7 With the fetal pig still on its back, identify these superficial and deep muscles of the *hindlimb* from the medial aspect:
- **Sartorius**
- **Gracilis**
- **Rectus femoris**
- **Vastus medialis**
- **Pectineus**
- **Iliopsoas**
- **Adductor longus**
- **Semimembranosus**
- **Semitendinosus**
- **Tensor fascia latae**

❑ 8 Place the fetal pig on its ventral surface and identify these muscles of the *hindlimb* from the lateral aspect:
- **Gluteus medius**
- **Tensor fasciae latae**
- **Biceps femoris**

HINT → Refer to LABORATORY REFERENCE, Plate 45 for a color photograph of dissected fetal pig musculature.

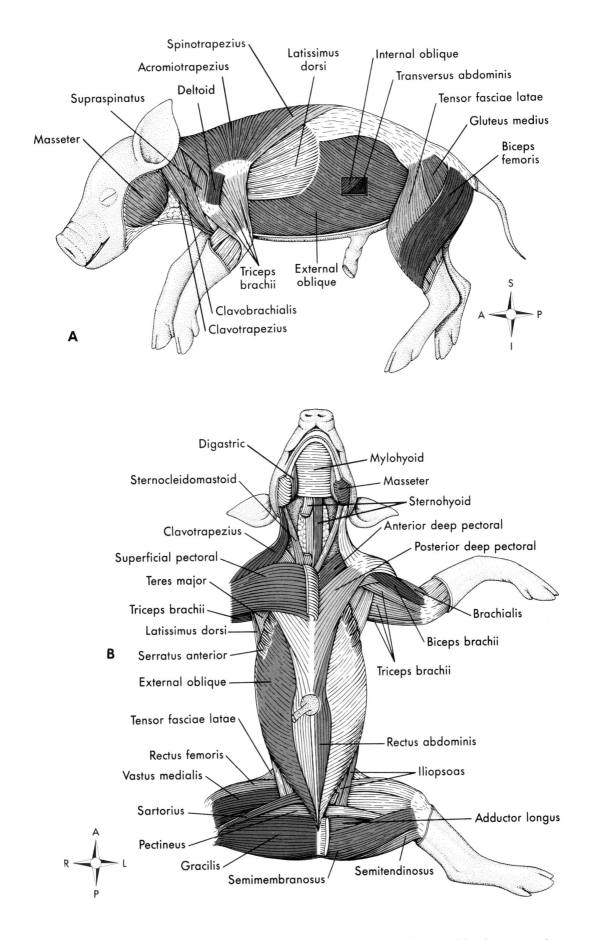

Figure 19-8 A, Lateral view of fetal pig musculature. **B,** Ventral view of fetal pig musclature.

Table 19-6 Muscles of the trunk and shoulder (fetal pig).

Muscle	Origin	Insertion	Action
Clavobrachialis	Clavicle	Ulna (proximal end)	Extends humerus, flexes elbow, rotates head
Superficial pectoral	Sternum	Humerus	Adducts humerus
Serratus anterior	Ribs (4 to 8)	Scapula (vertebral border)	Depresses scapula
External oblique	Lumbodorsal fascia, posterior ribs	Linea alba	Compresses abdomen; flexes, rotates vertebral column
Rectus abdominis	Coxal (pubis)	Sternum, costal cartilages	Compresses abdomen
Anterior deep pectoral	Sternum, costal cartilages	Scapula	Pulls scapula toward midline of body
Posterior deep pectoral	Sternum, costal cartilages	Humerus (proximal end)	Adducts humerus
Teres major	Scapula (lateral border)	Humerus (proximal end)	Rotates humerus
Internal oblique	Lumbodorsal fascia, pelvis	Linea alba	Compresses abdomen; flexes, rotates vertebral column
Transversus abdominis	Lumbar vertebra, costal cartilages	Linea alba	Compresses abdomen
Clavotrapezius	Back of skull, dorsal midline of neck	Clavicle	Moves head and clavicle toward each other
Supraspinatus	Scapula (above spine)	Humerus (greater tubercle)	Extends humerus; moves scapula toward head
Acromiotrapezius	Vertebrae C2 to T3 (spinous processes)	Scapula (spine)	Moves scapula toward midline of body
Spinotrapezius	Vertebrae C1 to C4 and T1 to T12 (spinous processes)	Scapula	Moves scapula toward midline of body and posteriorly
Deltoid	Scapula	Humerus (proximal end)	Adducts humerus
Latissimus dorsi	Lumbodorsal fascia and vertebrae T4 to L6	Humerus	Adducts, rotates, extends humerus

Table 19-7 Muscles of the forelimb (fetal pig).

Muscle	Origin	Insertion	Action
Triceps brachii	Long head: scapula (edge of glenoid cavity) Medial head: humerus (shaft) Lateral head: humerus (deltoid tuberosity)	Ulna (olecranon)	Extends distal forelimb
Biceps brachii	Scapula (edge of glenoid cavity)	Radius (radial tuberosity)	Flexes distal forelimb
Brachialis	Humerus (lateral surface)	Ulna (proximal end)	Flexes distal forelimb

Table 19-8 Muscles of the hindlimb (fetal pig).

Muscle	Origin	Insertion	Action
Sartorius	Coxal (iliac crest)	Tibia, patella	Adducts, rotates thigh and extends distal hindlimb
Gracilis	Coxal (pubis)	Tibia (medial surface)	Adducts hindlimb
Rectus femoris	Coxal (near acetabulum)	Patella	Extends hindlimb
Vastus medialis	Femur	Patella	Extends hindlimb
Pectineus	Coxal (pubis)	Femur (proximal end)	Adducts hindlimb
Iliopsoas	Vertebrae (lumbar), coxal (ilium)	Femur (lesser trochanter)	Flexes, rotates hindlimb
Adductor longus	Coxal (pubis)	Femur (proximal end)	Adducts hindlimb
Semi-membranosus	Coxal (ischium)	Femur (distal end)	Extends hindlimb
Semitendinosus	Coxal (ischium)	Tibia	Flexes distal hindlimb
Tensor fascia latae	Coxal (ilium)	Fascia of thigh	Extends hindlimb
Gluteus medius	Coxal (ilium), vertebrae (sacral and caudal)	Femur (proximal end)	Abducts hindlimb
Biceps femoris	Coxal (ischium)	Tibia, patella	Abducts hindlimb; extends distal hindlimb

NAME _____ DATE _____ SECTION _____

LAB REPORT 19

Dissection: Skeletal Muscles

A. Cat musculature dissection checklist

■ **Muscles of the head and neck**
- mylohyoid
- masseter
- digastric
- sternohyoid
- sternomastoid

■ **Muscles of the trunk and shoulder**
- clavobrachialis
- pectoantebrachialis
- pectoralis major
- pectoralis minor
- xiphihumeralis
- external oblique
- internal oblique
- transversus abdominis
- rectus abdominis
- clavotrapezius
- levator scapulae ventralis
- acromiotrapezius
- spinotrapezius
- acromiodeltoid
- spinodeltoid
- latissimus dorsi
- supraspinatus
- infraspinatus
- teres major
- rhomboideus major
- rhomboideus minor
- rhomboideus capitis
- splenius

■ **Muscles of the forelimb**
- triceps brachii
- epitrochlearis
- biceps brachii
- brachialis

■ **Muscles of the hindlimb**
- sartorius
- gracilis
- rectus femoris
- vastus medialis
- adductor longus
- semimembranosus
- semitendinosus
- tensor fascia latae
- flexor digitorum longus
- gastrocnemius
- soleus
- tibialis anterior
- gluteus medius
- gluteus maximus
- caudofemoralis
- biceps femoris
- peroneus

B. Fetal pig musculature dissection checklist

■ **Muscles of the head and neck**
- mylohyoid
- masseter
- digastric
- sternohyoid
- sternomastoid

■ **Muscles of the trunk and shoulder**
- clavobrachialis
- superficial pectoral
- serratus anterior
- external oblique
- rectus abdominis
- anterior deep pectoral
- posterior deep pectoral
- teres major
- internal oblique
- transversus abdominis
- clavotrapezius
- supraspinatus
- acromiotrapezius
- spinotrapezius
- deltoid
- latissimus dorsi

■ **Muscles of the forelimb**
- triceps brachii
- biceps brachii
- brachialis

■ **Muscles of the hindlimb**
- sartorius
- gracilis
- rectus femoris
- vastus medialis
- pectineus
- iliopsoas
- adductor longus
- semimembranosus
- semitendinosus
- tensor fascia latae
- gluteus medius
- biceps femoris

LAB EXERCISE 20

Skeletal Muscle Contractions

Now that you are familiar with the structure of the muscular system and its organs, it is time to take a closer look at the physiology of this system. The first activity demonstrates contraction at the cell level. The second and third activities challenge you to investigate concepts of muscle function at the organ level.

Before you begin

❏ Read the appropriate chapter in your textbook.

❏ Set your learning goals. When you finish this exercise, you should be able to
- describe some factors that influence the contraction of single skeletal muscle fibers
- interpret a myogram of a single twitch, determining its three phases
- understand the nature of the treppe phenomenon and tetanus
- demonstrate how muscle contraction can be studied in a laboratory setting

❏ Prepare your materials:
- glycerinated rabbit psoas muscle
- watch glass or Petri dish
- microscopes (dissecting and regular)
- fine glass probes
- glass slides and coverslips
- glycerine solution (50% in triple distilled water)
- metric ruler (mm)
- ATP + salt solution
- ATP solution
- salt solution

If Activity B is performed:
- frog (live)
- pithing needle
- dissection pan or board
- dissection tools
- Ringer's solution (amphibian)
- blunt glass probe or polished tube
- string
- kymograph or physiograph system
- stimulator apparatus

If Activity C is performed:
- PHYSIOGRIP system
- computer system (e.g., CPU, monitor, printer)
- electrode gel
- flat plate electrode (and tape)
- stimulator
- pen (washable ink)
- floppy disks (if data are to be stored)

❏ Read the directions and safety tips for this exercise **carefully** before starting any procedure.

A. Contraction of single muscle fibers

In this activity, you will observe contractions in single skeletal muscle fibers. Your specimen is a special glycerinated preparation of a rabbit psoas muscle.

> **SAFETY FIRST!** Take the usual precautions when using a microscope. Protect your eyes and skin from laboratory solutions. Be careful to avoid injury when using the glass probes.

❏ 1 Obtain a small piece (1 to 2 cm) of glycerinated muscle tissue in a watch glass or Petri dish.

❏ 2 Focus on the specimen after placing it on the stage of a dissecting microscope. (You may use a hand lens instead.)

❏ 3 Using fine glass probes or micropipette tips, try to separate some individual fibers from the rest of the bundle. *Avoid using your hands or metal instruments to manipulate the specimen.*

> **HINT** → Although it may seem difficult to separate fibers at first, care and patience will get the job done quickly and efficiently. Individual fibers are best, but a few in an unseparated bundle will do.

❏ 4 Remove one of the fibers and place it on a microscope slide with a coverslip. Observe the specimen on a compound microscope.

☐ 5 Observe the muscle fiber with low power, then high power. Make a sketch of your observation. Are the striations visible? Which parts of the muscle fiber form these stripes?

☐ 6 Put three or four separated fibers on a different microscope slide, straight and parallel to one another. Do not use a coverslip. With a metric ruler, measure the length of each fiber (in mm).

☐ 7 With the dropper built into the bottle, use the ATP + salt solution to bathe the fibers. Observe the fibers for 30 seconds and note any contraction.

☐ 8 Measure each fiber again and record the new length. Calculate the percentage of change in each fiber using the following formula:

$$\frac{(\text{Beginning length}) - (\text{Ending length})}{(\text{Beginning length})} \times 100 = \%$$

What is the *average* percentage of change in the fibers? How do you account for the change? If there was no change, why not?

☐ 9 Repeat steps 6 through 8 using ATP solution and again using salt solution. Based on your observations, what chemicals are essential for contraction?

B. Contraction in a muscle organ

Before beginning this activity, review these basic principles of muscle physiology:

Individual muscle fibers contract fully (under existing conditions) or not at all. This property is sometimes termed the *all-or-none* principle. The normal stimulus for a muscle fiber contraction is a *neurotransmitter* chemical from a nerve cell. The nerve-muscle fiber connection is called a **neuromuscular junction.** Electrical or mechanical stimulation can also cause a contraction. Regardless of whether the stimulus is chemical or electrical, it must be strong enough to pass the **threshold of stimulation**—the point at which it is just strong enough to initiate a contraction.

Fibers that compose a muscle organ are divided into functional teams, or **motor units.** Each fiber in a motor unit is innervated by the same nerve cell and so contracts at the same time. The relative strength of contraction in a whole organ depends in part on how many motor units are stimulated at one time. **Recruitment** of more units increases the organ's strength of contraction.

We will demonstrate muscle organ contraction by stimulating many motor units in a prepared muscle organ. To record our observations, we will use a **myograph.** Myograms, or graphic representations of muscle contractions, can be produced for **isotonic** contractions and for **isometric** contractions. Isotonic contractions decrease muscle length (without increasing tension), and isometric contractions increase tension (without decreasing length). On myograms, time is the horizontal axis, and strength (or length) of contraction is the vertical axis.

The myogram of a single **twitch** contraction shows these features:
- **Latent period**—Phase after stimulation has occurred but during which a contraction is not yet apparent
- **Contraction phase**—Period during which the myogram line rises, indicating that contraction is in progress
- **Relaxation phase**—Period when the line falls, indicating a return to the resting state
- **Refractory period**—Time during which the relaxed muscle cannot be stimulated to contract again (concurrent with the relaxation phase and, in some cases, shorter than the relaxation phase)

If restimulation of the muscle occurs soon enough after a twitch contraction, you may observe the **treppe** phenomenon, also called the *staircase effect*. The second in a series of contractions has a larger amplitude (size) than the first, the third an even greater amplitude, and so on. Eventually, a maximum is reached, and the waves reach a plateau. When the muscle can no longer sustain contractions, it is in a state of **muscle fatigue.**

When the refractory period is short and a muscle is continually restimulated before the relaxation phase is over, the myogram will not show the wave line returning to the base line. Instead, the twitch waves will seem to fuse together. Wave summation, or fusion, is often termed **tetanus.** *Incomplete tetanus* occurs when a slight relaxation can be seen between waves. *Complete tetanus* occurs when the waves fuse into a line.

In this activity, you (or your instructor) will use the gastrocnemius muscle of a frog to demonstrate some of the major principles of muscle contraction.

> **SAFETY FIRST!** Avoid cuts and punctures when using the dissection tools. Check for damaged or improper wiring on the electrical equipment. Do not touch the stimulator electrode or other bare wires directly.

Prepare a frog gastrocnemius muscle in the manner described in steps 1 through 7 (and Figure 20-2).

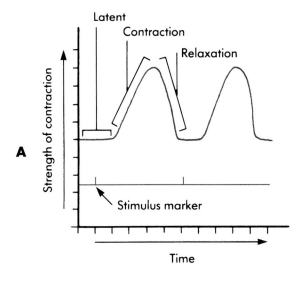
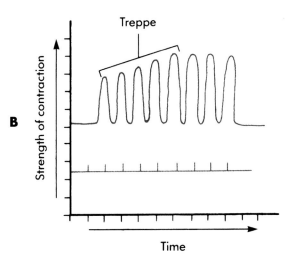
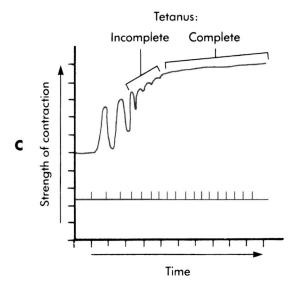

Figure 20-1 **A,** Twitch contractions on a myogram chart. **B,** The treppe phenomenon. **C,** Tetanus.

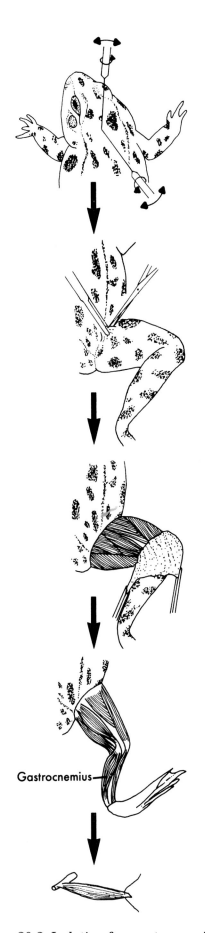

Figure 20-2 Isolating frog gastrocnemius muscle.

☐ 1 Holding the frog in your hand, locate the joint between the base of the skull and the first vertebra by bending the head down until a notch along the posterior edge of the skull can be felt or is visible.

☐ 2 Quickly push a pithing needle into the groove, and then into the cranial cavity. While rotating the needle, move it from side to side to destroy the brain tissue inside.

☐ 3 Pull out the needle partially so that you can insert the tip into the vertebral canal. Push the needle as far as you can, again rotating it to destroy the nerve tissue.

☐ 4 Place the frog on a dissection pan or board. Lift the skin around the hip joint with a forceps and puncture it with the tip of a scissors. Try not to damage the muscle tissue of the leg.

☐ 5 Using the scissors, cut the skin in a circle all the way around the base of the thigh. Grasp the cut edge of the thigh's skin with forceps and pull it toward the foot. The skin will pull away from the leg as if it were a stocking. Keep pulling until the ankle joint appears.

> **HINT** → Keep the muscle tissue moist with Ringer's solution so that it will remain functional.

☐ 6 Locate the *gastrocnemius muscle* (the fleshy part of the calf) and the *calcaneal* (Achilles) tendon, which attaches it to the ankle. Slip a blunt glass probe under the muscle and slide it back and forth to detach the muscle from underlying connective tissue. While lifting the muscle with the probe, tie a 10 to 15 cm piece of string to the calcaneal tendon. Finally, cut the tendon distal to the knot.

☐ 7 Cut the thigh (bone and muscle) just proximal to the gastrocnemius muscle's attachment to the femur. Break the femur away from the knee joint and remove all muscles but the gastrocnemius. You now have a muscle ready to use in an experiment.

> Animals used in research and education should be treated respectfully and humanely. Double pithing is an accepted method for destroying the central nervous system so that the animal will feel no pain. This method also assures that certain tissues will remain functional and will not be damaged by spinal reflexes. The frogs that you use have been raised for research and educational purposes.

☐ 8 Set up a kymograph or physiograph according to the instructions accompanying the instrument. You may be measuring the intensity of an isotonic contraction by using a movable lever attached to the muscle (as in a kymograph). Or, you may be measuring an isometric contraction by means of the force with which it pulls on a transducer (as in a physiograph).

> **HINT** → Different systems operate differently, but here is an easy way to calibrate just about any physiograph set up:
>
> ▪ suspend a 100 gram weight from the transducer
> ▪ adjust the signal amplifier SENSITIVITY or GAIN to produce a 5 cm deflection of the pen
>
> *Check the manufacturer's handbook for further details regarding your particular system.*

☐ 9 Set the stimulator *mode* to CONTINUOUS (REPEAT), *duration* (width) to 1 msec, *frequency* to 1 Hz (1 stimulus per second), and *voltage* to 0.1 volt (v). Stimulate the muscle with the stimulator probe. Continue increasing the voltage in 0.1 v steps until a response is observed. This voltage is the *threshold of stimulation*. Record it here for later reference: _____ volts.

☐ 10 Change the stimulator *mode* to SINGLE, then continue to increase voltage over a series of twitches a few seconds apart. You have to trigger each stimulus manually when in the SINGLE mode. Does the amplitude of the myogram curve remain the same? Why or why not? Record the voltage at which a moderately sized twitch is observed: _____ volts.

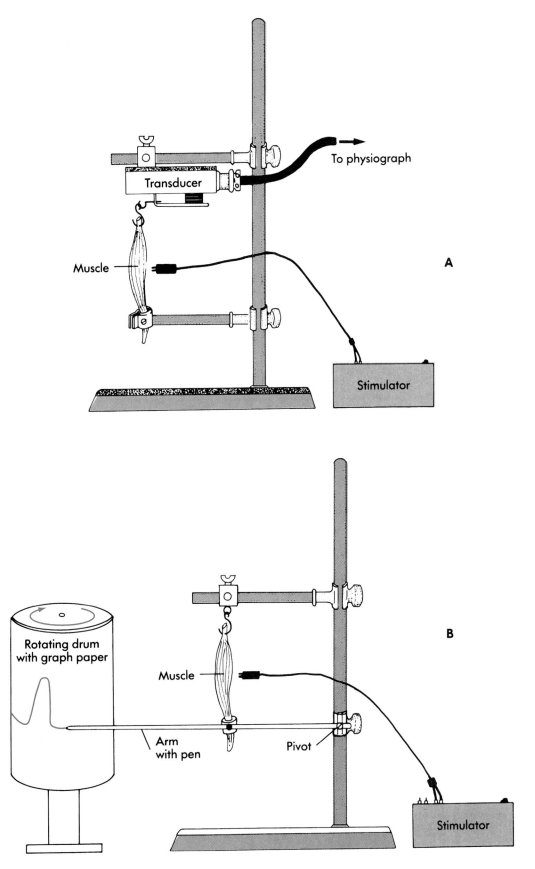

Figure 20-3 A, Typical physiograph setup. **B,** Typical kymograph apparatus. Refer to your system's owner's manual, or your instructor, for detailed instruction in setting up the apparatus.

☐ 11 Set the voltage to the level recorded in step 10, the *mode* to CONTINUOUS (REPEAT), and the *frequency* to 1 Hz. Keeping the voltage constant, record a series of twitches 1 second apart. Does the amplitude increase, decrease, or stay the same? Explain. Allow the muscle to rest for several minutes.

☐ 12 Keeping the voltage constant, increase the frequency slowly (from 1 Hz) until incomplete tetanus, then complete tetanus, is observed. Record the threshold frequencies for each phenomenon in Lab Report 20.

C. Human muscle contraction

☐ 1 Set up the PHYSIOGRIP apparatus according to the user's manual provided by the manufacturer (Intelitool).

☐ 2 Locate the **motor point** of the flexor digitorum superficialis muscle of the right forearm. The motor point is the spot most sensitive to external stimulation. Use this method to find it:
- Put electrode gel on the flat plate electrode and attach it to the dorsal surface of the right hand with a rubber strap (tape).
- With the power switched off, set the stimulator *frequency* to 1 Hz (stimulus/sec), *mode* to CONTINUOUS (REPEAT), *duration* to 1 msec, and *voltage* to 60 volts. Spread a tiny dab of gel on the stimulator probe, then turn the power on.
- Have the subject move the probe around the ventral surface of the forearm close to the belly of the flexor digitorum superficialis muscle until the area of strongest stimulation is found. When the strongest contraction of the third finger is observed, the subject has the correct spot. The subject may want to mark this spot with washable ink.

> **HINT** → Finding the correct motor point is critical to this activity. If you are having difficulty, try pressing rather hard with the stimulator. You may also try increasing the voltage gradually (but not exceeding 80 volts). Females, especially, may get better results with the alternate motor point (and resultant flexion of the fourth finger). See Figure 20-4.

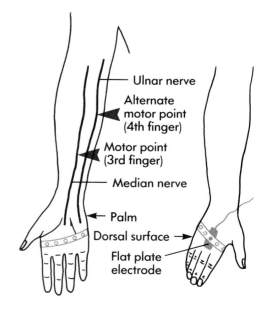

Figure 20-4 Forearm, showing placement of flat plate electrode and likely motor points.

☐ 3 Select CHANGE (ALTER) SWEEP SPEED from the main menu (Apple II) or experiment menu (DOS) and change the sweep speed to 15 seconds per sweep. Next, select RUN (CONTINUOUS) EXPERIMENT MODE. DOS users press Ⓐ to turn Autostop OFF. Have the subject place the right hand on the physiogrip pistol with the appropriate finger (third or fourth) around the trigger. Have the subject pull the trigger just enough to raise the monitor's graph line from the bottom of the screen.

This exercise may be done in addition to, or in place of, Activity B. Rather than using a laboratory animal as a subject, this activity calls for using *yourself* (or your lab partner) as an experimental subject. In this case, a muscle organ is used *in vivo* instead of being isolated from the subject's body. This activity is based on Intelitool's PHYSIOGRIP apparatus for use with Apple-compatible or DOS-based personal computers.

> **SAFETY FIRST!** Do not participate in this activity as a subject if you have health problems, such as a heart condition, that could be affected by this procedure. Be aware of electrical hazards. During the activity, the subject should not touch any person or object other than the pistol grip.

HINT → Press ⓘ in EXPERIMENT MODE to erase data and start over. Press Ⓟ in REVIEW/ANALYZE MODE to print data. A few other important keys:

Ⓢ = stop
Ⓕ = forward to next data frame
Ⓑ = backward to previous screen

Measuring time values:

→ = moves marker to right
← = moves marker to left

Apple II users:

Ⓓ = difference (as in subtracting time values; then press arrow keys to mark period to be measured)

DOS users:

Ⓞ = zero (creates a second interval marker that can be moved [by using arrow keys] to mark period to be measured)
Ⓙ = jump (moves marker much faster than arrow key)
Ⓛ = small jump (moves marker moderately faster than arrow key)

❏ 4 At a setting below 30 volts, stimulate the muscle at the motor point with the stimulator probe. The subject will feel a stimulation once per second. Increase the *voltage* gradually from 30 volts until a response is observed. This voltage is the *threshold of stimulation*. Record it here for later reference: _____ volts.

❏ 5 After the threshold has been reached, continue to increase the voltage slightly over a series of twitches. Does the amplitude of the myogram curve remain the same? Why or why not?

❏ 6 Select SINGLE TWITCH MODE from the main menu, then select RUN SINGLE TWITCH MODE. Set the voltage to the threshold level and the frequency to as low as possible.

❏ 7 Adjust the *voltage* and/or *duration* until the wave produced is 50% to 75% of the screen height. Apple II users press Ⓣ; DOS users press Ⓑ. New data are now being recorded but will not show up on the monitor until 15 data frames have been collected or until you stop the experiment (whichever comes first).

❏ 8 Select REVIEW/ANALYZE from the main menu. DOS users select DISPLACEMENT/TIME ANALYSIS. Measure the length of the latent period, contraction phase, and relaxation phase (in msec).

HINT → You may want to store your data on a floppy disk and do your measurements at a later time. Both the PHYSIOGRIP manual and the program screens instruct you in how to do this. You may also wish to take your measurements from a *hard copy* printout. Make a printout by pressing Ⓟ now (make sure your printer is ON LINE).

❏ 9 Select CHANGE SWEEP SPEED from the main menu and change the speed to 3 seconds per sweep. Select RUN (CONTINUOUS) EXPERIMENT MODE. DOS users press Ⓐ to turn Autostop OFF.

❏ 10 Set the stimulator *frequency* to 1 Hz, the *duration* to 1 msec, and adjust the *voltage* so that the wave is about 25% of the screen height. Gradually increase the *frequency* until incomplete tetanus, then complete tetanus, is observed. Sustain complete tetanus for only 1 or 2 seconds.

❏ 11 Select REVIEW/ANALYZE from the main menu. DOS users select DISPLACEMENT/TIME ANALYSIS. Make any measurements or observations that your instructor may suggest. At what frequency does incomplete tetanus occur? At what frequency does complete tetanus occur? Print out your results and attach them to Lab Report 20.

HINT → The PHYSIOGRIP manual contains additional experiments, as well as troubleshooting tips.

EXERCISE

The term **exercise** has many definitions, depending on the context in which it is used. For some *exercise physiologists,* the term refers to any significant use of skeletal muscles. Exercise physiology has many important applications in athletic training, injury prevention, *ergonomics* (study of body movement as it relates to work activities), physical therapy, and everyday health issues.

An interesting aspect of exercise physiology is that it involves the study of nearly every system and organ in the body, not just the skeletal muscles. When skeletal muscles contract, the sudden increase in metabolism affects many of the processes throughout the body.

For example, the increased use of oxygen in muscles usually triggers an increase in respiratory rate. Why? Because the rate of oxygen extraction from the blood by muscles increases. Respiratory control centers increase the respiratory rate in response to the low blood concentration of oxygen.

Answer the following questions. They will help you appreciate the whole-body aspect of exercise physiology. Read all the questions before answering. You may want to put your answers in the form of a table on separate sheet.

1. List as many metabolic needs of skeletal muscles during exercise as you can. One has already been given to you: *oxygen.* (Think of things from the extracellular environment that are needed by cells as they work.)

2. For each need listed in question 1, name the body process that supplies that need. Think about *all* the processes that fulfill a need. In our example of *oxygen,* breathing is cited as a process that supports oxygen availability. However, blood flow is also required to deliver the oxygen.

3. Go back over each process listed in question 2. Next to each one, indicate how the process may change during exercise. Does its rate increase, decrease?

4. During exercise, some processes slow or stop because metabolic necessities such as oxygen are diverted to the skeletal muscles. What processes must be "put on hold" until exercise is over?

LAB REPORT 20

Skeletal Muscle Contractions

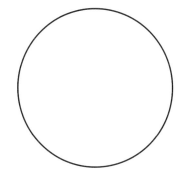

Specimen: *psoas muscle (rabbit)*

Total Magnification: _____

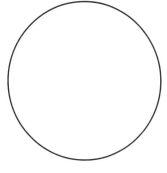

Specimen: *psoas muscle (rabbit)*

Total Magnification: _____

CONTRACTION OF SINGLE MUSCLE FIBERS			
Treatment	Beginning length	Ending length	Percent change
ATP + salt solution			
	Average		
ATP solution			
	Average		
Salt solution			
	Average		
Explanation/interpretation:			

\multicolumn{3}{c}{**CONTRACTION IN A MUSCLE ORGAN**}
Phenomenon	**Value**	**Explanation**
Threshold level of stimulation	volts	What does this value represent?
Level of stimulation required for a standard twitch	volts	What does this value represent? How does knowing this value help you do the rest of the experiment?
Series of twitch contractions		Does the amplitude remain the same for each twitch in the series? Why or why not?
Threshold frequency or incomplete tetanus	Hz	What does this value represent? A common finding in the frog gastrocnemius is a threshold of about 70 Hz. Is your result higher or lower?
Threshold frequency or complete tetanus	Hz	What does this value represent? A common result is around 200 Hz. How does your result compare?
Duration of latent period (of one twitch)	msec	What does this value represent? Can you explain it in terms of the physiology of contraction?
Duration of contraction phase (of one twitch)	msec	What does this value represent?
Duration of relaxation phase (of one twitch)	msec	What does this value represent?
Total duration of one twitch contraction	msec	Looking at all your data, does this value seem to be constant or variable?

Note: Do not forget to attach copies of your myogram charts (with your measurements and notations) to this report.

LAB EXERCISE 21

Nerve Tissue

This exercise is the first of eleven that challenge you to investigate the human **nervous system.** Before going any further, it is best to lay a foundation by discussing the overall organization of this system.

The nervous system is composed of two major divisions: the **central nervous system (CNS)** and the **peripheral nervous system (PNS).** The CNS includes the brain and spinal cord; the PNS includes all the **nerves** that conduct impulses to and from the CNS. Often, the nervous system is instead subdivided into the **afferent division** and the **efferent division.** The afferent division includes nerves and tracts leading *toward* the CNS—the *sensory nerves* and *sensory tracts.* The efferent division includes nerves and tracts leading *away from* the CNS—the *motor nerves* and *motor tracts.* The efferent division, in turn, can be subdivided into the **somatic motor nervous system** and the **autonomic nervous system (ANS).** Although both efferent divisions carry nerve signals away from the brain and spinal cord, they differ in their final destination. Somatic motor nerves end at skeletal muscles, whereas autonomic nerves innervate cardiac muscle, smooth muscle, and glands. (The autonomic nervous system also includes sensory pathways that provide feedback information about autonomic receptors.)

This exercise is an investigation of the cells of the nervous system. Exercise 22 will continue the study of the organization of the nervous system with an exploration of concepts of the *reflex arc.*

Before you begin

❑ Read the appropriate chapter in your textbook and Exercise 9 in this Lab Manual.

❑ Set your learning goals. When you finish this exercise, you should be able to
- describe the structural components of a typical neuron and identify them on a model or chart
- name the three structural categories of neurons and identify their principal locations
- identify specimens of *multipolar* and *bipolar* neurons
- identify a figure of *unipolar* neurons
- name the major categories of neuroglia and identify their principal locations
- identify specimens of figures of neuroglia
- describe the structure of a typical peripheral nerve

❑ Prepare your materials:
- models or charts:
  Multipolar neuron
  Unipolar neuron
  Nerve c.s.
- microscope
- prepared microslides:
  Spinal cord smear (multipolar neurons)
  Retina c.s. (bipolar neurons)
  Peripheral nerve (myelinated) c.s.
  Spinal cord c.s.

❑ Read the directions and safety tips for this exercise **carefully** before starting any procedure.

> **SAFETY FIRST!** Observe the usual precautions when using the microscope.

A. The neuron

As you already know from your study of nervous tissue histology (Exercise 9), the **neuron** is the cell type that conducts impulses, or **action potentials.** The **neuroglia,** on the other hand, support the neurons in any number of ways. Neurons can be *unipolar,* having a single projection from the cell body; *bipolar,* having two projections from the cell body; or *multipolar,* having many projections (see Activity B). This activity asks you to find as many of the listed neuron structures as possible on a multipolar neuron model (or chart) and in a slide of multipolar neurons (spinal cord smear).

❑ 1 Locate the cell body, or **soma.** It is an enlarged area filled with cytoplasm and containing the nucleus and organelles called **Nissl bodies.**

❑ 2 The soma forms a cone-shaped projection, or **axon hillock,** as it projects to become the **axon.** The axon is one of two types of neuron projections (fibers). The axon usually conducts action potentials away from the cell body.

☐ 3 The axon may be wrapped with a series of neuroglial cells called **Schwann cells.** Schwann cells wrap around the axons of some peripheral nerves like tape, each spiraling around a fiber to form a multilayered coating. The inner layers of the Schwann cell are filled with the fatty white substance, **myelin.** Because the Schwann cells are found in series, they form a segmented *sheath of Schwann,* or **myelin sheath.** The gaps between the Schwann cells are termed **nodes of Ranvier.** The outer wrapping of each Schwann cell is normal cytoplasm, with organelles and a nucleus. In the case described, the axon is called a **white fiber,** or **myelinated fiber.** A group of white fibers together is called **white matter.** Schwann cells occur only in the PNS. Within the CNS, myelinated axons are wrapped with extensions of **oligodendrocytes,** another type of neuroglial cell.

☐ 4 Schwann cells or oligodendrocytes are also associated with **unmyelinated axons,** which together with cell bodies and dendrites form **gray matter.** However, in this case, the neuroglia do not form multiple wrappings and are not partially filled with myelin.

☐ 5 **Collateral axons,** or axon branches, can be observed in some cells. Also, the distal ends of axons are often branched. These smaller, distal branches are termed **telodendria.**

☐ 6 Multipolar neurons have many projections from the soma called **dendrites.** Dendrites are branched extensions that are sensitive to stimuli from other cells. Other neurons form an association, or **synapse,** at a bump on the dendrite or cell body (soma). Stimulation of a dendrite or the cell body results in a local change in potential.

B. Structural classification of neurons

There are three categories of neurons based on their shape (see Figure 21-2):

☐ 1 **Multipolar neurons**—Multipolar neurons have multiple projections from the cell *body.* Multipolar neurons comprise most of the neurons in the central nervous system (CNS). Almost all motor neurons are multipolar. Examine a multipolar neuron in a prepared spinal cord smear. Note the many projections. See LABORATORY REFERENCE, Plate 58.

☐ 2 **Bipolar neurons**—Bipolar neurons have exactly two projections from the cell body. Normally, one process conducts impulses toward the cell body. Such a process is called the *dendrite.* The other process conducts impulses away from the cell body and is called the *axon.* Bipolar neurons are most commonly found in the sensory tracts of the eye (for vision) and nasal epithelium (for smell). Examine a prepared slide of the retina (the sensitive lining of the eyeball). Can you distinguish any bipolar cells? A color micrograph of a retina cross section is shown in LABORATORY REFERENCE, Plate 50.

☐ 3 **Unipolar neurons**—Unipolar neurons have one extension from the cell body. However, the single process splits near the body to become two long branches. One branch has dendritelike endings that receive stimulation for an impulse. The remainder of that branch and all of the other branch then conduct the impulse along. The two branches together function as a single axon. Most sensory neurons are unipolar. Examine a figure or model of a unipolar neuron.

C. Types of neuroglia

Neuroglia, or simply **glia,** are the most numerous kinds of cells in nerve tissue. The conducting cells are called **neurons.** Glia, which are often associated with particular neurons, support the structure or function of the neurons. A brief exploration of glia in the human nervous system is outlined in this activity.

☐ 1 **Astrocytes**—Astrocytes look like tiny stars in many preparations. They have a cell body with many elongated projections. Astrocyte projections have broad "feet" that cover blood vessels in the brain. Together with the vessel wall itself, the layer formed by the astrocyte feet form the **blood-brain barrier.** Substances that pass in and out of the blood vessel wall must also move across the wall formed by astrocyte projections. Locate astrocytes in Figure 21-3.

☐ 2 **Microglia**—Microglia are small glial cells that develop from different tissue than other nervous tissue cells. Because of their location, they are considered along with other glial cells. Microglia are able to move to sites of damage or infection to phagocytize harmful substances. Locate microglia in Figure 21-4.

☐ 3 **Oligodendrocytes**—Like astrocytes, oligodendrocytes have many processes. Oligodendrocyte processes form a layer over axons in the central nervous system. This layer, called a **myelin sheath,** affects the speed of conduction in an axon. Locate oligodendrocytes in Figure 21-4.

☐ 4 **Schwann cells**—Schwann cells form myelin sheaths around axons in the peripheral

COLORING EXERCISE Using colored pens or pencils, shade in the figure and accompanying labels in contrasting colors of your choice as indicated by the red numerals.

The Multipolar Neuron

SOMA₁
   NUCLEUS₂
   NUCLEOLUS₃
   MITOCHONDRION₄
   GOLGI APPARATUS₅
   NISSL BODY₆
   AXON HILLOCK₇
DENDRITE₈
AXON₉
SCHWANN CELL₁₀

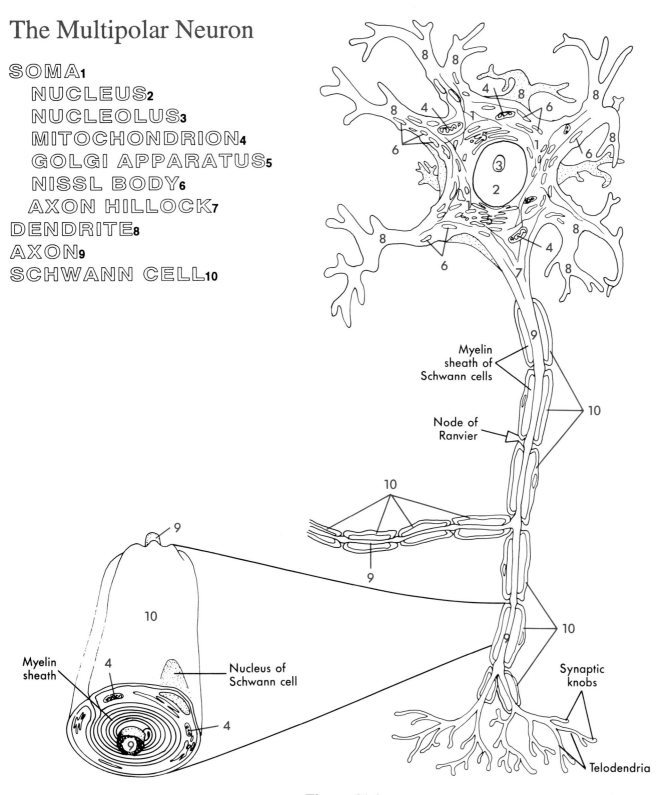

Figure 21-1

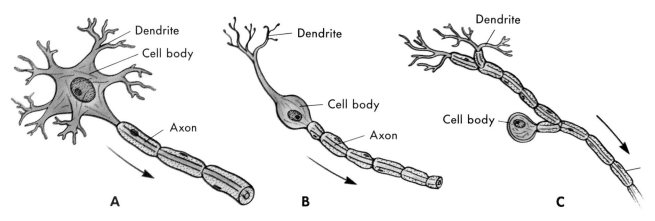

Figure 21-2 Structural classification of neurons. **A,** Multipolar neuron. **B,** Biopolar neuron. **C,** Unipolar neuron.

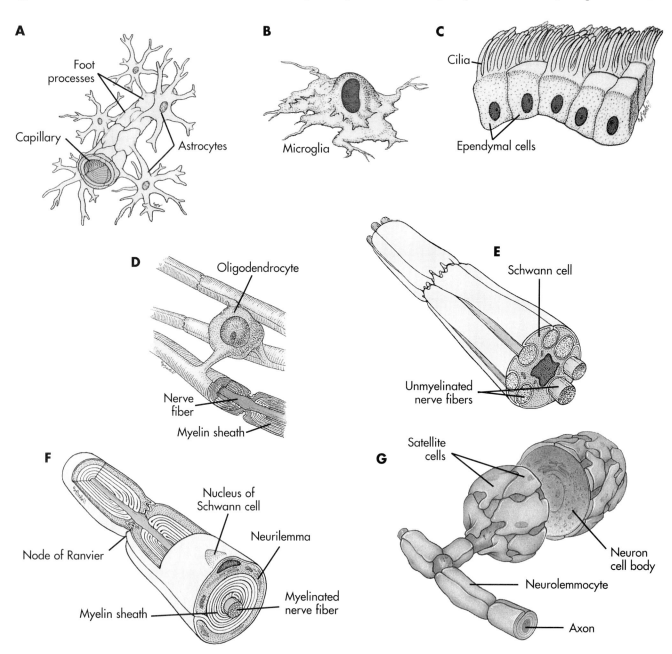

Figure 21-3 Types of glia. **A,** Astrocytes attached to the outside of a capillary blood vessel in the brain. **B,** A phagocytic microglial cell. **C,** Ciliated ependymal cells forming a sheet that usually lines fluid cavities in the brain. **D,** An oligodendrocyte with processes that wrap around nerve fibers in the CNS to form myelin sheaths. **E,** A Schwann cell supporting a bundle of nerve fibers in the PNS. **F,** Another type of Schwann cell wrapping around a peripheral nerve fiber to form a thick myelin sheath. **G,** Satellite cells, another type of Schwann cell, surround and support cell bodies of neurons in the PNS.

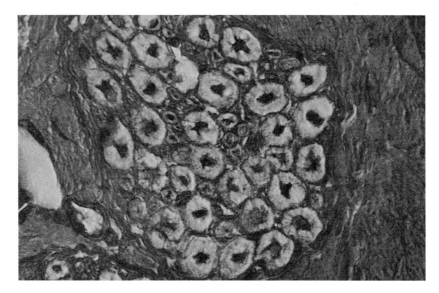

Figure 21-4 Peripheral nerve cross section. Myelin appears light gray in this micrograph (250×).

nervous system. Each Schwann cell wraps entirely around a single axon, rather than projecting extensions to several axons. Locate Schwann cells in Figure 21-3. Examine a prepared slide of a peripheral nerve cross section. Locate the myelin that surrounds each nerve fiber. Figure 21-4 shows a cross section of a peripheral nerve with the myelin surrounding each nerve fiber clearly visible.

❏ 5 **Ependymal cells**—Ependymal cells are found lining the fluid spaces of the brain and spinal cord. Some form *cerebrospinal fluid* (*CSF*) and some are ciliated and assist the movement of CSF through the fluid spaces. Locate ependymal cells in Figure 21-4. Examine a prepared slide of a spinal cord cross section. Locate small ciliated ependymal cells that form the lining of the central canal. LABORATORY REFERENCE, Plate 58 shows such a cross section, but at too low a magnification to see ependymal cells clearly.

D. Nerve Structure

The **nerve** is a bundle of axons that lies in the peripheral nervous system (PNS). The coloring exercise in Figure 21-5 and the steps outlined as follows describe the essential structure of the peripheral nerve. Locate these features in a model, diagram, and/or a prepared microslide of a peripheral nerve cross section (see Figure 21-3).

❏ 1 **Epineurium**—The epineurium is a fibrous membrane that forms a sheath around the entire bundle of structures that make up the nerve.

❏ 2 **Perineurium**—The perineurium is a fibrous structure that is essentially an inward continuation of the epineurium. Acting as a sort of "inner wrapping" or "packing material," the perineurium segregates the nerve fibers (axons) within the nerve into groupings called **fascicles.**

❏ 3 **Endoneurium**—Just as the perineurium is an inward extension of the epineurium, the endoneurium is an inward extension of the perineurium. The endoneurium is a fibrous membrane that covers each individual axon (and its sheath of Schwann) within each fascicle.

The Nerve

COLORING EXERCISE Using colored pens or pencils, shade in the figure and accompanying labels in contrasting colors of your choice as indicated by the red numerals.

EPINEURIUM₁
PERINEURIUM₂
ENDONEURIUM₃
AXON₄
FASCICLE₅

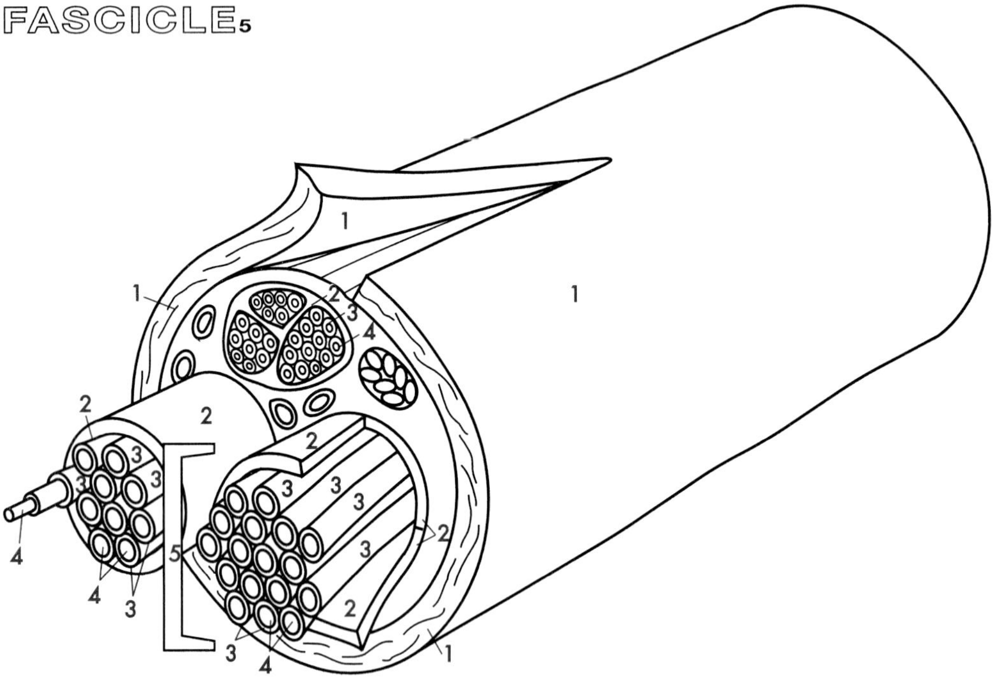

Figure 21-5

LAB REPORT 21

Nerve Tissue

Fill in this table summarizing your examination of a multipolar neuron model. Check off each structure as it is identified. For functions, refer to a reference book or your textbook.

Identification	Structure	Function(s)
☐	soma	
☐	Nissl bodies	
☐	axon hillock	
☐	axon	
☐	Schwann cells	
☐	nodes of Ranvier	
☐	collateral axon	
☐	telodendria	
☐	dendrite	

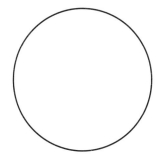

Specimen: *multipolar neurons*

Total Magnification: _____

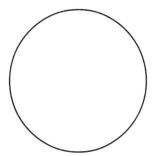

Specimen: *bipolar neurons*

Total Magnification: _____

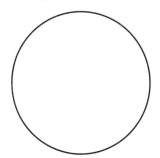

Specimen: *unipolar neurons*

Total Magnification: _____

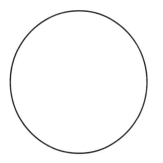

Specimen: *Schwann cells*

Total Magnification: _____

217

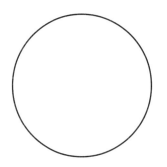

Specimen: *ependymal cells*

Total Magnification: _____

Fill-in

_____ 1
_____ 2
_____ 3
_____ 4
_____ 5
_____ 6
_____ 7
_____ 8
_____ 9
_____ 10
_____ 11
_____ 12
_____ 13
_____ 14
_____ 15
_____ 16
_____ 17
_____ 18
_____ 19
_____ 20

Fill-in (complete each statement with the correct term)

1. The soma forms a cone-shaped __?__ as it projects to form an axon.
2. __?__ are small, distal branches of an axon.
3. Either Schwann cells or extensions of __?__ can form myelin sheaths.
4. The gaps between segments of a myelin sheath are called __?__.
5. A group of myelinated fibers may form a region of nerve tissue called __?__ matter.
6. The __?__ neuron connects an afferent neuron to an efferent neuron.
7. A junction between two neurons, or between a neuron and effector, is called a(n) __?__.
8. A bundle of parallel neurons encased in fibrous connective tissue is called a(n) __?__ (in the PNS).
9. A bundle of parallel neurons in the CNS is called a(n) __?__.
10. An action potential traveling down a myelinated axon travels __?__ (faster/slower) than in an unmyelinated axon.
11. Unmyelinated nerve tissue is called __?__ matter.
12. __?__ neurons have exactly two projections from the cell body.
13. __?__ neurons have multiple dendrites and a single axon extending from the cell body.
14. In __?__ neurons, a single process from the cell body diverges to form two long branches—one acting as a dendrite, the other as an axon.
15. Schwann cells form myelin sheaths in the __?__ nervous system.
16. Oligodendrocytes form myelin sheaths in the __?__ nervous system.
17. Neuroglial cells called __?__ cells line the fluid spaces of the brain.
18. Small glial cells that phagocytize harmful matter are called __?__.
19. __?__ are ciliated neuroglia that assist the circulation of cerebrospinal fluid.
20. The most numerous type of cell in nerve tissue is the __?__ (neuron/glial cell).

LAB EXERCISE 22

Nerve Reflexes

This exercise describes the organization of the human nervous system and relates it to the concept of the *reflex arc*. The first activity describes the nature of the reflex arc and the manner in which it elicits a response called a *reflex* in an effector. The remainder of the exercise challenges you to demonstrate some simple reflexes that are often used in clinical situations to assess the condition of the nervous system.

Before you begin

❏ Read the appropriate chapter in your textbook.

❏ Set your learning goals. When you finish this exercise, you should be able to
- outline the features of a reflex arc and be able to apply this model to specific nerve pathways
- demonstrate several nerve reflexes in a human subject

❏ Prepare your materials:
- rubber reflex mallet
- sterile cotton balls
- penlight

❏ Read the directions and safety tips for this exercise **carefully** before starting any procedure

A. Organization of nerve pathways

An often-used model of nerve pathways is the **reflex arc.** A reflex arc is a way of visualizing the direction of transmission of nerve signals (action potentials). For each step that follows, find the reflex arc featured in the diagram of a simple reflex arc in Figure 22-1.

❏ 1 The arc begins with a **receptor,** a specialized cell or cell projection, which is stimulated by a change in its environment. For example, some receptors in the skin are sensitive to heat, others to pressure, and so on.

❏ 2 If stimulation of the receptor was significant enough to initiate an action potential in the **afferent neuron,** the signal is transmitted toward the CNS. The afferent, or sensory, neuron brings the signal into the brain or spinal cord.

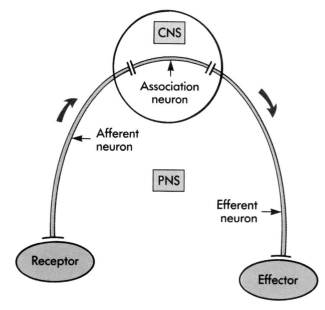

Figure 22-1 A simple three-neuron reflex arc. Dark arrows indicate direction of nerve transmission.

❏ 3 At the peak of the arc, where the signal is "turned around," the afferent neuron may synapse directly with an **efferent neuron** (forming a **two-neuron arc**). Often, an **association neuron** synapses with the afferent neuron and transmits the signal to an efferent neuron. This is a **three-neuron arc.**

❏ 4 The efferent, or motor, neuron then proceeds to an **effector.** An effector is a muscle or gland innervated by a motor nerve. The effector responds in some way to nerve signals, perhaps by contracting or secreting a chemical.

Synthesizing the component parts of the reflex arc, we see that a stimulus at the receptor results in a reaction by an effector. A simple example is a pin prick that causes a reflexive withdrawal of the pricked limb. Reflexes may be more complex, with sensory information being relayed to several different points in the CNS before triggering a motor response. Also, more than one sensory neuron may **converge** on a single association neuron, or multiple motor neurons may **diverge** from a single association neuron. Autonomic reflexes typically involve

219

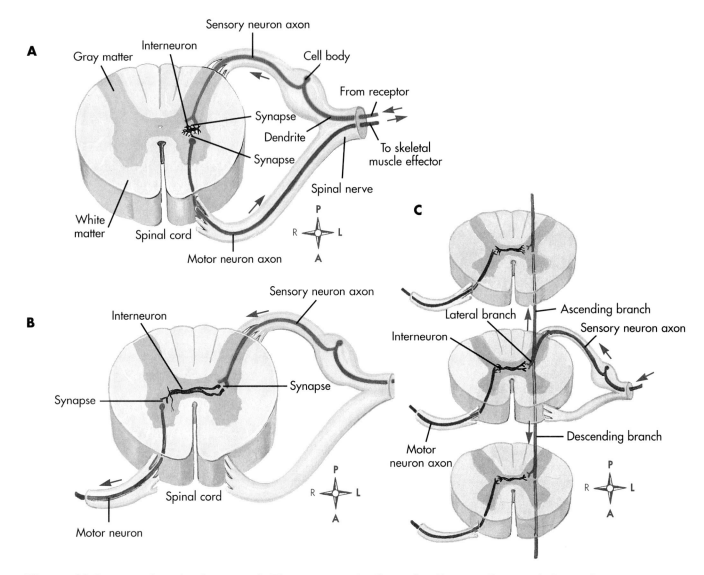

Figure 22-2 Examples of reflex arcs. **A,** Three-neuron ipsilateral reflex arc. Sensory information enters on the same side of the CNS as the motor information leaves the CNS. **B,** Three-neuron contralateral reflex arc. Sensory information enters on the opposite side of the CNS from the side that motor information exits the CNS. **C,** Intersegmental contralateral reflex arc. Divergent branches of a sensory neuron bring information to several segments of the CNS at the same time. Motor information leaves each segment on the opposite side of the CNS.

two efferent neurons, whereas somatomotor reflexes involve only one efferent neuron.

Figure 22-2 shows that reflex arcs can become quite complicated. Study this figure (and the descriptive legend) carefully, noting the different pathways efferent and afferent neurons may take in the human nervous system.

Afferent and efferent neurons generally form bundles of parallel fibers called **nerves** (in the PNS) or **tracts** (in the CNS). Nerves are supported by fibrous connective tissue membranes similar to those found in skeletal muscle organs. See Lab Exercise 21 for more details of nerve structure.

B. Stretch reflex demonstrations

Examples of reflexes can be seen in clinical tests performed in patients suspected of having some type of nerve damage. If the reflex shows an abnormal reaction or no reaction at all, damage to some component of the reflex arc is suspected. This activity demonstrates several reflexes that are categorized as *stretch* reflexes. Other types of simple reflexes are demonstrated in later activities.

❑ 1 The **patellar reflex,** or *knee-jerk reflex,* is mediated by a two-neuron arc centered in the spinal cord. Have the subject sit on a table with the legs dangling above the floor. Tap the knee sharply with a reflex mallet at the ligament just inferior to the patella (kneecap). The tap stretches a quadriceps muscle in the anterior thigh, stimulating *stretch receptors* located in the muscle. In response to the increased stretch, which normally would only occur when the muscle load has suddenly increased, the muscle contracts and extends the lower leg. In this demonstration, such a reaction seems strange, but in normal circumstances the patellar reflex allows the quadriceps muscle to reflexively increase its strength of contraction in response to increased load.

> HINT → Possible abnormal spinal reflex results:
>
> - **hyperflexia** (exaggerated response) results from a damaged or diseased motor areas in the CNS.
> - **hypoflexia** (inhibited response) results from degeneration of nerve pathways, voluntary motor control, and other factors.

❏ 2 The **biceps reflex** is a spinal reflex that involves nerves C5 and C6. Have the subject sit with the elbow flexed at about 90 degrees and palm facing downward. Put your thumb on the biceps (brachii) tendon at the inside angle of the elbow and press gently. Tap your thumb with a reflex mallet. Flexion of the elbow is the normal response.

❏ 3 The **achilles reflex** is a spinal reflex that results in plantar flexion of the foot. Have the subject kneel on a chair, facing away from you, and the toes pointing toward the floor. Tap the middle of the achilles (calcaneal) tendon with a reflex mallet. Is the response normal?

❏ 4 The **triceps reflex** is another stretch reflex. Have the subject lie on a table, with an arm across the abdomen. Supporting the subject's arm with the elbow flexed at a 90-degree angle, sharply tap the posterior surface of the upper arm just proximal to the olecranon. The lower arm should extend as the triceps brachii muscle reflexively contracts. How do you explain this response?

C. Cutaneous reflexes

Cutaneous reflexes are those that result from stimulation of cutaneous (skin) receptors.

❏ 1 The **plantar reflex** involves cutaneous receptors rather than deep receptors in muscles or tendons. Position the subject's bare foot with the lateral surface resting on a table or chair. Demonstrate this reflex by firmly sweeping the handle of the mallet along the lateral region of the sole. In a normal adult, the toes flex. In a **Babinski response,** the toes extend and move apart. The Babinski response is normal in infants in whom the nerves have not fully myelinated but is abnormal in adults.

❏ 2 Another cutaneous reflex is the **abdominal reflex.** Begin the abdominal reflex demonstration by having the subject lie on a clean, flat surface. Expose the skin of the abdomen. Use the handle of the reflex mallet to sweep gently along the anterior, lateral aspect of the skin. Move in a superior-to-inferior direction. What is the response of the abdominal muscles?

D. Cranial reflexes

Another type of reflex is mediated by the brain. For example, the *pupillary reflexes* demonstrated in Step 1 involve cranial nerves and autonomic reflex centers.

❏ 1 In a dimly lit area, demonstrate the **pupillary reflexes,** which are centered in the brainstem. Have the subject stare straight ahead, with a hand held vertically between the eyes. Shine a penlight in the subject's left eye (from 5 to 7 cm away). Normally, the light receptors in the eye receive the bright light and trigger a reflexive response by the muscles in the iris. Does the pupil increase or decrease in diameter? What is the advantage of this response? The right pupil may also exhibit a reflexive response, even though light is not shown on it. Does the right pupil increase or decrease in diameter? What advantage is gained by this *consensual* reflex?

❏ 2 Demonstrate the **corneal reflex,** which involves the fifth cranial nerve. Begin the corneal reflex demonstration by having the subject stand facing at a 90-degree angle to you. While the subject stares straight ahead, quickly move a clean, sterile cotton ball toward the surface of one eye. *Do not actually touch the eye, however.* What is the reaction? What survival advantage does this reflex have?

Reflex Arc

📝COLORING EXERCISE Using colored pens or pencils, shade in the figure and accompanying labels in contrasting colors of your choice as indicated by the red numerals.

RECEPTOR₁
AFFERENT NEURON₂
INTERNEURON₃
EFFERENT NEURON₄
EFFECTOR₅
DORSAL SPINAL NERVE ROOT₆
DORSAL ROOT GANGLION₇
　VENTRAL SPINAL NERVE ROOT₈
SPINAL NERVE₉
BRANCH OF SPINAL NERVE₁₀
SPINAL CORD
WHITE MATTER₁₁
GRAY MATTER₁₂

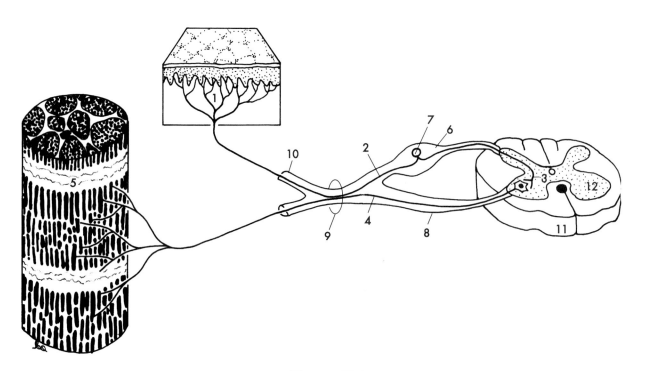

Figure 22-3

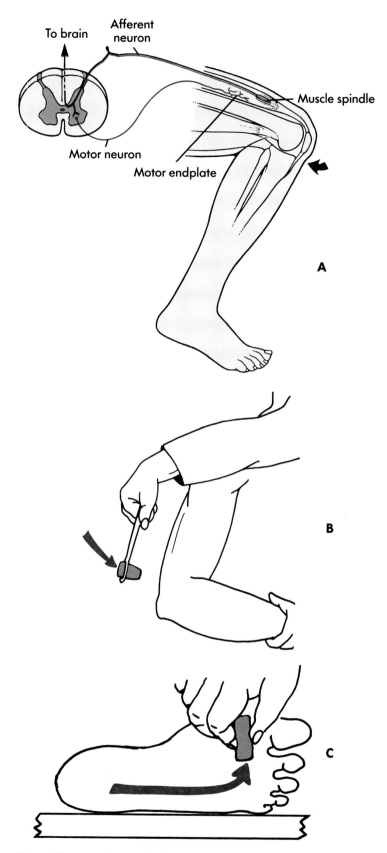

Figure 22-4 A, The patellar reflex arc. Arrow indicates tapping point. **B,** Tapping point for triceps reflex. **C,** Demonstrating plantar reflex.

NAME_____ DATE_____ SECTION_____

LAB REPORT 22

Nerve Reflexes

Record your data and interpretations of the reflex tests. Be sure to indicate the function each may have.

Reflex	Result		Explanation/discussion
	Left	Right	
Patellar	❏ normal ❏ hyperflexia ❏ hypoflexia ❏ no response	❏ normal ❏ hyperflexia ❏ hypoflexia ❏ no response	
Biceps	❏ normal ❏ hyperflexia ❏ hypoflexia ❏ no response	❏ normal ❏ hyperflexia ❏ hypoflexia ❏ no response	
Achilles	❏ normal ❏ hyperflexia ❏ hypoflexia ❏ no response	❏ normal ❏ hyperflexia ❏ hypoflexia ❏ no response	
Triceps	❏ normal ❏ hyperflexia ❏ hypoflexia ❏ no response	❏ normal ❏ hyperflexia ❏ hypoflexia ❏ no response	
Plantar	❏ normal ❏ Babinski response ❏ no response	❏ normal ❏ Babinski response ❏ no response	
Abdominal reflex	❏ normal ❏ hyperflexia ❏ hypoflexia ❏ no response	❏ normal ❏ hyperflexia ❏ hypoflexia ❏ no response	
Pupillary	❏ dilation ❏ constriction ❏ no response	❏ dilation ❏ constriction ❏ no response	
Corneal	❏ response ❏ no response	❏ response ❏ no response	

Put in order

1. _____
2. _____
3. _____
4. _____
5. _____

Fill-in

1. _____
2. _____
3. _____
4. _____
5. _____
6. _____
7. _____
8. _____
9. _____
10. _____

Put in order (put these components of the reflex arc in the order in which nerve signals pass through them)

association neuron
effector
motor neuron
receptor
sensory neuron

Fill-in (complete each statement with the correct term)

1. In infants, the __?__ response is normal when testing the plantar reflex.
2. When testing the plantar reflex in an adult, extension of the toes may indicate damage to the __?__ somewhere along the reflex arc.
3. In the __?__ reflex, a muscle contracts when the load increases.
4. The center of the patellar reflex is in the __?__.
5. A stretch reflex involving the biceps brachii muscle causes __?__ of the elbow if the biceps muscle is stretched.
6. In the triceps reflex demonstration, the triceps muscle was the effector, and one or more __?__ were the receptors.
7. If the Achilles (calcaneal) tendon is tapped, one would expect the ankle to __?__.
8. When suddenly illuminated with a penlight, the pupil of the eye normally __?__ (dilates/constricts).
9. __?__ reflexes result from the stimulation of sensory receptors in the skin.
10. __?__ reflexes are centered in the brain and involve cranial nerves.

LAB EXERCISE 23

The Spinal Cord and Spinal Nerves

As you know, the spinal cord is one of two principal organs of the central nervous system (CNS). The brain is the other. This exercise presents the spinal cord and its accessory structures on the gross level and the microscopic level. This exercise also includes a peek into the peripheral nervous system (PNS) by means of an introduction to the **spinal nerves.**

Before you begin

❏ Read the appropriate chapter in your textbook.

❏ Set your learning goals. When you finish this exercise, you should be able to
  ■ name and describe the coverings of the spinal cord
  ■ identify principal structures of the spinal cord in figures and specimens
  ■ define *spinal nerve, nerve root,* and *plexus*
  ■ name the spinal nerves and locate them in a model or figure

❏ Prepare your materials:
  ■ preserved specimen: *ox spinal cord segment*
  ■ dissection tools and tray
  ■ model or chart of the human spinal cord
  ■ microscope
  ■ prepared microslide: *mammalian spinal cord c.s.*

❏ Read the directions and safety tips for this exercise **carefully** before starting any procedure.

A. Spinal cord dissection

Use both the preserved spinal cord specimen and a model or chart to explore the features described in this activity.

> **SAFETY FIRST!** Heed the precautions that accompany the chemical preservative used with your specimen, especially those regarding ventilation and skin protection. Use your dissection tools carefully.

The spinal cord is nerve tissue continuous with that of the brain, extending inferiorly from the brain at the level of the foramen magnum. The spinal cord is covered by several layers of protective or nourishing tissue:

❏ 1 The *bony covering* of the spinal cord is formed by the vertebral column. The spinal cord lies within the spinal cavity formed by the vertebral foramina of the stacked vertebrae. What is the bony covering of the brain?

❏ 2 The outermost of three *membranous coverings,* or **meninges** (singular, *meninx*), of the spinal cord is the tough, skinlike **dura mater.** The spinal meninges are continuous with the meninges surrounding the brain.

❏ 3 Deep to the dura mater is a much thinner meninx, the spidery **arachnoid.** The space deep to the arachnoid, the **subarachnoid space,** is filled with circulating **cerebrospinal fluid (CSF)** in life. The CSF has been drained from your specimen.

❏ 4 The deepest meninx, the delicate **pia mater,** is a thin, vascular membrane adhering to the surface of the spinal cord.

❏ 5 The **anterior median fissure** is a deep groove on the anterior surface of the human spinal cord (along the midline). This fissure is analogous to that on the ventral surface of the ox spinal cord.

❏ 6 The **posterior median sulcus** is a somewhat shallower groove along the dorsal surface of the spinal cord.

❏ 7 A **cervical enlargement** is seen in the cervical region of the spinal cord, and a similar **lumbar enlargement** is found in the lower thoracic region. The lumbar enlargement tapers to a cone-shaped end called the **medullary cone.**

☐ 8 The spinal cord proper does not extend below vertebra L1. Instead, separate nerves trail inferiorly until each pair has exited the spinal cavity. This group of nerves, resembling a horse's tail, is called the **cauda equina.**

☐ 9 **Ventral** and **dorsal spinal nerve roots** may be intact in your specimen. These structures are discussed more fully in a later activity.

> **HINT** → The following structures may or may not be easily seen in the preserved specimen, depending on the specimen's condition. A freshly cut, even cross section usually gives the best results.

☐ 10 A cross-sectional view of the spinal cord shows a distinct H-shaped area of gray matter surrounded by areas of white matter. The two lateral sections of gray matter are joined by a transverse **gray commissure.** In the center of the gray commissure, you may see the **central canal.** This canal contains CSF in life but contains preservative in prepared specimens.

☐ 11 Each lateral mass of the gray matter exhibits extensions termed the **anterior gray horn** and the **posterior gray horn.** The thoracic and lumbar regions have additional lateral extensions called **lateral gray horns.**

☐ 12 Columns of white matter surround the central gray matter area. These include the **anterior white column, lateral white column,** and **posterior white column.** Columns are sometimes called **funiculi.**

B. Microscopic exploration

> **SAFETY FIRST!** Observe the usual precautions when using the microscope and slide.

☐ 1 Obtain a stained cross section preparation of a mammalian spinal cord and observe it at a low total magnification.

☐ 2 Try to find as many features of the spinal cord (as described in Activity A) as you can. White matter and gray matter are often distinguished by differential staining (contrasting colors) as in LABORATORY REFERENCE, Plate 58.

C. Spinal nerves

Afferent neurons extending through the spinal cord enter as part of a **spinal nerve.** Efferent neurons in the spinal cord exit in spinal nerves. Although not really part of the spinal cord proper (they are in the PNS rather than the CNS), we will consider them in this exercise. As each structure is presented, try to locate it on a model or chart of the human nervous system and in Figure 23-2.

☐ 1 There are 31 pairs of spinal nerves communicating with the spinal cord by way of the intervertebral foramina formed as the vertebrae stack on one another. The spinal nerves are named for the vertebral region with which they are associated:
- **Cervical spinal nerves**—C1 through C8 (one pair between the skull and vertebra C1 and one pair inferior to each of seven cervical vertebrae)
- **Thoracic spinal nerves**—T1 through T12
- **Lumbar spinal nerves**—L1 through L5
- **Sacral spinal nerves**—S1 through S5 (branches of these nerves communicate via the dorsal and pelvic foramina of the sacrum)
- **Coccygeal spinal nerves**—C or Cx

☐ 2 Each spinal nerve communicates with the cord by means of two separate **spinal nerve roots,** each formed by the combination of six to eight tiny *rootlets*. The **ventral (anterior) spinal nerve root** includes efferent fibers (somatomotor and autonomic axons). The **dorsal (posterior) spinal nerve root** contains afferent fibers. Nuclei of afferent neurons form an area of gray matter in the dorsal root called the **dorsal root ganglion.** The spinal nerve roots are found within the spinal cavity. The spinal nerve proper passes through the intervertebral foramina.

☐ 3 Outside the spinal cavity, each spinal nerve branches, forming a **ventral ramus** and a **dorsal ramus.** In most of the thoracic and lumbar nerves, the ventral ramus branches into autonomic and somatic pathways. In other segments, the ventral rami's fibers branch to form **plexi.** Plexi are complex nerve networks in which fibers from several different spinal nerves are reorganized to form specific peripheral nerves. Identify each plexus on a model or chart:
- **Cervical plexus**
- **Brachial plexus**
- **Lumbosacral plexus**

The Spinal Cord

COLORING EXERCISE Using colored pens or pencils, shade in the figure and accompanying labels in contrasting colors of your choice as indicated by the red numerals.

MENINGES
DURA MATER₁
ARACHNOID₂
  SUBARACHNOID SPACE₃
PIA MATER₄

SPINAL CORD
CENTRAL CANAL₅
GRAY COMMISSURE₆
ANTERIOR GRAY HORN₇
POSTERIOR GRAY HORN₈
ANTERIOR WHITE
  COLUMN₉
POSTERIOR WHITE
  COLUMN₁₀
LATERAL WHITE COLUMN₁₁

SPINAL NERVE
VENTRAL NERVE ROOT₁₂
DORSAL NERVE ROOT₁₃
  DORSAL ROOT
    GANGLION₁₄
SPINAL NERVE **PROPER** 15
VENTRAL RAMUS₁₆
DORSAL RAMUS₁₇

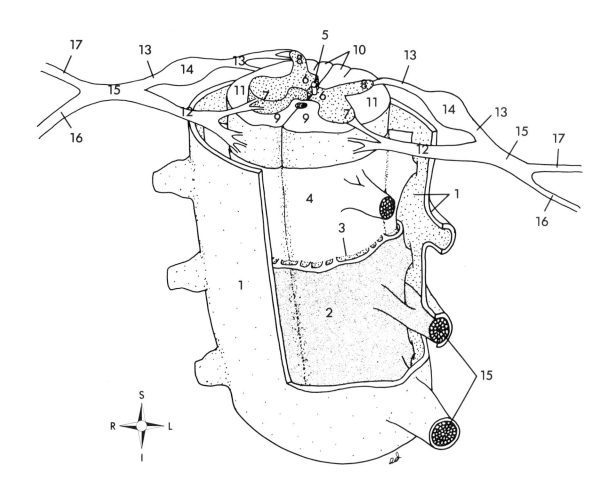

229

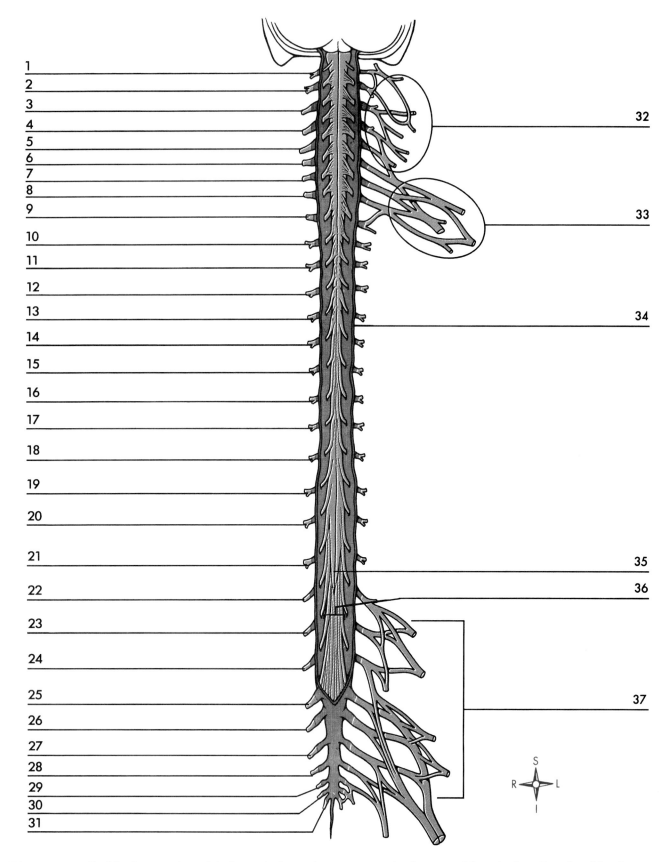

Figure 23-2 In blanks numbered 1 through 31, write correct spinal nerve abbreviation. In other blanks, write name of cord or nerve feature indicated.

LAB REPORT 23

The Spinal Cord and Spinal Nerves

Figure 23-2

_____ 1
_____ 2
_____ 3
_____ 4
_____ 5
_____ 6
_____ 7
_____ 8
_____ 9
_____ 10
_____ 11
_____ 12
_____ 13
_____ 14
_____ 15
_____ 16
_____ 17
_____ 18
_____ 19
_____ 20
_____ 21
_____ 22
_____ 23
_____ 24
_____ 25

continued on next page

Summarize your exploration of the spinal cord and spinal nerves by using this checklist.

Structure	Notes
❏ dura mater	
❏ arachnoid	
❏ subarachnoid space	
❏ pia mater	
❏ anterior median fissure	
❏ posterior median sulcus	
❏ cervical enlargement	
❏ lumbar enlargement	
❏ medullary cone	
❏ cauda equina	
❏ ventral spinal nerve root	
❏ dorsal spinal nerve root	
❏ dorsal root ganglion	
❏ gray commissure	
❏ central canal	
gray horns: ❏ anterior ❏ lateral ❏ posterior	
white columns: ❏ anterior ❏ lateral ❏ posterior	
spinal nerves: ❏ cervical ❏ thoracic ❏ lumbar ❏ sacral ❏ coccygeal	
❏ ventral ramus of spinal nerve	
❏ dorsal ramus of spinal nerve	

Figure 23-2 (continued)

_____ 26
_____ 27
_____ 28
_____ 29
_____ 30
_____ 31
_____ 32
_____ 33
_____ 34
_____ 35
_____ 36
_____ 37

Put in order I

_____ 1
_____ 2
_____ 3
_____ 4
_____ 5
_____ 6
_____ 7

Put in order II

_____ 1
_____ 2
_____ 3
_____ 4
_____ 5
_____ 6
_____ 7

Multiple choice

____ 1
____ 2
____ 3
____ 4
____ 5

Put in order I (put these structures in the order in which a sensory signal might pass)

    cervical plexus
    dorsal root ganglion
    posterior gray horn
    receptor
    sensory area of the brain
    spinal nerve C2
    ventral ramus

Put in order II (put these structures in the order in which a motor signal might pass)

    anterior gray horn
    effector
    lateral gray horn
    motor area of the brain
    spinal nerve T6
    ventral ramus
    ventral nerve root

Multiple choice (only one response is correct in each item that follows)

1. The medial deep groove on the ventral surface of the spinal cord is called the
    a. posterior median sulcus
    b. medial enlargement
    c. anterior lateral sulcus
    d. anterior median fissure
    e. ventral lateral column

2. The correct order for the meninges of the brain (from superficial to deep):
    a. pia mater, dura mater, arachnoid
    b. dura mater, arachnoid, pia mater
    c. dura mater, pia mater, arachnoid
    d. dura mater, subarachnoid, pia mater
    e. the brain does not have meninges

3. The cauda equina is
    a. not present in humans
    b. a cone-shaped end of the spinal cord proper
    c. a bundle of separate nerves
    d. a bundle of cranial nerves
    e. an enlargement of the spinal cord proper

4. Spinal nerve branches from several spinal segments (vertical levels) may exchange fibers. Such a network of nerve fibers is called a
    a. braid
    b. ramus
    c. plexus
    d. ganglion
    e. spinal nerve root

5. Large groups of neuron cell bodies are likely to be found in
    a. the lateral gray horn
    b. the gray commissure
    c. a dorsal root ganglion
    d. the central canal
    e. a, b, and c are correct

LAB EXERCISE 24

The Brain and Cranial Nerves

Having already explored the spinal cord in the previous exercise, it is time to move on to an exploration of the other major organ of the central nervous system: the **brain.** The brain occupies most of the cranial cavity within the skull. This rather large, complex organ has been partitioned, for convenience of study, into numerous divisions and subdivisions. Nerves that enter and exit the brain directly are called **cranial nerves.** In this exercise, you are challenged to explore the organization of the brain and the structure and function of the cranial nerves.

Before you begin

❑ Read the appropriate chapter in your textbook.

❑ Set your learning goals. When you finish this exercise, you should be able to
- list and describe the principal structures of the brain
- identify structures of the human brain in a chart, model, or preserved specimen

❑ Prepare your materials:
- chart, model, or preserved specimen of a human brain

❑ Read the directions and safety tips for this exercise **carefully** before starting any procedure.

A. External aspect of the brain

Using a model or preserved specimen of the human brain, locate the structures of the brain listed in the following steps.

> HINT → Your textbook has descriptions and figures that will help you identify structures of the brain. You may wish to use your textbook as a supplemental resource for this and the following activities.

❑ 1 Determine whether any of the **meninges,** or membranous coverings of the brain, are intact. See Figure 24-1. If they are present, locate the following:
- **Dura mater**
- **Sagittal sinus**
- **Subdural space**
- **Arachnoid**
- **Subarachnoid space**
- **Pia mater**

❑ 2 Locate the largest part of the brain, the **cerebrum.** The cerebrum is divided into nearly symmetrical left and right **cerebral hemispheres** by a deep **longitudinal fissure.** Joining the left and right cerebral hemisphere is a band of white matter called the **corpus callosum,** which may be seen at the bottom of the longitudinal fissure. The surface of the cerebrum, or **cerebral cortex,** is characterized by large convoluted folds of gray matter called **gyri.** The grooves between the gyri are called **sulci** if they are shallow and **fissures** if they are deep. Some fissures serve as landmarks to divide the cerebrum into regions called **lobes** that correspond to the overlying skull bones:
- **Frontal lobe** (left and right)
- **Parietal lobe**
- **Occipital lobe**
- **Temporal lobe**
- **Insula** (also called the *island of Reil,* a region hidden within a fissure that separates the temporal lobe from the parietal lobe)

In addition to using lobes as a means of identifying different regions of the cerebral cortex, neurobiologists often prefer to identify **functional regions** of the cortex. Figure 24-2 is a coloring exercise that illustrates some of the major functional areas of the cerebral cortex. How many of these areas can you identify on a specimen or model of the human brain?

❑ 3 The smaller, rounded structure under the posterior portion of the cerebrum that possesses somewhat parallel gyri is the **cerebellum.**

233

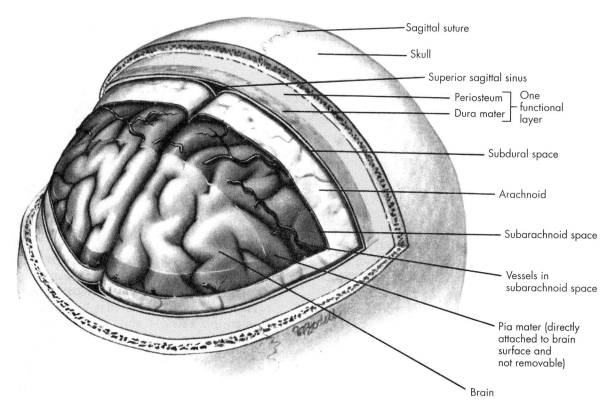

Figure 24-1 External aspect of human brain and its coverings.

❏ 4 The **diencephalon** is the brain region just under the central portion of the cerebrum, superior and somewhat anterior to the brainstem. One major region of the diencephalon, the **thalamus,** is internal and cannot be seen from the outside of the brain. From the external aspect one can see some of these features of the **hypothalamus** of the diencephalon:
- **Optic chiasma**
- **Optic tracts**
- **Mammillary bodies**
- **Infundibulum**

One can also see the **pituitary,** or *hypophysis,* which is not part of the diencephalon proper but is attached to its ventral surface by the stalklike infundibulum.

❏ 5 The **brainstem** is the roughly cylindrical region of the brain that projects inferiorly from the diencephalon, anterior to the cerebellum. Locate these features if you can:
- **Midbrain** (upper portion, may not be visible in some models or specimens)
- **Pons** (bulging middle portion)
- **Medulla,** or *medulla oblongata* (cylindrical, inferior portion)

B. Cranial nerves

The cranial nerves are bundles of nerve fibers and their coverings that project from the brain, mostly from the brainstem. There are twelve pairs, each known by both name and number (usually expressed as a Roman numeral). Cranial nerves are numbered in the sequence of their origin from the ventral brain surface, from anterior to posterior. *Sensory nerves* contain only sensory (afferent) fibers. *Motor nerves* contain primarily motor (efferent fibers) but also have a few sensory fibers that carry feedback proprioceptive information. *Mixed nerves* have significant numbers of both sensory and motor fibers. Locate each of the twelve pairs of cranial nerves listed here in your model. Table 24-1 will help you identify their types and functions. Figure 24-3 will help you find them.

HINT → The first letter of each word in the following sentence, or one like it, helps in memorizing the names and numbers of the cranial nerves. It is, *"On Old Olympus' Tiny Tops, A Friendly Viking Grew Vines And Hops."* The functional type of each cranial nerve can be remembered by using this sentence: *"Some Say 'Marry Money,' But My Brothers Say 'Bad Business, Marry Money."* In this sentence, *S* indicates *sensory, M* indicates *motor,* and *B* indicates *both* sensory and motor (mixed).

Cerebral Cortex

COLORING EXERCISE Using colored pens or pencils, shade in the figure and accompanying labels in contrasting colors of your choice as indicated by the red numerals.

LOBES
FRONTAL LOBE₁
PARIETAL LOBE₂
OCCIPITAL LOBE₃
TEMPORAL LOBE₄
INSULA₅

FUNCTIONAL AREAS
PREFRONTAL AREA₆
PREMOTOR AREA₇
MOTOR SPEECH AREA₈
PRIMARY SOMATIC MOTOR AREA₉
PRIMARY SOMATIC SENSORY AREA₁₀
PRIMARY TASTE AREA₁₁
SOMATIC SENSORY ASSOCIATION AREA₁₂
SENSORY SPEECH AREA₁₃
VISUAL ASSOCIATION AREA₁₄
VISUAL CORTEX₁₅
PRIMARY AUDITORY AREA₁₆
AUDITORY ASSOCIATION AREA₁₇

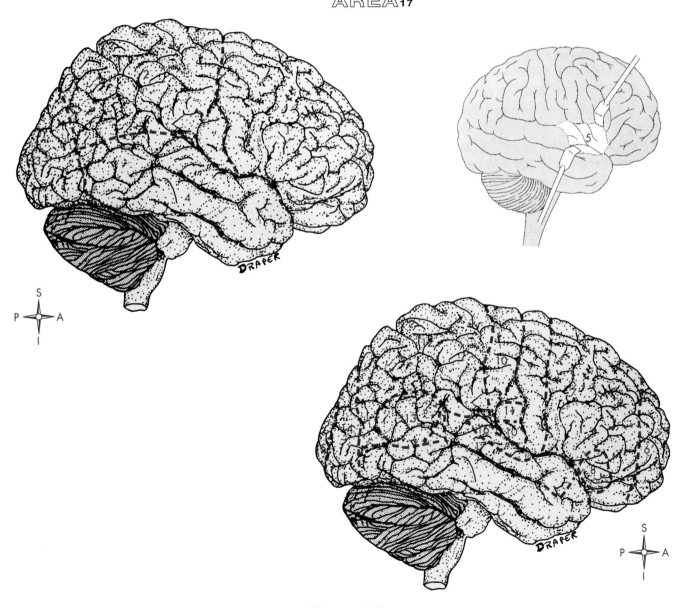

Figure 24-2

Cranial Nerves

COLORING EXERCISE Using colored pens or pencils, shade in the figure and accompanying labels in contrasting colors of your choice as indicated by the red numerals.

OLFACTORY NERVE (I)₁
OPTIC NERVE (II)₂
OCULOMOTOR NERVE (III)₃
TROCHLEAR NERVE (IV)₄
TRIGEMINAL NERVE (V)₅
ABDUCENS NERVE (VI)₆
FACIAL NERVE (VII)₇
VESTIBULOCOCHLEAR NERVE (VIII)₈
GLOSSOPHARYNGEAL NERVE (IX)₉
VAGUS NERVE (X)₁₀
ACCESSORY NERVE (XI)₁₁
HYPOGLOSSAL NERVE (XII)₁₂

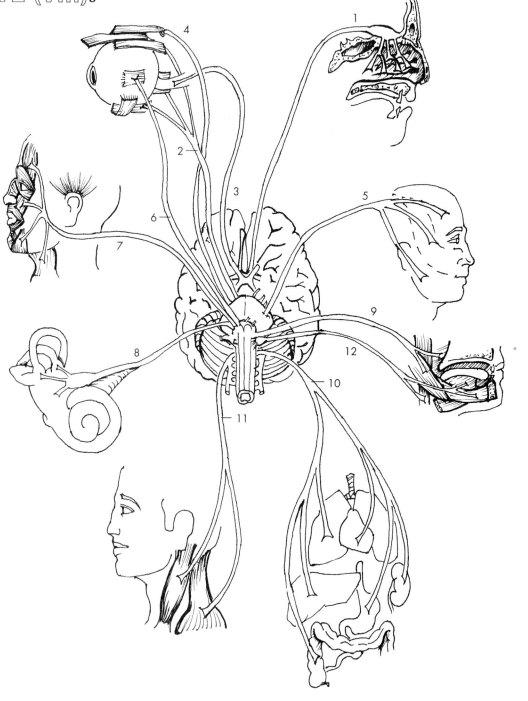

Figure 24-3

Table 24-1 Cranial nerves.

Name	Number	Type	General function(s)
Olfactory	I	Sensory	Olfaction (sense of smell)
Optic	II	Sensory	Vision
Oculomotor	III	Motor	Controls upper eyelid muscles; controls superior rectus, medial rectus, inferior rectus, and inferior oblique muscles of the eye; controls ciliary muscle of the eye and sphincter in the iris
Trochlear	IV	Motor	Controls superior oblique muscle of the eye
Trigeminal	V	Mixed	Controls chewing movements *Ophthalmic branch:* sensation around the eye *Maxillary branch:* sensation from eye to upper jaw and throat *Mandibular branch:* sensation in mandibular region
Abducens	V	Motor	Controls lateral rectus muscle of the eye
Facial	VII	Mixed	Controls facial muscles; controls secretion of tears and saliva; taste (anterior two thirds of tongue)
Vestibulocochlear	VIII	Sensory	*Vestibular branch:* senses of equilibrium *Cochlear branch:* hearing
Glossopharyngeal	IX	Mixed	Controls salivation; controls swallowing muscles; taste (posterior third of tongue); blood pressure sensation
Vagus	X	Mixed	Controls swallowing muscles; control and sensation in various visceral effectors
Accessory	XI	Motor	*Cranial branch:* controls some swallowing movements *Spinal branch:* controls some head movements
Hypoglossal	XII	Motor	Controls tongue muscles (swallowing and speech)

C. Internal aspect of the brain

If your brain specimen is dissectible, separate the two halves of the model along the midsagittal plane to reveal the inner structures of the brain. If you are using a preserved or mounted specimen, you will need to use one that is cut in a midsagittal section to see the structures described in the following steps. Locate each structure listed.

❏ 1 The **corpus callosum** had been connecting the two cerebral hemispheres and can now be clearly seen in the midsagittal section. The **fornix,** part of the diencephalon, is ventral to the corpus callosum and is also composed of white fibers. You may be able to see a hollow cavity just ventral to the corpus callosum in each brain half. These cavities are the right and left lateral ventricles. In the whole brain, they are separated by a thin membrane, the **septum pellucidum.** In some models and specimens, the septum may be visible as a drumlike covering to one ventricle or may partially cover both ventricles. The **choroid plexus,** a tiny mass of capillaries within each ventricle, produces the *cerebrospinal fluid* (*CSF*) that fills each ventricle.

❏ 2 The lateral ventricles can also be seen in a frontal section. If your model can be dissected along a frontal plane, do so. If you are using preserved specimens, you will need one that is cut in a frontal section. Locate the lateral ventricle(s). Just lateral to each lateral ventricle is a small circle of gray matter called the *caudate nucleus,* part of the cerebrum's basal ganglia. Separated from the caudate nucleus

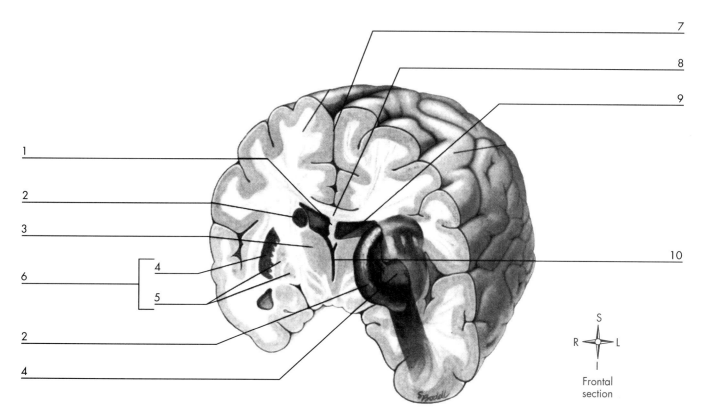

Figure 24-4 Label structures of brain (frontal section) indicated.

by a band of white matter called the *internal capsule* is another part of the basal ganglia. This portion of the basal ganglia, inferior and lateral to the caudate nucleus, is called the **lentiform nucleus.** The lentiform nucleus is comprised of a lateral *putamen* and a medial *globus pallidus,* both made of gray matter. Just inferior to the lateral ventricles, along the midline, is a single third ventricle. Connecting each lateral ventricle to the third ventricle is a foramen of Monro, which is often not visible in any of the brain sections. The lateral walls of the third ventricle are formed by a mass of gray matter called the **thalamus,** a region of the diencephalon. The two *lateral masses* of the thalamus are joined by the *intermediate mass* that passes through the third ventricle. Note the distribution of gray and white matter in the frontal section. What do you notice about its pattern?

❏ 3 Return your attention to the midsagittal section. Identify the white matter pattern, or arbor vitae, of the cerebellum. Ventral to the cerebellum is the fourth ventricle, which is connected to the third ventricle by the cerebral aqueduct (aqueduct of Sylvius).

❏ 4 Many features of the diencephalon already seen on the external aspect or in the frontal section can be seen in the midsagittal section. Locate the **optic chiasma, mammillary bodies,** and **infundibulum.** A small, rounded body on the midline, nearest the cerebrum, is the **pineal body** (or *pineal gland*) of the diencephalon. In the midsagittal section, only the intermediate mass of the thalamus is visible. It appears as a circle of gray matter surrounded by the shallow section of the third ventricle.

❏ 5 The **corpora quadrigemina** is easily seen in the midsagittal section, along the dorsal surface of the midbrain, just posterior to the pineal body. The corpora quadrigemina is comprised of four rounded bumps, arranged as two pairs. The larger pair is called the **superior colliculi** and the slightly smaller pair is called the **inferior colliculi.** The white matter tracts in the ventral portion of the midbrain constitute the **cerebral peduncles.**

❏ 6 Identify the pons and medulla in the midsagittal section.

The Human Brain

COLORING EXERCISE Using colored pens or pencils, shade in the figure and accompanying labels in contrasting colors of your choice as indicated by the red numerals.

CEREBRUM
CORPUS CALLOSUM₁
LATERAL VENTRICLE₂

CEREBELLUM
GRAY MATTER₃
ARBOR VITAE₄
FOURTH VENTRICLE₅

DIENCEPHALON
OPTIC CHIASMA₆
INFUNDIBULUM₇
MAMMILLARY BODIES₈
THALAMUS **INTERMEDIATE MASS** ₉

PINEAL BODY₁₀
FORNIX₁₁
THIRD VENTRICLE₁₂

MIDBRAIN
CEREBRAL PEDUNCLES₁₃
CEREBRAL AQUEDUCT₁₄
CORPORA QUADRIGEMINA₁₅

PONS₁₆

MEDULLA₁₇

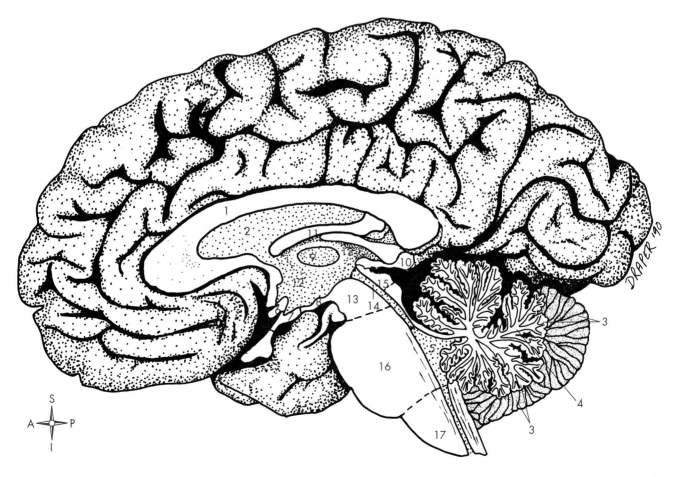

Figure 24-5

MAGNETIC RESONANCE IMAGING

As you recall, traditional radiographic imaging relies on x-rays that pass through soft tissues easily but cast shadows when they try to pass through dense tissues. Thus, soft tissues can only be seen when they are coated or filled either with a substance that blocks radiation or with a substance that emits its own radiation.

A newer technology that produces finely detailed images of soft tissue without relying on x-rays is **nuclear magnetic resonance (NMR).** Called **magnetic resonance imaging (MRI)** in many clinical settings, this technique relies on the magnetic properties of the nuclei of hydrogen molecules in human tissues.

When placed in a strong magnetic field, the hydrogen nuclei in a living tissue all orient themselves in one direction. At different intervals, depending on the type of tissue, the nuclei spin back to their original position and give off a burst of RF (radio frequency) energy. A computer with sensors that can detect the relative position and rate of energy bursts then constructs a three-dimensional image of the tissues. Thus, a subject is placed in a magnetic field, and the MRI monitor displays sections similar to, but clearer than, CT images.

The brain is a particularly interesting subject for MRI because it is composed entirely of soft tissue. Although CT scanning technology can produce images of brain structure, MRI pictures are much more detailed. That patients scanned with MRI receive no harmful nuclear radiation makes this technology even more attractive.

Label the brain parts indicated in the magnetic resonance image that follows.

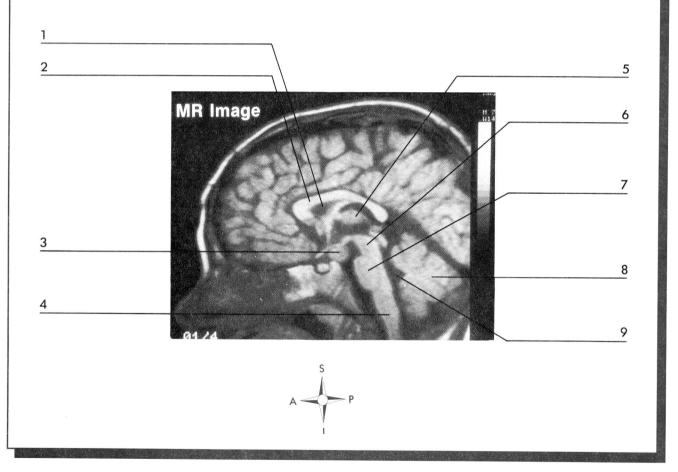

LAB REPORT 24

The Brain and Cranial Nerves

Identify (write the names of the structures described as follows)

1. A mass of white fibers connecting the left and right cerebral hemispheres
2. A cerebral lobe not visible in an ordinary external inspection
3. A fold of cortical gray matter on the surface of the cerebrum
4. A deep sulcus
5. A vein in the dura mater roughly parallel to the longitudinal fissure
6. A rounded structure dorsal to the brainstem
7. A distinct, branched pattern of white matter in the cerebellum
8. The middle of three divisions of the brainstem
9. A vascular structure that produces CSF and is present in all the fluid ventricles
10. The fluid ventricle associated with the cerebellum
11. The membrane that separates the left and right lateral ventricle
12. The portion of the thalamus that passes through the third ventricle
13. This structure is composed of four colliculi
14. The gland is not technically a part of the brain but is attached to the brain via the infundibulum
15. The gray matter on the surface of the cerebrum is called the cerebral __?__
16. The white matter tracts in the ventral portion of the midbrain
17. The most caudal structure of the brain

Put in order (arrange these in the order in which CSF passes through them, beginning with the place in which CSF could first be formed)

cerebral aqueduct
foramen of Monro
fourth ventricle
lateral ventricle
third ventricle

Figure 24-4

_____ 1
_____ 2
_____ 3
_____ 4
_____ 5
_____ 6
_____ 7
_____ 8
_____ 9
_____ 10

MR image: human brain

_____ 1
_____ 2
_____ 3
_____ 4
_____ 5
_____ 6
_____ 7
_____ 8
_____ 9

LAB EXERCISE 25

The Dissection: The Brain

In this exercise, you are challenged to dissect a mammalian brain (the sheep brain). This activity will strengthen your understanding of the structure of the human brain, which is very similar.

Before you begin

❏ Read the appropriate chapter in your textbook.

❏ Set your learning goals. When you finish this exercise, you should be able to
- list and describe the principal structures of the sheep brain
- identify important parts of the sheep brain in a preserved specimen

❏ Prepare your materials:
- dissection tools and trays
- preserved specimen: *sheep brain (whole)*

❏ Read the directions and safety tips for this exercise **carefully** before starting any procedure.

> **SAFETY FIRST!** Heed the safety instructions that accompany the preservative used with your specimen(s). Avoid cuts and other injuries when using the dissection tools.

A. The sheep brain: external

The sheep brain is remarkably similar to the human brain. One major difference is in proportion. For example, the sheep has a proportionally smaller cerebrum. Another difference is in orientation. The sheep brain is oriented anterior to posterior, as in any four-legged animal. The human brain is oriented superior to inferior. Examine the external features of the sheep brain as outlined in these steps:

❏ 1 Determine whether the **meninges** of your specimen are still intact. If so, find the following:
- **Dura mater**—A tough, leathery outer covering (the dura forms a large cavity for blood along the top of the brain called the **sagittal sinus**)
- **Arachnoid**—The thin, translucent middle meninx (recall that the *subarachnoid space* contains cerebrospinal fluid, CSF, in living specimens)
- **Pia mater**—Thin, vascular membrane covering the brain (preserved blood in the vessels may give it a dark brown appearance)

❏ 2 The most prominent feature of the brain is the **cerebrum.** This region is divided into nearly symmetrical left and right **hemispheres** by a deep **longitudinal fissure.** It is along this fissure that the sagittal sinus of the dura lies. Bend the two hemispheres away from each other and observe the **corpus callosum,** a mass of white fibers that connects the two hemispheres. The surface of the cerebrum is covered with large folds of tissue called **gyri.** The grooves between the gyri are **sulci.** The deeper sulci are often termed **fissures.** The fissures are used by anatomists as landmarks to divide the surface of the cerebrum (the **cerebral cortex**) into regions that roughly correspond to the overlying skull bones:
- **Frontal lobes**
- **Parietal lobes**
- **Occipital lobes**
- **Temporal lobes**
- **Insula** (buried within a fissure between the temporal and parietal lobes)

❏ 3 The smaller, rounded structure caudal to the cerebrum (and dorsal to the brainstem) but still possessing gyri is the **cerebellum.** The cerebellum has smaller gyri that are roughly parallel to one another, unlike the convoluted gyri of the cerebrum.

❏ 4 The **diencephalon** is the brain region ventral to the cerebrum and cranial to the brainstem. One can see some of its features in a ventral view of the sheep brain. The **hypothalamus** includes the **optic chiasma,** an X-shaped junction of fibers at the bases of the **optic nerves.** The legs of the chiasma attached to the brain are the beginnings of the **optic tracts.** The **mammillary bodies,** also part of the hypothalamus, are just posterior to the optic chiasma. The stalklike **infundibulum** projects ventrally from between the chiasma and the mammillary bodies. The infundibulum connects the hypothalamus to the **pituitary gland,** or *hypophysis.* Another major part of the diencephalon is the **thalamus,** which is not visible in an external examination.

> **HINT** → If your specimen has an intact pituitary (not all do), it will be in a mass of tissue on the ventral aspect. To see the infundibulum and adjacent structures, it is necessary to pull the pituitary away from the ventral surface of the brain. You may be able to keep the infundibulum in place as you do the separation, as if it were a hinge on a flap of tissue. Some cranial nerves may be pulled away from the brain, so cut them in a way that leaves as much of the nerve attached to the brain as possible.

❑ 5   Holding the specimen in gloved hands, dorsal aspect toward you, carefully bend it to expose the tissue between the cerebrum and cerebellum. You may have to tease some connecting tissue apart with a probe. You now see the dorsal surface of the **midbrain.** A small, rounded body on the midline, nearest the cerebrum, is the **pineal body** of the diencephalon. The four rounded bumps caudal to the pineal body constitute the **corpora quadrigemina** of the midbrain. The four bumps are distinguished as the **superior colliculi** (the slightly larger pair) and **inferior colliculi** in the human. The midbrain is part of the brain division called the **brainstem,** which is found between the diencephalon and the spinal cord.

❑ 6   The portion of the brainstem just caudal to the midbrain is the **pons.** The cerebellum covers the dorsal surface of the pons, but the ventral surface of the pons can be clearly seen as a large bulge.

❑ 7   The **medulla,** or *medulla oblongata,* is caudal to the pons. The medulla looks as if it were a swollen region of the spinal cord, to which it is cranial. Together, the midbrain, pons, and medulla constitute the brainstem.

❑ 8   Because of their smallness and the manner in which specimens are usually prepared, some of the **cranial nerves** may be impossible to locate on your specimen. Normally, there would be twelve pairs of nerves extending from various parts of the brain. These nerves, of course, are part of the peripheral nervous system (PNS) rather than part of the brain proper. Using Figure 25-1, *A* as a guide, try to locate as many cranial nerves as you can.

> **HINT** → LABORATORY REFERENCE Plates 46 and 47 show color photographs of the external surface of the sheep brain.

B. The sheep brain: internal

❑ 1   Use a knife or long-bladed scalpel to cut the specimen along a midsagittal plane. Use the longitudinal fissure as a cutting guide.

❑ 2   The corpus callosum had been connecting the two cerebral hemispheres and can now be clearly seen in the brain section. The **fornix,** part of the diencephalon, is ventral to the corpus callosum and is also composed of white fibers. You may be able to see a hollow cavity just ventral to the corpus callosum in each brain half. These cavities are the right and left **lateral ventricles.** In the whole brain, they are separated by a thin membrane, the **septum pellucidum.** Because of variations in the plane of sectioning, the septum may be visible as a drumlike covering to one ventricle or may partially cover both ventricles. The **choroid plexus,** a tiny mass of capillaries within each ventricle, produces the cerebrospinal fluid that fills each ventricle.

❑ 3   The lateral ventricles can also be seen in a frontal section. Use one of the sheep brain halves or a new whole brain to cut a frontal section. The infundibulum is a good landmark for the correct position of the frontal cut. Locate the lateral ventricle(s). Just lateral to each lateral ventricle is a small circle of gray matter called the **caudate nucleus,** part of the cerebrum's *basal ganglia.* Just inferior to the lateral ventricles, along the midline, is a single **third ventricle.** Connecting each lateral ventricle to the third ventricle is a **foramen of Monro,** which is not visible in any of the brain sections. The lateral walls of the third ventricle are formed by a mass of gray matter called the **thalamus,** a region of the diencephalon. The two *lateral masses* of the thalamus are joined by the *intermediate mass* that passes through the third ventricle. Take note of the distribution of gray and white

ANATOMICAL ATLAS OF THE SHEEP BRAIN

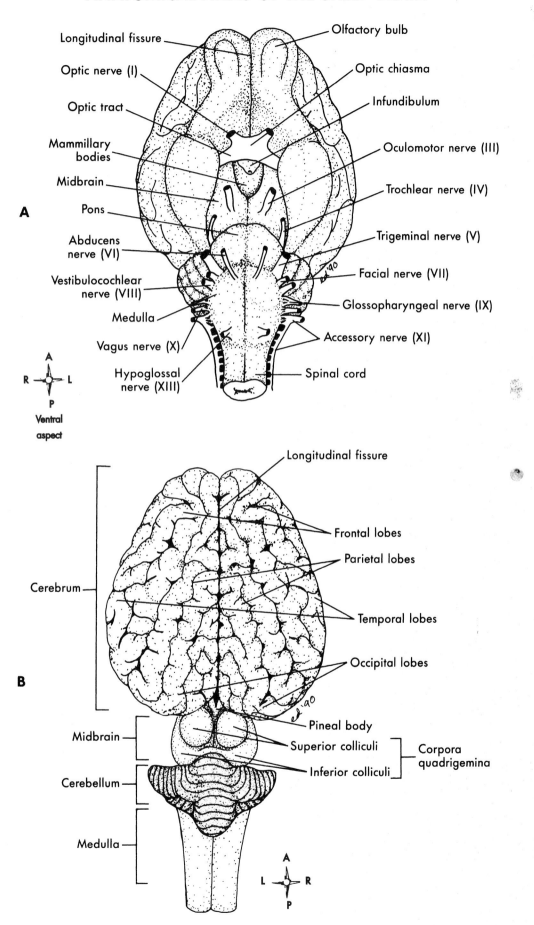

Figure 25-1 External view of sheep brain. **A,** Ventral aspect. **B,** Dorsal aspect.

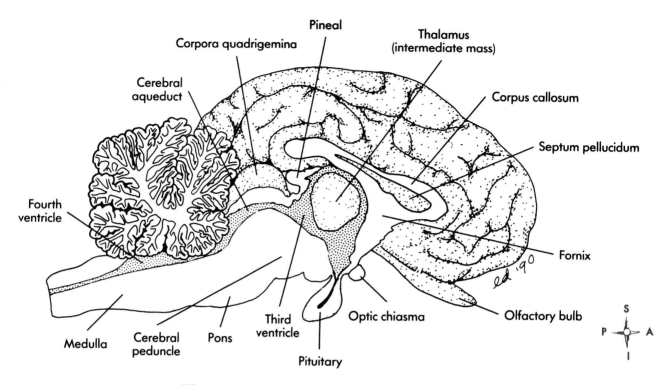

Figure 25-2 Midsagittal section of sheep brain.

matter in the frontal section. What do you notice about its pattern?

❏ 4 Return your attention to the midsagittal section. Identify the white matter pattern, or **arbor vitae,** of the cerebellum. Ventral to the cerebellum is the **fourth ventricle,** which is connected to the third ventricle by the **cerebral aqueduct** (aqueduct of Sylvius).

❏ 5 Many features of the diencephalon already seen on the external aspect or in the frontal section can be seen in the midsagittal section. Locate the pineal body, optic chiasma, mammillary bodies, and infundibulum. In this section, only the intermediate mass of the thalamus is visible. It appears as a circle of gray matter surrounded by the shallow section of the third ventricle.

❏ 6 The corpora quadrigemina is easily seen in the midsagittal section, along the dorsal surface of the midbrain, just posterior to the pineal body. The white matter tracts in the ventral portion of the midbrain constitute the **cerebral peduncles.**

❏ 7 Identify the pons and medulla in the midsagittal section.

> **HINT** → LABORATORY REFERENCE Plate 49 shows a color photograph of some of the internal features of the sheep brain.

NAME _____ DATE _____ SECTION _____

LAB REPORT 25

The Dissection: The Brain

Use this table as a checklist to keep track of your study of the sheep brain specimen.

Brain region	Structure	Sheep	Notes
Meninges	Dura mater	☐	
	Arachnoid	☐	
	Pia mater	☐	
Cerebrum	Hemispheres	☐	
	Longitudinal fissure	☐	
	Corpus callosum	☐	
	Frontal lobe	☐	
	Parietal lobe	☐	
	Occipital lobe	☐	
	Temporal lobe	☐	
	Insula	☐	
	Lateral ventricle	☐	
	Septum pellucidum	☐	
	Caudate nucleus	☐	
Cerebellum	Gray matter	☐	
	Arbor vitae	☐	
	Fourth ventricle	☐	

Continued

Brain region	Structure	Sheep	Notes
Diencephalon	Optic chiasma	☐	
	Infundibulum	☐	
	Mammillary bodies	☐	
	Thalamus	☐	
	Pineal body	☐	
	Fornix	☐	
	Third ventricle	☐	
Midbrain	Cerebral peduncles	☐	
	Cerebral aqueduct	☐	
	Corpora quadrigemina	☐	
Pons	Pons	☐	
Medulla	Medulla	☐	

LAB EXERCISE 26

Somatic Senses

Sensations occur in the central nervous system, after sensory stimuli in the internal or external environment generate nerve impulses, which in turn convey information regarding the environmental change to the CNS. Sensory receptors can be classified a number of ways. For example, they can be classified by **mode:** *photoreceptors* are sensitive to light, *chemoreceptors* are sensitive to certain chemicals, *mechanoreceptors* are sensitive to mechanical changes, and so on. Sensory receptors can also be classified according to their **location:** *exteroceptors* are located in the skin, *proprioceptors* are located in the muscles, and *visceroceptors* are located in deeper tissues (the viscera). Senses mediated by the sensory receptors of the body can be categorized into two major groups: the **somatic senses** and the **special senses.** The special senses involve elaborate sensory organs such as the eye or ear. Somatic senses involve less elaborate sensory mechanisms—often just single receptors embedded in skin or muscle tissue. This exercise briefly explores some of the somatic senses. Later exercises discuss the special senses.

Before you begin

❏ Read the appropriate chapter in your textbook.

❏ Set your learning goals. When you finish this exercise, you should be able to
- explain the distinction between *somatic senses* and *special senses*
- demonstrate that exteroceptors are not evenly distributed in the skin
- demonstrate the use of the proprioceptive sense in controlling body movement

❏ Prepare your materials:
- compass
- metric ruler (mm)
- nontoxic, washable marking pen
- horse hair
- blunt metal probe
- hot water tap
- ice water
- chalkboard and chalk

❏ Read the directions and safety tips for this exercise **carefully** before starting any procedure.

A. Cutaneous senses

Cutaneous senses are somatic senses that are mediated by exteroceptors—sensory receptors in the skin. Cutaneous senses include sensitivity to touch, pressure, heat, cold, vibration, and so on. In this activity, you will explore the distribution pattern of some of the cutaneous senses.

❏ 1 Perform the two-point discrimination test on your lab partner by following the directions given here.
- Adjust a geometry compass so that its two points are touching.
- With your partner's eyes closed and palm upward on the lab table, gently place the compass points on the surface of your partner's index finger.
- Move the compass points apart (by about 1 mm) and touch the finger again.
- Repeat this sequence, each time increasing the distance slightly, until your partner tells you that two distinct points are felt. Record the distance between compass points in the lab report.
- Repeat the test on the back of the arm and compare the density of cutaneous touch receptors in these two external body regions.

SAFETY FIRST! Do not press hard with the compass. Pressing hard ruins the results of the test and may cause injury to the subject.

❏ 2 Measure the density of touch receptors on the back of the hand using a different method:
- Draw a 2.5 cm (1 in) square on the back of your lab partner's hand with a nontoxic, washable marker (or rubber stamp). Divide the square into 16 blocks by drawing a grid within the box (use a ruler).
- With your partner's eyes closed, touch the tip of a horse hair to a point on the grid just hard enough to bend it slightly.
- Mark the position on the matching grid in the Lab Report. Record "0" if the subject does not feel the touch; record "+" if the subject does feel the touch. Repeat this for about 25 to 35 different points within the grid.

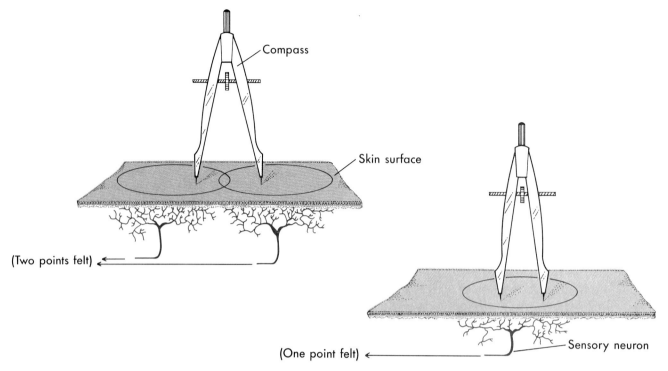

Figure 26-1 Two-point discrimination test. As soon as two points can be distinguished, two separate receptors are being stimulated.

> **HINT** → Touch your partner at random intervals—not with a noticeable rhythm—to avoid anticipation that might alter your results. Also, do not move from block to block in an orderly pattern. This might also cause the subject to anticipate your next move and indicate that he or she felt the hair even though he or she did not really feel it.

❏ 3 Cold receptors are thought to outnumber heat receptors. These two types of *thermoreceptors*, or "temperature receptors," are also thought to have different distribution patterns. Determine the density of these two types of somatic receptors:

- Draw a 2.5 cm (1 in) square on the palmar surface of your lab partner's wrist with a nontoxic, washable marker (or rubber stamp). Divide the square into 16 blocks by drawing a grid within the box (use a ruler).
- With your partner's eyes closed, gently but firmly touch the tip of blunt metal probe that has been cooled in ice water to a point on the grid.
- Mark the position on the matching grid in the Lab Report. Record "0" if the subject does not feel the coolness; record "+" if the subject does feel the coolness. Repeat this for about 25 to 35 different points within the grid.
- Repeat the test, using a probe that has been warmed in hot tap water.

> **SAFETY FIRST!** Do not heat the probe in a flame. Make sure the probe is not dangerously hot or dangerously cold.

> **HINT** → Make sure the probe is kept dry. Make sure that only sensations of *heat* and *cold* are reported—not merely the sensation of touch. This is a test for thermoreceptors, not touch receptors.

B. Proprioception

Also known as "muscle sense," **proprioception** is the ability to sense the contraction or tension of a muscle organ. Receptors for proprioception include stretch receptors called *muscle spindles* and *Golgi tendon organs*. Can you think of any advantages to survival that this sense gives us? Demonstrate the use of the proprioceptive sense:

❑ 1 On the first signature line in the Lab Report, write your name—with your eyes open, as you normally would. On the second signature line in the Lab Report, write your name again—only this time keep your eyes closed. You are using mainly the proprioceptive sense this time. Is your signature significantly different the second time? Can you explain your result?

❑ 2 With your eyes open, draw a cross or addition symbol on a chalkboard at about eye level. Now close your eyes, extend your hand out to the side, then touch the center of the cross with the chalk. Make a small mark at that spot. Now open your eyes and measure the distance from your spot to the center of the cross. Can you relate your result to the operation of the proprioceptive sense?

NAME_____ DATE_____ SECTION_____

LAB REPORT 26

Somatic Senses

Two-point discrimination test

Location	Result
Finger	
Arm	

Density of touch receptors

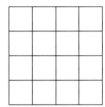
Finger

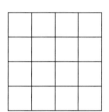
Arm

Density of temperature receptors

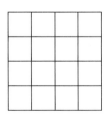
Cold

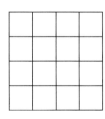
Heat

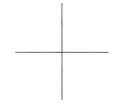

Proprioception
Signature (eyes open) _____
Signature (eyes closed) _____

Distance from center:____cm

Multiple choice

____ 1
____ 2
____ 3
____ 4
____ 5
____ 6
____ 7
____ 8
____ 9
____ 10

Multiple choice (only one response is correct)

1. Somatic senses may include
    a. proprioception
    b. sense of touch
    c. sense of vision
    d. sense of hearing
    e. a and b are correct
    f. all of the above are correct

2. Mechanoreceptors are sensitive to changes in
    a. chemistry
    b. position, pressure, shape
    c. light intensity
    d. temperature
    e. all of the above are correct

3. Exteroceptors may be sensitive to changes in
    a. temperature
    b. pressure
    c. sound
    d. visual images
    e. a and b are correct

4. The two-point discrimination test determines the density of
    a. cutaneous touch receptors
    b. cutaneous thermoreceptors
    c. proprioceptors
    d. chemoreceptors
    e. a and b are correct

5. Thermoreceptors are sensitive to
    a. heat
    b. cold
    c. changes in temperature
    d. external stimuli
    e. any of the above may be correct

6. The sense of vision is an example of a
    a. special sense
    b. somatic sense
    c. neither type of sense

7. Cold receptors and heat receptors are probably distributed evenly over the skin of the body.
    a. true
    b. false

8. Touch receptors are probably distributed evenly over the skin's surface.
    a. true
    b. false

9. Proprioceptors may include
    a. thermoreceptors
    b. Golgi tendon organs
    c. muscle spindles
    d. a and b are correct
    e. b and c are correct

10. The proprioceptive sense involves exteroceptors.
    a. true
    b. false

LAB EXERCISE 27

Senses of Taste and Smell

Sensations of smell and taste are "chemical senses" because both use **chemoreceptors** to detect specific molecules, or types of molecules, in the external environment. Being able to determine the chemical nature of the atmosphere, and of various substances in our external environment, better equips us to find nutritious food and essential minerals, avoid poisonous substances, and otherwise make decisions that affect our overall survival. In this exercise, we will briefly explore the sense of **olfaction** (smell) and the **gustatory sense** (taste).

Before you begin

❏ Read the appropriate chapter in your textbook.

❏ Set your learning goals. When you finish this exercise, you should be able to
- briefly describe the sense of smell, identifying the modality, location of receptors, and the physiological role of smell
- explain the term *adaptation* as it relates to the physiology of sensory reception
- briefly describe the sense of taste, identifying the modality, location and structure of taste buds, and the physiological role of taste

❏ Prepare your materials:
- model of the human head (midsagittal section)
- six small vials, each containing a different recognizable scent
- stopwatch or timer
- microscope
- prepared slide: *human tongue surface c.s.*
- sterile, disposable cotton swabs
- waste container for used swabs
- small paper cup containing *fresh* salt solution
- small paper cup containing *fresh* sugar solution
- small paper cup containing *fresh* lemon juice
- small paper cup containing *fresh* bitter almond solution

❏ Read the directions and safety tips for this exercise **carefully** before starting any procedure.

A. Sense of smell

The olfactory sense, or the sense of smell, is a sense mediated by chemoreceptors in the patch of mucous membrane called the **olfactory epithelium** (see Figure 27-1). The olfactory epithelium is located along the lining of the upper one third of the nasal cavity. Sensory olfactory cells possess hairlike cilia that detect the presence of specific molecules when those molecules react with specialized receptor molecules in the plasma membrane covering the cilia. Because air is continually being drawn into the nasal cavity through the nostrils (and sometimes even through the mouth), the olfactory epithelium can analyze the chemical content of the atmosphere. This allows us to find nutritious foods, avoid dangerous situations (such as fire), and otherwise maintain the stability of our internal environment.

❏ **1** In a model of the head, locate the *olfactory epithelium* in the nasal cavity. Next, locate the *olfactory bulb*. The olfactory bulb has branches of cranial nerve I (the *olfactory nerve*) that project through the *cribriform plate of the* ethmoid bone.

> **HINT** → Use Figure 27-1 to help you find the olfactory epithelium and the olfactory bulb. In many models, a midsagittal section reveals these structures—but they may not be highlighted or labeled in any special way.

❏ **2** The olfactory sense can detect many different kinds of odors, many of which are instantly recognizable to the conscious mind. Demonstrate this fact by opening each of six vials (marked A, B, C, D, E, and F) in front of your partner. Record whether your partner can detect the odor and what he or she thinks the substance in the vial is. Can you identify the odors? Check your results with the lab instructor's key.

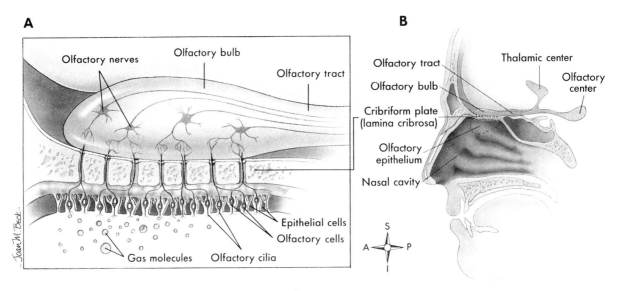

Figure 27-1 Olfactory epithelium. **A,** Enlarged view of cross section of olfactory epithelium and olfactory bulb that supplies sensory fibers to it. **B,** Location of olfactory epithelium in nasal cavity.

> **SAFETY FIRST!** Smell the substances in the vials by waving your hand over the top of each vial, driving the air toward your nose. *Do not hold your nose directly over the vial and sniff!*

☐ 3 **Adaptation** can occur in many types of sensory receptors, including olfactory receptors. In adaptation, prolonged exposure to a particular stimulus causes a receptor to fatigue and thus cease sending information to the central nervous system. Olfactory adaptation occurs frequently; for example, when we put on cologne or use a scented soap and soon cannot smell it on ourselves. Demonstrate the concept of sensory adaptation by performing the following activity:
■ Hold one of the vials used in Step 2 near your nose, breathing normally, but still able to smell the substance.
■ Have your partner begin timing (using a stopwatch or timer) as soon as you begin smelling the substance.
■ Continue to hold the vial in place for several minutes, indicating to your partner the moment you can no longer smell the odor distinctly.
■ Record your adaptation time in the Lab Report.

B. Sense of taste

The sense of taste includes at least four modalities of chemical sensation: *salt, sweet, sour,* and *bitter.* The receptor organs that mediate these sensations are the **taste buds,** found primarily along the sides of the papillae of the tongue. **Gustatory hairs** projecting from sensory cells possess receptors that can be triggered by the molecules detected by the sense of taste.

☐ 1 Locate taste buds embedded within the papillae of a prepared microscopic specimen of the human tongue. Sketch your observations in Lab Report 27.

> **HINT** → Color micrographs of the tongue papillae and taste buds can be found in the LABORATORY REFERENCE, Plates 59 and 60.

☐ 2 Determine the location of the various types of taste buds (salt, sweet, sour, and bitter) by following this procedure:
■ Blot the saliva from the surface of your tongue with a *clean* paper towel.
■ Dip a sterile, disposable cotton swab in one of the four taste solutions provided in a clean vial and place it on various areas of the tongue. Record the areas of the tongue where you can perceive the taste by mapping them on the diagrams provided in Lab Report 27.
■ Repeat this procedure for all four taste solutions:
salt: salt (NaCl) solution
sweet: sugar solution
sour: lemon juice
bitter: bitter almond solution

> **SAFETY FIRST!** Use only *fresh*, uncontaminated solutions. Do not reuse the cotton swabs, even with the same person. Never dip a used swab into the solution. Dispose of used swabs properly. Check the contents of each solution and avoid contact with substances that your physician has advised you to avoid.

NAME _____ DATE _____ SECTION _____

LAB REPORT 27

Senses of Taste and Smell

Olfactory recognition

Vial	Unrecognized	Recognized	Substance
A	☐	☐	
B	☐	☐	
C	☐	☐	
D	☐	☐	
E	☐	☐	
F	☐	☐	

Olfactory adaptation

Trial	Substance	Adaptation time
1		
2		

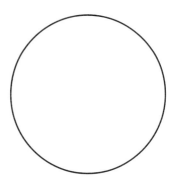

Specimen: *human tongue surface c.s.*

Total Magnification: _____

Location of taste buds by taste modality

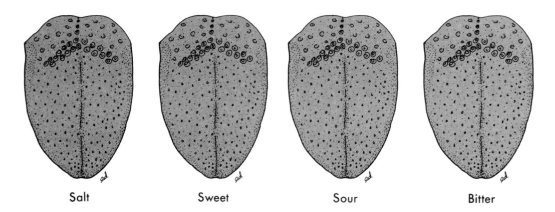

Essay: Write a few brief paragraphs outlining how the senses of smell and taste operate in ways that help the body maintain a homeostatic balance. Include at least two examples of how each might be involved in a homeostatic feedback loop. Use another sheet of paper if you need more room.

LAB EXERCISE 28

The Ear

The **ear** is a complex sensory organ located on each side of the skull, mostly buried within each temporal bone. The ear serves as a receptor organ for at least three special senses: *hearing, static equilibrium,* and *dynamic equilibrium.* This exercise challenges you to learn the basic divisions and structures of the human ear. Once you are familiar with the structure of the ear, you will be ready to explore the functional aspects of hearing and equilibrium in Exercise 29.

Before you begin

❑ Read the appropriate chapter in your textbook.

❑ Set your learning goals. When you finish this exercise, you should be able to
- distinguish the three main divisions of the ear
- identify the principal structures of the human ear in models and figures
- locate features of the cochlea in a prepared microscopic specimen

❑ Prepare your materials:
- model or chart of the human ear
- microscope
- prepared microslide: *cochlea c.s.*

❑ Read the directions and safety tips for this exercise **carefully** before starting any procedure.

Use a chart or model of the human ear to locate and study the ear structures listed in the three activities of this exercise.

A. External ear

You may want to use your textbook as an additional aid in this activity. The **external (outer) ear** is the ear division comprising the elements described in the following steps.

❑ 1 **Auricle (pinna)**—This external ear flap protects the auditory opening, directing sound waves toward it. It also functions as a "radiator" in thermoregulation.

❑ 2 **External auditory meatus**—This tubelike passage carries airborne sound waves further into the ear apparatus; also called the *ear canal.*

❑ 3 **Tympanic membrane**—Also called the *eardrum,* it covers the end of the external auditory meatus to form a boundary with the middle ear. It vibrates when struck by airborne sound waves, carrying the sound energy into the middle ear.

B. Middle Ear

The **middle ear** begins where the external ear ends, with the tympanic membrane. The middle ear is an air-filled cavity lined with mucous membrane. Identify the important parts of the middle ear described in the following steps.

❑ 1 **Malleus**—It is one of the three *auditory ossicles* in each ear. Also called the *hammer,* it is a tiny club-shaped bone attached to the eardrum. It vibrates when sound waves pass to it from the eardrum.

❑ 2 **Incus**—Also called the *anvil,* this tiny bone forms a synovial joint with the malleus. The incus vibrates when it receives energy from the malleus.

❑ 3 **Stapes**—Called the *stirrup* because of its shape, this ossicle is joined to the incus, from which it receives vibrations. A flat portion of the stapes fits into the **oval window,** a passage into the inner ear. Because of the structural relationship of this chain of ossicles, sound waves are carried from the tympanic membrane to the oval window (that is, from the external ear to the inner ear). Along the way, the vibrations are amplified for better reception.

❑ 4 **Auditory (eustachian) tube**—It is a collapsible tube running between the middle ear and pharynx. It allows internal air pressure to equalize with atmospheric air pressure so that high pressure on one side does not distort or muffle the eardrum.

❑ 5 The middle ear also has an opening to the mastoid air cells in the mastoid process of the temporal bone. What significance might this have if the middle ear becomes infected?

> **SAFETY FIRST!** Observe the usual precautions when using the microscope and prepared slides.

C. Inner ear

The **inner ear** is the third division of the ear apparatus. The receptors for hearing and equilibrium are located here. The inner ear is within a hollow area in the petrous portion of the temporal bone. A mazelike **bone labyrinth** contains a similarly shaped but smaller **membranous labyrinth.** The fluid inside the membranous labyrinth is called **endolymph,** and the fluid outside the membranous labyrinth is called **perilymph.** The bony labyrinth, and the membranous labyrinth inside it, is composed of the three main regions described in the following steps.

❑ 1 **Cochlea**—It is a long passage coiled like a snail. A cross section reveals that the passage is divided into three chambers by a Y-shaped partition. The base of the Y is a projection of the bone called the **spiral lamina.** The two branches are pieces of membrane called the **vestibular membrane** and the **basilar membrane.** The space between the two membranes is the endolymph-filled **cochlear duct.** The space outside the vestibular membrane is the **scala vestibuli,** whereas the **scala tympani** is the space outside the basilar membrane. Sound waves move into the perilymph of the scala vestibuli as the stapes vibrates in the oval window. The vestibular membrane vibrates, causing the endolymph in the cochlear duct to vibrate. This, in turn, causes the flaplike **tectorial membrane** to vibrate and bend the hairs projecting from the **organ of Corti** on the basilar membrane (which also vibrates). The bending of hairs induces receptor potentials in the sensory neurons. The energy dissipates as it moves through the scala tympani to the round window. A branch of cranial nerve VIII called the **cochlear nerve** carries information to auditory areas in the brain. Observe a cross section of a mammalian cochlea in the microscope and sketch your observations. Label as many parts as you can identify.

> **HINT** → LABORATORY REFERENCE Plate 51 shows a light micrograph of the mammalian cochlea and features a good view of the organ of Corti in cross section.

❑ 2 **Vestibule**—This is the central area of the inner ear. The vestibule contains saclike portions of the membranous labyrinth called the **utricle** and the **saccule.** Each has a patch of sensory hair cells called a **macula.** The macula's hair cells are covered by a gelatin coating embedded with hard, tiny crystals called **otoliths.** When the head is tilted, gravity pulls on the heavy otoliths and the hairs bend. This induces a receptor potential. Sensory information regarding the effects of gravity (static equilibrium) is transmitted to the brain through the **vestibular nerve,** another branch of cranial nerve VIII.

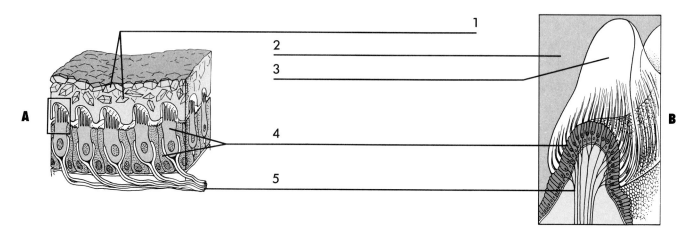

Figure 28-1 Identify **A,** Structure of macula. **B,** Structure of ampulla.

- ☐ 3 **Semicircular canals**—These are three round passages, each on a different plane. The semicircular canals have bubbles at their bases called **ampullae.** Within each ampulla is a crest of tissue called the **crista ampullaris.** Each crista is a patch of sensory hair cells covered with a gelatinous mass (without otoliths) called the **cupula.** When the speed or direction of movement of the head changes, the inertia of the perilymph within the semicircular canals causes it to circulate. As the perilymph circulates, it pushes the cupula and generates receptor potentials in the crista's sensory cells. The vestibular nerve carries the signal to the brain, where it is interpreted as kinetic (dynamic) equilibrium.

The Ear

COLORING EXERCISE Using colored pens or pencils, shade in the figure and accompanying labels in contrasting colors of your choice as indicated by the red numerals.

EXTERNAL EAR
AURICLE 1
EXTERNAL AUDITORY MEATUS 2
TYMPANIC MEMBRANE 3

MIDDLE EAR
MALLEUS 4
INCUS 5
STAPES 6
AUDITORY TUBE 7

INNER EAR
COCHLEA 8
 SPIRAL LAMINA 9
 VESTIBULAR MEMBRANE 10
 BASILAR MEMBRANE 11
 COCHLEAR DUCT 12
 SCALA VESTIBULI 13
 SCALA TYMPANI 14
 TECTORIAL MEMBRANE 15
 ORGAN OF CORTI 16
 COCHLEAR NERVE 17
VESTIBULE 18
 VESTIBULAR NERVE 19
SEMICIRCULAR CANALS 20

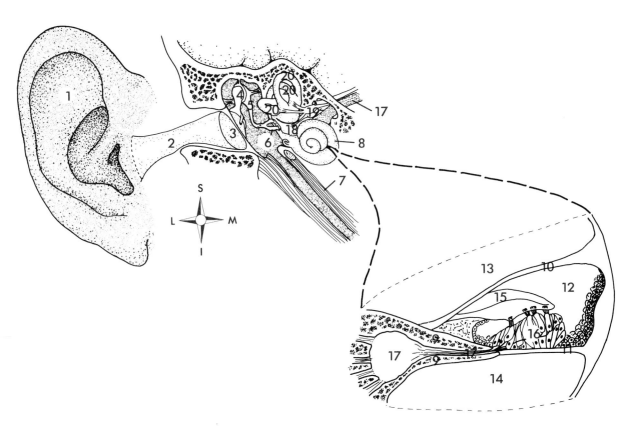

Figure 28-2

NAME _____ DATE _____ SECTION _____

LAB REPORT 28

The Ear

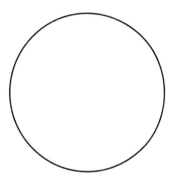

Specimen: *cochlea c.s.*

Total Magnification: _____

Figure 28-1

1. _____
2. _____
3. _____
4. _____
5. _____

Put in order

1. _____
2. _____
3. _____
4. _____
5. _____
6. _____
7. _____
8. _____
9. _____

Put in order (arrange these structures in the order in which sound waves pass through them)

auricle
endolymph
external auditory meatus
incus
malleus
perilymph
stapes
tectorial membrane/organ of Corti
tympanum

263

Ear divisions

Ear divisions (for each structure, identify whether it is part of the *external ear*, *middle ear*, or *inner ear*)

1. auricle
2. bony labyrinth
3. cochlea
4. external auditory meatus
5. incus
6. malleus
7. membranous labyrinth
8. organ of Corti
9. saccule
10. semicircular canal
11. stapes
12. tympanic membrane
13. utricle
14. vestibule

Fill-in (complete each statement with the correct term)

1. The __?__ are three passages, each forming a circle and each in a different plane.
2. When traveling in an automobile, you sense that you have suddenly turned a corner. The sensation you have is mainly an aspect of the sense of __?__ equilibrium.
3. In a weightless environment, as in deep space, the sense of __?__ equilibrium would not work well, if at all.
4. The vestibular membrane and basilar membrane are walls of the __?__ labyrinth.
5. The sensory patch in the utricle's lining is called the __?__.
6. The sensory patch in the utricle's lining has receptors for the sense of __?__.
7. The __?__ is a passage that allows air pressure in the middle ear to reach equilibrium with atmospheric air pressure.
8. The auditory ossicles are joined to one another with __?__ (type) joints.
9. The __?__ has a flat, pluglike portion that fits into the oval window.
10. Both the oval window and the __?__ window are on the boundary of the middle and inner ear.
11. During sound reception, the tectorial membrane vibrates and stimulates the hair cells in the __?__.
12. The __?__ is the most external portion of the external ear.
13. The __?__ amplify sound waves by as much as twenty times their original intensity.
14. The semicircular canals have sensory structures called __?__ within the ampullae.
15. The semicircular canals have sensory structures for the sense of __?__.
16. The maculae of the vestibular sensory regions are covered with gelatinous material. __?__ are embedded in this material.

LAB EXERCISE 29

Hearing and Equilibrium

Hearing, like vision, is considered to be one of the *special senses*. Hearing, the ability to detect a range of sound **frequencies** (pitches) and **intensities**, is only one of the senses mediated by the ear apparatus. Two types of **equilibrium** are also mediated by ear receptors. One type, **static equilibrium**, allows you to determine your position relative to a center of gravity. In other words, it tells you whether you are right-side-up, upside-down, or something in between. Your sense of **kinetic (dynamic) equilibrium** gives information regarding the speed and direction of your body's motion.

This exercise challenges you to perform tests that demonstrate the basic functions of hearing and equilibrium.

Before you begin

❑ Read the appropriate chapter in your textbook.

❑ Set your learning goals. When you finish this exercise, you should be able to
- demonstrate some clinical screening tests for hearing and equilibrium

❑ Prepare your materials:
- tuning fork (256 Hz)
- swimmer's earplugs (optional)
- chalkboard and chalk
- bright desk lamp
- metric ruler (30 to 100 cm)
- watch or clock
- swivel chair

❑ Read the directions and safety tips for this exercise **carefully** before starting any procedure.

A. Hearing tests

Conduction impairment refers to a blocking of sound waves as they are conducted through the external and middle ear to the sensory areas of the inner ear (the *conduction pathway*). **Nerve impairment** implies insensitivity to sound because of inherited or acquired nerve damage. The two hearing tests described are often used in clinical settings to screen patients with suspected hearing impairments.

❑ 1 Conduct the **Rinne test** on a subject:
- Strike a tuning fork on the base of your palm and place the handle against the subject's mastoid process.
- When the subject no longer hears the tuning fork's hum, quickly move the prongs to the opening of the external auditory meatus (but do not touch it).
- Normally, a hum will be heard again. If not, a conductive impairment is suspected. Repeat on the other side of the head. You may want to try simulating conductive impairment in a normal subject by using swimmer's earplugs.

❑ 2 Conduct the **Weber test** on a subject:
- Strike a tuning fork on the base of your palm and place the handle against the middle of the forehead.
- Ask the subject on which, if any, side the sound seems louder.
- A conduction problem is indicated on a side that is noticeably louder than the other. Again, you may want to simulate a conduction impairment by having the subject wear one earplug.

B. Equilibrium tests

❑ 1 The **Romberg test** demonstrates a subject's ability to use information sensed by the utricle and saccule alone, without any help from the sense of vision.

> **SAFETY FIRST!** If an equilibrium impairment exists, the Romberg test may result in the subject toppling over! One or two people should stand ready to catch the subject if necessary.

- Position the subject standing near a chalkboard or other vertical writing surface, facing away from the board.
- Use a desk lamp or other light source to cast a shadow of the subject onto the chalkboard.
- With feet together and eyes open, have the subject stand for 2 minutes.

- Mark the farthest edges of the subject's shadow as it sways.
- After 2 minutes, measure the longest distance of sway.
- Repeat with the eyes closed.
- Repeat with the subject facing perpendicular to the board.
- Results showing excessive sway, almost to the point of falling, indicate an equilibrium problem.

❏ 2 The **Barany test** evaluates the function of the semicircular canals. Be sure to record your results in Lab Report 29 as you proceed.

> **SAFETY FIRST!** Hold onto the chair tightly and surround the subject so that no one topples over and becomes injured. Students prone to motion sickness or related problems should not participate in this activity.

- Have your lab partner sit in a rotating chair with eyes closed and head tilted 30 degrees forward.
- Rotate the chair in one direction for about 10 seconds.
- When you stop the rotation, look into your lab partner's eyes. The side-to-side movement that should be apparent is called *nystagmus*. Side-to-side nystagmus is evidence that the lateral semicircular canals are being stimulated by the rotation.
- Allowing time for rest, repeat the demonstration with your lab partner's head tilted toward one side about 90 degrees. This stimulates the superior canals, causing up-and-down nystagmus.
- Allowing time for rest, repeat the demonstration with your lab partner's head tilted as far forward as possible. This arrangement stimulates the posterior canals maximally, producing rotary nystagmus.

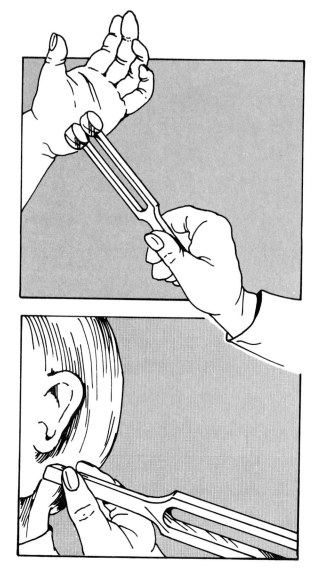
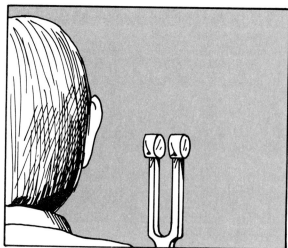

Figure 29-1 Using tuning fork for Rinne test.

NAME_____ DATE_____ SECTION_____

LAB REPORT 29

Hearing and Equilibrium

Hearing tests

Test	Left		Right	
	Normal	Impaired	Normal	Impaired
Rinne test (normal)	☐	☐	☐	☐
Rinne test (simulated impairment)	☐	☐	☐	☐
Weber test (normal)	☐	☐	☐	☐
Weber test (simulated impairment)	☐	☐	☐	☐

Explain the difference in results between the normal hearing tests and the hearing tests performed with a simulated impairment.

Romberg test
    Observations:

    Result: ❏ normal ❏ impaired

Barany test

Position	Observations	Explanation
Head forward 30 degrees		
Head to side 90 degrees		
Head full forward		

Explain how the results of the Barany test relate to the function of the semicircular canals.

LAB EXERCISE 30

The Eye

Vision, one of the *special senses,* relies on a very complex receptor apparatus, the **eye.** Vision depends on a variety of positioning and focusing mechanisms to form an image in the correct spot on the light-sensitive receptor cells inside the eye. These mechanisms involve muscles, lenses, and other structures that are all a part of the visual apparatus. The complexity of structure allows for complexity of function. The visual image perceived by humans has the qualities of **resolution, brightness, color,** and **depth.** These and other aspects of visual structure and function are presented in this exercise and in Exercise 31.

Before you begin

❏ Read the appropriate chapter in your textbook.

❏ Set your learning goals. When you finish this exercise, you should be able to
- describe the major features of the eye and identify them in specimens, models, and figures
- explain the basic function of the eye and its structures

❏ Prepare your materials:
- sheep or beef eye (fresh or preserved)
- dissection tools and trays
- chart or dissectible model of the eye
- microscope
- prepared microslide: *retina c.s.*

❏ Read the directions and safety tips for this exercise **carefully** before starting any procedure.

A. The human eye

The human eye and its accessory structures provide the structural apparatus required for processing and receiving visual images. Using charts or dissectible models of the human eye and its accessories, locate each of the parts listed and describe its function in Lab Report 30.

❏ 1 Locate these features of the **lacrimal apparatus:**
- **Lacrimal gland**—Exocrine "tear" gland in the superior lateral corner of the orbit.
- **Nasolacrimal duct**—Duct in the inferior medial corner of the orbit, drains tears from **lacrimal canals** toward the nasal cavity.

❏ 2 Locate these *extrinsic eye muscles,* which are skeletal muscles that control eye movement:
- **Rectus muscles**—(four) *Superior, inferior, medial, lateral*
- **Oblique muscles**—(two) *Superior, inferior*
What is the action of each of these muscles?

❏ 3 Identify these additional accessory structures:
- **Eyelids**
- **Conjunctiva**—Thin, transparent mucous membrane adhering to the anterior surface of the eye and lining the eyelids.
- **Eyebrows**

❏ 4 Distinguish the three *coats* of the eyeball:
- **Outer coat**—The **cornea** is the anterior, transparent portion. The **sclera** is the white, fibrous portion.
- **Middle coat**—The posterior portion is the thin, heavily pigmented **choroid.** The anterior portions include the circular **ciliary body,** in which are incorporated the **ciliary muscles** (*intrinsic eye muscles*). The ciliary muscles are attached to the rim of the **lens** by means of **suspensory ligaments.** Attached to the anterior edge of the ciliary body is the colored **iris,** which has an opening called the **pupil.**
- **Inner coat**—The **retina** is divided into two layers: the outer *pigmented retina* and the inner *sensory retina.* The sensory retina contains photoreceptor cells of two types: **rods** and **cones,** as well as association neurons. In the posterior retina is a small, yellow **macula lutea** with a pit called the **fovea centralis.** The fovea is normally the center of the visual field and contains many cones. Rods become more dominant farther away from the fovea. Medial to the macula lutea is the white **optic disc.** Here, blood vessels, as well as the nerve fibers that exit as the **optic nerve,** pass out of the eyeball.

> **SAFETY FIRST!** Do not forget the rules for safe use of the microscope.

Observe a cross section of the retina in a prepared microscopic specimen. Can you identify the features of the retina shown in Figure 30-1? Are either of the other two coats of the eyeball visible in your specimen?

> **HINT** → LABORATORY REFERENCE Plate 50 shows a light micrograph of the retina.

❑ 5 Identify these cavities and chambers of the eye:
- **Anterior cavity**—Filled with **aqueous humor,** a watery filtrate produced by the ciliary body. Aqueous humor circulates from the **posterior chamber** behind the iris, through the pupil, to the **anterior chamber,** where it is reabsorbed.
- **Posterior cavity**—Filled with transparent, jellylike **vitreous humor.** This cavity is posterior to the lens.

B. Eye dissection

Apply your knowledge of mammalian eye anatomy by carefully dissecting a beef or sheep eye.

> **SAFETY FIRST!** Observe the usual precautions while dissecting the preserved or fresh tissue of the eye specimen.

❑ 1 Trim away any excess adipose tissue, leaving the stub of the optic nerve and extrinsic eye muscles intact.

❑ 2 Locate these structures on the external aspect:
- **Optic nerve**—The nerve bundle projecting from the posterior of the eyeball.
- **Extrinsic eye muscles** (six)—These may have been cut from your specimen during preparation.
- **Sclera**—The white portion of the outer coat.
- **Cornea**—The clear, anterior portion of the outer coat.
- **Iris**—The pigmented region under the cornea.
- **Pupil**—The iris's hole (it may be oblong in your specimen rather than round as in the human).
- **Conjunctiva**—The thin mucous membrane over the anterior portion of the eye, extending to line the inner eyelid.
- Any other accessory structures that may be present in your specimen

❑ 3 Puncture the sclera with the tip of your scissors (or a scalpel) about 1 cm posterior to the edge of the cornea. Use the scissors to cut a circle around the eye, staying 1 cm from the cornea's margin. Pull the anterior portion away from the posterior portion.

❑ 4 Identify these features:
- **Vitreous humor**—The thick fluid of the **posterior cavity.**
- **Aqueous humor**—A watery liquid in the **anterior cavity.**
- **Lens**—A plasticlike ball of translucent tissue (in fresh specimens it is very clear; hold it up and try to look through it).
- **Ciliary body**—A ring of ridges around the outside of the iris's margin (the ridges are formed by the ciliary muscles within).
- **Retina**—A thin film of gray matter loosely associated with the inside posterior wall of the eyeball (it may have fallen away, appearing as a crumpled mass still attached at the **optic disc**).
- **Choroid**—The pigmented region of the **vascular tunic** in the posterior region of the eyeball's wall, deep to the sclera, superficial to the retina (in some mammals, but not in humans, an iridescent **tapetum lucidum** can be seen in the choroid).

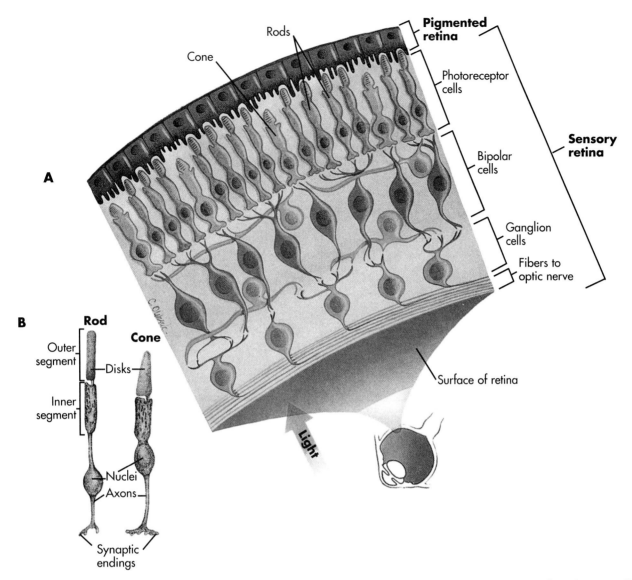

Figure 30-1 Cell layers of the retina. **A,** Pigmented and sensory layers of the retina. **B,** Rod and cone cells. Note their variation in the general structure of a neuron.

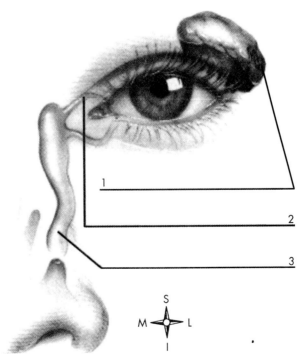

Figure 30-2 Label these parts of the lacrimal apparatus.

The Eye

> **COLORING EXERCISE** Using colored pens or pencils, shade in the figure and accompanying labels in contrasting colors of your choice as indicated by the red numerals.

OUTER COAT₁
   CORNEA₂
   SCLERA₃
MIDDLE COAT₄
   CHOROID₅
   CILIARY BODY₆
   IRIS₇
   LENS₈
INNER COAT₉
   RETINA₁₀
   AQUEOUS HUMOR₁₁
   VITREOUS HUMOR₁₂
   CONJUNCTIVA₁₃

EXTRINSIC MUSCLES
SUPERIOR RECTUS₁₄
INFERIOR RECTUS₁₅
LATERAL RECTUS₁₆
SUPERIOR OBLIQUE₁₇
INFERIOR OBLIQUE₁₈

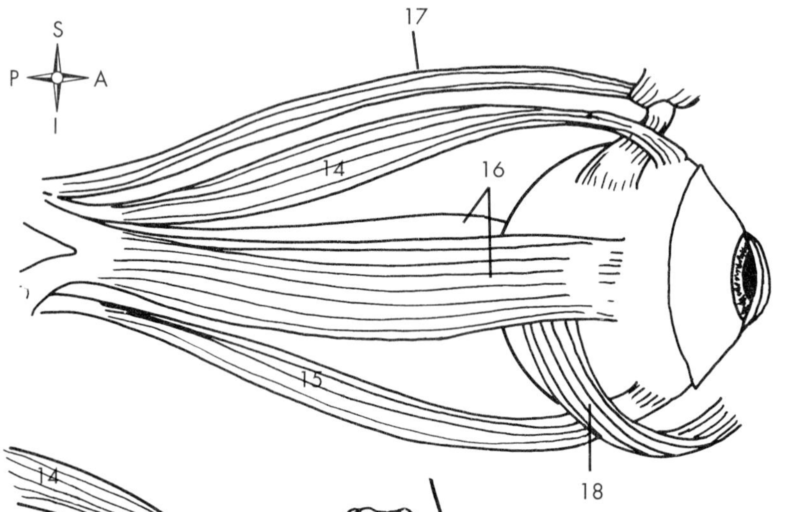

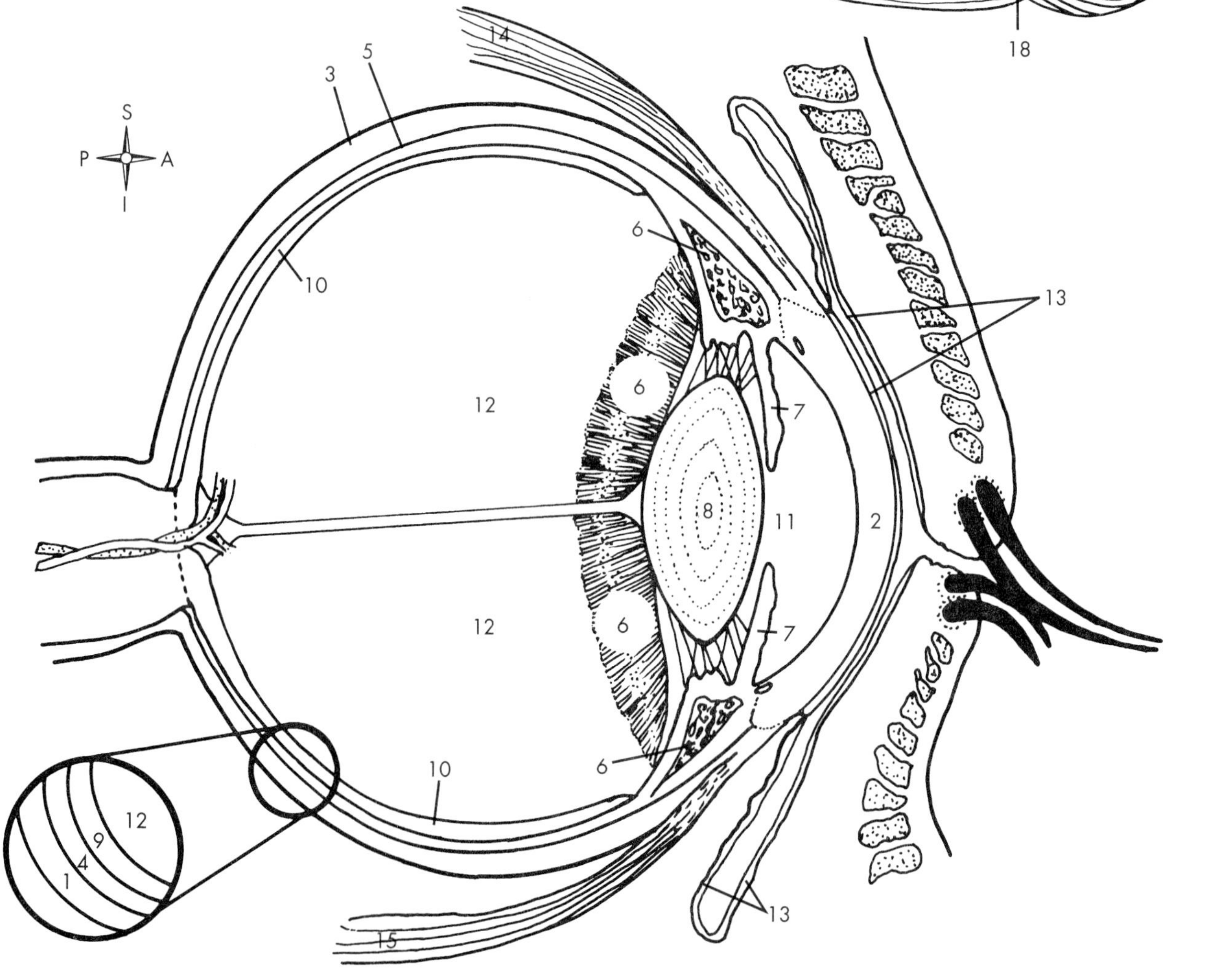

Figure 30-3

ANATOMICAL ATLAS OF THE EYE

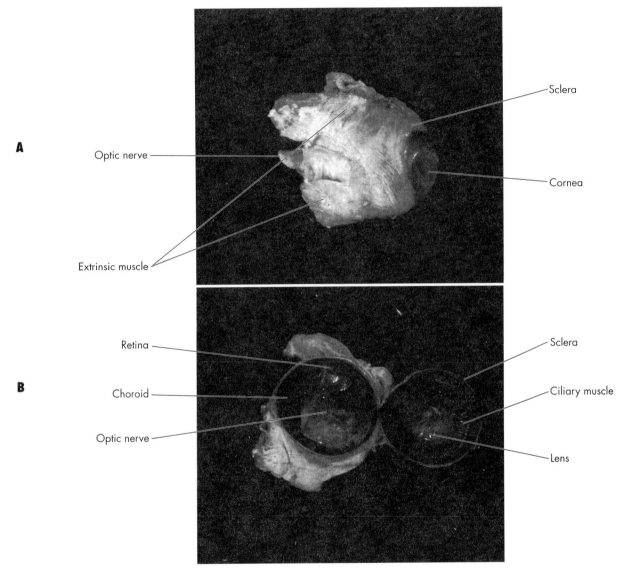

Figure 30-4 Mammalian eye dissection. **A,** External aspect of beef eye. **B,** Internal aspect of beef eye that has been cut according to instructions given in this exercise. Notice grayish retina resting loosely over dark choroid. Ridges of ciliary body can also be seen in this photograph.

LAB REPORT 30

The Eye

Use this table as a checklist for your study of the human and sheep (or beef) eye. Do not forget to fill in the function column.

Tunic or region	Structure	Sheep	Human	Function(s)
Lacrimal apparatus	Lacrimal gland	☐	☐	
	Nasolacrimal duct	☐	☐	
Extrinsic eye muscles	Rectus: superior, inferior, lateral, medial	☐	☐	
	Oblique: superior, inferior	☐	☐	
External accessories	Conjunctiva	☐	☐	
	Eyelids, eyebrows	☐	☐	
Outer coat	Cornea	☐	☐	
	Sclera	☐	☐	
Middle coat	Choroid	☐	☐	
	Ciliary body	☐	☐	
	Lens	☐	☐	
	Suspensory ligaments	☐	☐	
	Iris	☐	☐	
	Pupil	☐	☐	
	Tapetum lucidum	☐	☐	
Inner coat	Retina	☐	☐	
	Macula lutea	☐	☐	
	Fovea	☐	☐	
	Optic disc	☐	☐	
Anterior cavity	Aqueous humor	☐	☐	
Posterior cavity	Vitreous humor	☐	☐	
	Optic nerve	☐	☐	

Figure 30-2

1. _____
2. _____
3. _____

Fill-in

1. _____
2. _____
3. _____
4. _____
5. _____
6. _____

Put in order

1. _____
2. _____
3. _____
4. _____
5. _____
6. _____

Fill-in (complete each statement with the correct term)

1. __?__ is a type of photoreceptor specialized for vision in dimly lit environments.
2. Tears are produced by the __?__ glands.
3. When you try to keep your eyes focused on a rocket as it ascends, you contract the __?__ muscle attached to your eyeballs.
4. Muscles in the __?__ relax or contract to alter the diameter of the pupil.
5. The redness seen in one's eyes in a smoke-filled environment is a result of temporary inflammation of the __?__.
6. On a hot day, sweat collects on your forehead. It is prevented from dripping directly toward your eye by the __?__.

Put in order (arrange these in the order in which light passes through them to form an image inside the eye)

conjunctiva
cornea
lens
pupil
retina
vitreous humor

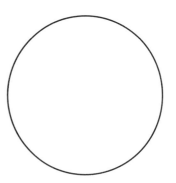

Specimen: *retina c.s.*

Total Magnification: _____

LAB EXERCISE 31

Visual Function

Human vision is a complex function involving **image formation, reception** by photoreceptors, **transmission** of sensory information to appropriate brain centers, and **perception** of the image in the mind. The activities described in this exercise challenge you to demonstrate just a few of the many functional characteristics of the human sense of vision.

Before you begin

❑ Read the appropriate chapter in your textbook.

❑ Set your learning goals. When you finish this exercise, you should be able to
- demonstrate visual function by means of standard tests and demonstrations

❑ Prepare your materials:
- meter stick
- wall chart: astigmatism test
- wall chart: Snellen eye test
- pack of assorted brightly colored papers
- color blindness test plates (Ichikawa Test Book or similar)

❑ Read the directions and safety tips for this exercise **carefully** before starting any procedure.

A. Acuity

Acuity is the sharpness of the visual image and is often tested in clinics with the *Snellen eye test*.

❑ 1   Place a Snellen chart on the wall and have the subject stand 20 feet (6.1 m) away, facing the chart.

❑ 2   Have the subject cover one eye and read the letters on the chart as you point to them. Begin at the top and work your way down.

❑ 3   Note the lowest row that can be read accurately. Record the number printed next to that row in Lab Report 31. That number is the farthest distance (in feet) that a person with normal acuity can see the letters in that row. For example, if the number is 20, the subject has 20-20 vision (meaning the subject can see at 20 feet what a person with normal vision can see at 20 feet). If the number is 100, the subject has 20-100 vision, meaning that the subject sees at 20 feet what a person with normal vision can see at 100 feet). Test both eyes; if the subject wears eyeglasses, try the test corrected (with glasses) and uncorrected (without glasses).

B. Astigmatism

Astigmatism is abnormal curvature of the cornea or lens.

❑ 1   Test for a curvature flaw by facing the subject toward an *astigmatism test chart* 20 feet (6.1 m) away.

> **HINT** → If you don't have a wall chart, you can use Figure 31-2 at about 1 foot away.

❑ 2   Cover one of the subject's eyes and ask him or her to stare at the circle in center of the chart.

❑ 3   If all the lines radiating from the center appear to the subject to be straight and of equal darkness, no astigmatism is present. If the lines appear wavy, curved, or of unequal darkness, astigmatism is suspected. Test the other eye.

C. Accommodation

Accommodation is the ability to adjust the focusing apparatus to account for changes in distance from the viewed object. Your lens changes shape to accommodate near vision. Likewise, your pupil changes diameter to help accommodation.

❑ 1   Position the subject's chin on the lab table and put a meter stick on the table perpendicular to the subject's face (with 0 cm at the chin).

❑ 2   Have the subject cover one eye and ask him or her to focus on the word TEST in Figure 31-3 as you hold it 1 meter from the face.

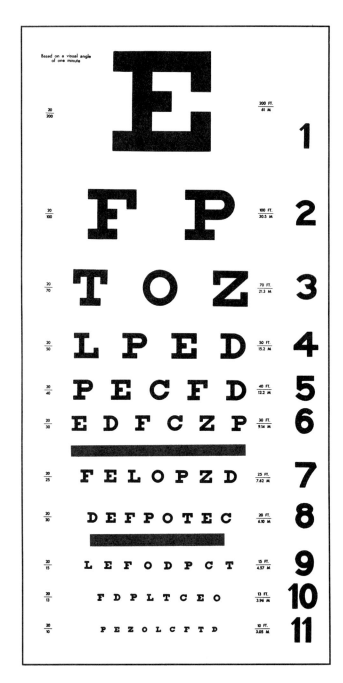

Figure 31-1 Snellen chart for testing visual acuity.

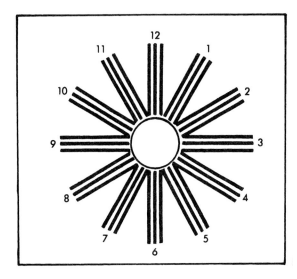

Figure 31-2 Example of astigmatism chart.

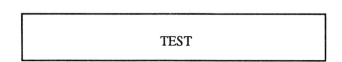

Figure 31-3 Near-point accommodation test figure.

❑ 3 Slowly move the figure closer and record the nearest point at which the subject can focus. Does the diameter of the pupil change as you approach the subject's *near point*? Repeat with the other eye.

D. Blind spot

The **blind spot** is actually the optic disc, a point where there are no photoreceptors in the retina. The part of the *visual field* focused on the blind spot cannot be seen. We normally do not notice it, because the brain "fills in" the blank area for us. Demonstrate the location of the blind spot in your visual field by covering one eye and looking at Figure 31-4. Stare at the object (square or circle) *medial* to the other object in the figure. Starting at about 35 cm, slowly bring the figure closer to your eye. At one point, the *lateral* object will seem to disappear because its image has fallen on the blind spot.

E. Peripheral monochrome vision

Rods are photoreceptors adapted to reception of dim light of a broad color range (that is, *dim light–black-and-white vision*). Cones are adapted for *bright light–color vision* because they require higher light intensity, but different cones can perceive different wavelengths (colors) of light. The cones are

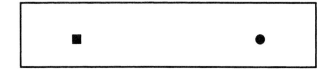

Figure 31-4 The blind spot test objects.

densest at the fovea and the rods in the peripheral portion of the visual field. You can demonstrate the *monochrome* nature of the peripheral field by means of this demonstration:

❑ 1 While the subject stares forward, slowly bring a brightly colored piece of paper into the visual field from behind the subject's head.

❑ 2 Stop when the subject indicates that the object has just entered the visual field.

❑ 3 Ask what color the object is. If you did the test properly and the subject isn't cheating, the color is difficult or impossible to determine. Why?

F. Color vision

Color blindness, or color vision deficiency, can be the result of brain or nerve damage but is usually an inherited condition. In the more common forms, *red-green deficiency*, red- or green-sensitive cones are missing or defective. Individuals with no green-sensitive cones have *deuteranopia;* those with no red-sensitive cones have *protanopia*. Test for red-green deficiency in your subject by following these steps (for the Ichikawa Test Book) or the steps with your test kit.

❑ 1 Hold the book of plates about 75 cm from the subject in bright, natural light, avoiding glare.

❑ 2 *Without touching the figures,* ask the subject to name the numeral seen in each mosaic. If more than one numeral can be seen, ask for the one that appears most distinct. Record the numeral for each plate on the form provided with the test kit.

❑ 3 Add up the marks in the two columns of the form's first section. Using the formula given there, determine whether the subject is normal or color deficient.

❑ 4 *Only if the subject is color deficient,* complete the second section of the form. Determine whether the subject has protanopia or deuteranopia.

NAME _____ DATE _____ SECTION _____

LAB REPORT 31

Visual Function

Test or demonstration	Results		Remarks
	Left eye	Right eye	
Snellen (acuity) test	corrected: 20-_____	corrected: 20-_____	
	uncorrected: 20-_____	uncorrected: 20-_____	
Astigmatism	corrected: ❏ normal ❏ not normal	corrected: ❏ normal ❏ not normal	
	uncorrected: ❏ normal ❏ not normal	uncorrected: ❏ normal ❏ not normal	
Accommodation (near point)	corrected: _____ cm	corrected: _____ cm	
	uncorrected: _____ cm	uncorrected: _____ cm	
Blind spot	❏ found	❏ found	
Peripheral vision test	❏ correct color identification ❏ incorrect color identification	❏ correct color identification ❏ incorrect color identification	
Color blindness screening	❏ normal ❏ protanopia ❏ deuteranopia	❏ normal ❏ protanopia ❏ deuteranopia	

Fill-in

1. _____
2. _____
3. _____
4. _____
5. _____
6. _____
7. _____
8. _____
9. _____
10. _____
11. _____
12. _____
13. _____
14. _____
15. _____

Fill-in (complete each statement with the correct term)

1-2. __?__ is the sharpness of the perceived visual image and is often evaluated in clinical situations with the __?__ eye test.

3. A person with 20-100 vision has (*normal/abnormal*) vision.

4. Abnormal curvature of the cornea or lens is called __?__.

5. The ability to adjust the focusing apparatus of the eye to account for changes in distance from the viewed object is called __?__.

6. When accommodating, the pupil and the __?__ may change shape.

7. The blind spot is actually the __?__.

8. Black-and-white vision is also called __?__ vision.

9. Rods are involved in (*bright-light/dim-light*) vision.

10. Cones are involved in (*bright-light/dim-light*) vision.

11. Rods are involved in (*black-and-white/color*) vision.

12. Cones are involved in (*black-and-white/color*) vision.

13. Individuals with a lack of green-sensitive cones are said to have __?__, a form of red-green color blindness.

14. Individuals with a lack of red-sensitive cones are said to have __?__, a different form of red-green color blindness.

15. Color blindness *usually* results from __?__.

LAB EXERCISE 32

Endocrine Glands

The **endocrine system** includes a number of glands that secrete regulatory chemicals called **hormones** into the blood for distribution throughout the body. The hormones interact only with their **target cells,** cells with receptors for those particular hormones. Endocrinology, the study of endocrine glands and hormones, is advancing rapidly as research continues. Biologists now realize that practically every tissue in the body produces at least one hormone. However, only a handful of glands are large and specialized enough and secrete enough hormone to be considered here as part of the endocrine system. The gross and microscopic structure of some of the major endocrine glands are presented in this exercise. The next exercise presents topics related to endocrine hormones.

Before you begin

❑ Read the appropriate chapter in your textbook.

❑ Set your learning goals. When you finish this exercise, you should be able to
- describe the principal endocrine glands and locate each in a model and in figures
- identify histological features of some of the important endocrine glands

❑ Prepare your materials:
- dissectible model of the human torso and head
- microscope
- prepared microslides:
  Pituitary gland
  Thyroid/parathyroid glands
  Adrenal gland
  Pancreas
  Thymus
  Pineal

❑ Read the directions and safety tips for this exercise **carefully** before starting any procedure.

A. Gross anatomy

Using a dissectible model of the human body, locate each endocrine organ described. As you identify each gland, investigate its gross structure.

❑ 1 **Pituitary**—This tiny, pea-shaped gland is located on the inferior aspect of the brain. It is cradled in the sella turcica of the sphenoid bone. The pituitary, or *hypophysis cerebri,* is actually two glands that are fused together: the **anterior pituitary (adenohypophysis)** and the **posterior pituitary (neurohypophysis).**

❑ 2 **Thyroid**—This single gland is located on the anterior aspect of the trachea (windpipe), near the larynx (voice box). It resembles a bow tie in that it has two **lateral lobes** and a narrow medial **isthmus** that joins them in the middle.

❑ 3 **Parathyroids**—The parathyroids are tiny masses of tissue embedded in the posterior surface of the thyroid. There are usually two superior and two inferior parathyroid glands. (Many models do not represent these glands.)

❑ 4 **Adrenals**—The adrenals are a pair of glands, each on the superior surface of a kidney. Each adrenal resembles a cone-shaped hat on top of a kidney. Many models distinguish between the outer *adrenal cortex* and the inner *adrenal medulla.*

❑ 5 **Pancreas**—Both an endocrine and exocrine gland, the pancreas is a long, narrow mass of glandular tissue. Resembling a fish, the pancreas is cradled in the bend of the C-shaped duodenum (first part of the small intestine, just inferior to the stomach).

❑ 6 **Thymus**—This two-lobed gland is located in the anterior mediastinum. The thymus grows until puberty, then degenerates through adulthood. It is not shown in models of the adult human body.

❑ 7 **Pineal**—The tiny pineal body has already been identified as a nervous structure in the diencephalon of the brain. Because it secretes several hormones, it is also considered an endocrine gland.

☐ 8 **Testis**—The testes, the pair of male *gonads* (primary sex organs), are ovoid organs in the saclike *scrotum* external to the lower anterior trunk.

☐ 9 **Ovaries**—The pair of female gonads called the *ovaries* are ovoid organs within the pelvic cavity. Each is located at the distal end of a tube leading to the uterus (womb) called a **uterine (fallopian) tube.**

☐ 10 **Other**—Many other organs secrete endocrine hormones and so are considered by some biologists to be components of the endocrine system. For example, the digestive tract, the placenta, the kidneys, and the skin all produce endocrine hormones.

B. Endocrine histology

In this activity, you are asked to identify some of the histological characteristics of endocrine glands. For each, locate the structures described in a prepared specimen and sketch your observations in Lab Report 32.

> **SAFETY FIRST!** Do not forget to be cautious when using the microscope and when handling prepared slides.

☐ 1 **Pituitary**—As already stated, the pituitary (hypophysis cerebri) is actually two glands in one. The anterior pituitary (adenohypophysis) is modified epithelium, which is typical for endocrine glands. The posterior pituitary (neurohypophysis) is composed of **neurosecretory tissue,** nerve tissue whose neurons secrete hormones rather than neurotransmitters. Refer to LABORATORY REFERENCE, Plate 61.

> **LANDMARK CHARACTERISTICS**
> The anterior pituitary is usually more darkly stained than the posterior pituitary. Remember that you should be able to find both in a single cross section of the pituitary. The anterior pituitary has two regions: the *pars distalis* and the *pars intermedia.* Of the two, the pars intermedia is more lightly stained. The posterior pituitary has many dark, nonsecreting cells and only a few secretory cells (which are normally not distinguishable).

☐ 2 **Thyroid**—In thyroid tissue, the glandular epithelium forms **thyroid follicles,** which contain **thyroid colloid** (the stored form of thyroid hormones). The **follicular cells,** which form the walls of each follicle, secrete the colloid. The cells between the follicles, called **parafollicular cells,** also secrete a hormone. See LABORATORY REFERENCE, Plate 62.

> **LANDMARK CHARACTERISTICS**
> One of the easiest specimens to identify, thyroid tissue has circles of cuboidal cells with uniformly stained centers. These are cross-sectional views of follicles filled with colloid. Some parafollicular cells should be visible between the circles.

☐ 3 **Parathyroids**—Parathyroid tissue can often be found in the thyroid tissue specimen because of the close association of these two gland types. Parathyroid tissue does not have any follicles but rather a dense packing of glandular cells. **Chief cells** form cords, or rows, and are thought to be primary producers of hormone. **Oxyphil cells** are scattered among the chief cells and are considered by some to be backup hormone producers. See LABORATORY REFERENCE, Plate 63.

> **LANDMARK CHARACTERISTICS**
> Chief cells are smaller, more densely packed cells than oxyphil cells. Oxyphil cells have more cytoplasm per cell, giving them a lighter overall appearance. As is typical in prepared histology specimens, the nuclei are darkly stained, and the cytoplasm is lightly stained.

☐ 4 **Adrenals**—Each adrenal gland is composed of an inner **medulla** and an outer **cortex.** The neurosecretory **chromaffin cells** form the medullary tissue. Cortical tissue is composed of three distinct regions of glandular epithelium:
  ▪ **Zona glomerulosa**—The outermost region, it is just deep to the *capsule* of the organ. Here, cells are arranged in balls and loops.
  ▪ **Zona fasciculata**—It is the middle, and thickest, region of the cortex. The cells in this zone form rather straight rows that are perpendicular to the gland's surface.
  ▪ **Zona reticularis**—This is the innermost zone. Cells in this region form branched rows. See LABORATORY REFERENCE, Plate 52.

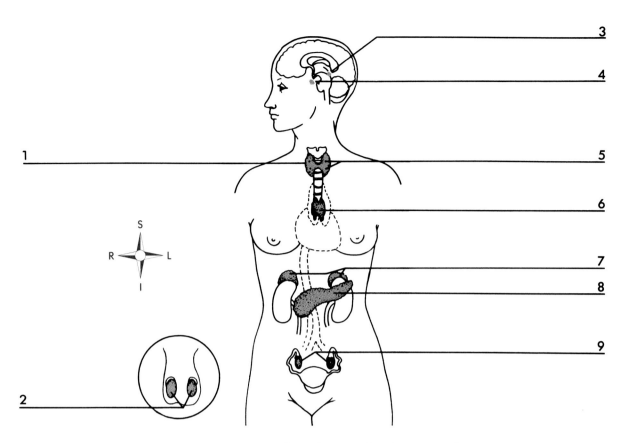

Figure 32-1 Identify endocrine glands indicated.

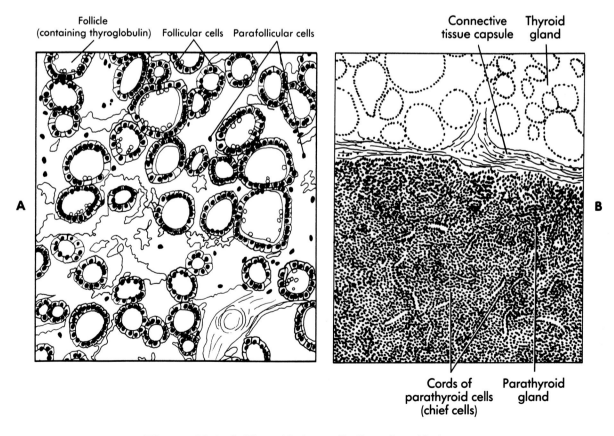

Figure 32-2 A, Thyroid tissue. **B,** Parathyroid tissue.

285

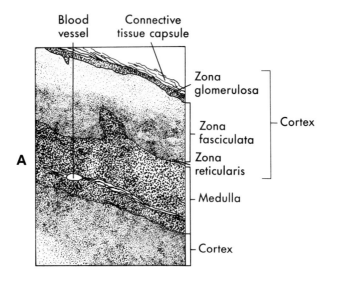
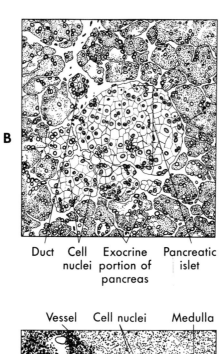
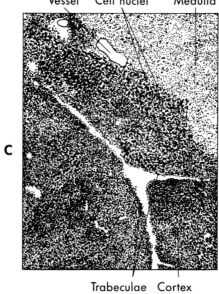

Figure 32-3 A, Adrenal tissue. **B,** Pancreatic tissue. **C,** Thymic tissue.

> **LANDMARK CHARACTERISTICS**
>
> With low-power magnification, you may be able to see at least part of the medulla and all the cortical zones. The medullary chromaffin cells stain more darkly than cells in the adrenal cortex. Switching to high-power magnification, you may be able to distinguish the arrangements of cells in the three zones of the cortex. The cortex appears lighter in color than the medulla because the cortical cells do not accept the stain well.

❑ 5 **Pancreas**—The pancreas can be thought of as an exocrine gland with bits of endocrine tissue scattered throughout. The exocrine **acinar cells** secrete digestive *pancreatic juice.* The endocrine **pancreatic islets (of Langerhans)** are spherical masses of cells that secrete pancreatic hormones. See LABORATORY REFERENCE, Plate 53.

> **LANDMARK CHARACTERISTICS**
>
> Pancreatic islets are usually easy to locate because their endocrine cells are more lightly stained than the surrounding acinar cells. Often, several islets are visible in a single, low-power field.

❑ 6 **Thymus**—The thymus has a dual role in that it functions as both a lymphatic organ and an endocrine gland. It serves as a site for the development of T-lymphocytes (immune cells) and secretes a hormone that influences lymphocyte development. The gland is surrounded by a *capsule* that extends inward, forming *trabeculae* that divide the thymus into **lobules.** Each lobule has a **cortex** largely filled with lymphocytes and a **medulla** with few lymphocytes. In the medulla, small spheres called **thymic (Hassall's) corpuscles,** whose function is unclear, can be found. See LABORATORY REFERENCE, Plate 54.

> **LANDMARK CHARACTERISTICS**
>
> The capsule and its inward extensions (trabeculae) are normally pale, easily found structures. Near these walls, the cortex can be seen as a region of darkly stained lymphocytes. Farther away from the walls, the more lightly stained medullary region can be seen. Occasional thymic corpuscles (circles with a nonuniform interior) can sometimes be seen.

❑ 7 **Pineal gland**—The pineal gland is composed of neuroglia and cells sometimes called *pinealocytes,* which are thought to secrete a number of different hormones. Around puberty, calcifications called **pineal (brain) sand** may begin developing. Although several hypotheses have been proposed, the significance of brain sand is still unknown.

> **LANDMARK CHARACTERISTICS**
>
> Pineal tissue appears as a rather random hodge-podge of different cells but may have several distinct dark spots that represent calcifications ("sand").

NAME_____ DATE_____ SECTION_____

LAB REPORT 32

Endocrine Glands

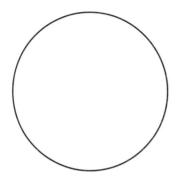

Specimen: *pituitary gland*
Total Magnification: _____

Pituitary gland

Structure:

Location:

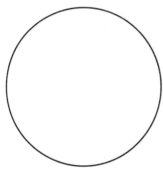

Specimen: *thyroid gland*
Total Magnification: _____

Thyroid gland

Structure:

Location:

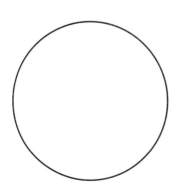

Specimen: *parathyroid gland*
Total Magnification: _____

Parathyroid gland

Structure:

Location:

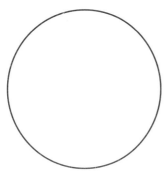

Specimen: *adrenal gland*

Total Magnification: _____

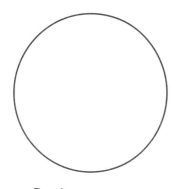

Specimen: *pancreas*

Total Magnification: _____

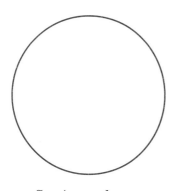

Specimen: *thymus*

Total Magnification: _____

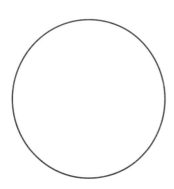

Specimen: *pineal gland*

Total Magnification: _____

Adrenal gland

Structure:

Location:

Pancreas

Structure:

Location:

Thymus

Structure:

Location:

Pineal gland

Structure:

Location:

LAB EXERCISE 33

Hormones

Hormones are organic molecules derived from lipids or proteins. Hormones are secreted by endocrine tissues and travel throughout the body by means of the blood stream. Once dispersed, the hormone molecules have a regulatory effect on cells that have **receptors** for those particular hormones. The affected cells are **target** cells, affected tissues are target tissues, and so on.

Generally, secretion of hormones is regulated by **feedback mechanisms.** *Negative feedback* mechanisms slow or stop secretion when the metabolic changes triggered in the target cells become evident or when blood hormone concentration gets too high. Negative feedback is the most commonly observed type of regulation. For example, **insulin** tends to decrease blood glucose concentration. When blood glucose is high, insulin secretion increases. When blood glucose drops, insulin secretion decreases. *Positive feedback,* less commonly observed, results in an increased hormone secretion when the metabolic changes become evident or blood hormone levels increase. For example, high levels of labor-stimulating **oxytocin** trigger uterine contractions, which push the baby and stretch the birth canal. The stretch sensation feeds back to cause more oxytocin secretion, and so on. Another mechanism of regulating hormone secretion is the **biological clock.** Endocrinologists have long observed cyclic rises and falls in the secretion of several hormones related to reproduction, but more recent evidence suggests that most, if not all, hormones may be regulated to some degree by a biological timekeeping mechanism.

This activity challenges you to learn the names, sources, targets, and actions of some of the principal endocrine hormones and presents some demonstrations of hormone detection tests.

Before you begin

❑ Read the appropriate chapter in your textbook.

❑ Set your learning goals. When you finish this exercise, you should be able to
- name the major hormones of the endocrine system and identify their sources, targets, and principal actions
- appreciate the noncontinuous pattern of secretion in hormones
- explain the hormonal basis of pregnancy testing
- explain how hormone testing can predict ovulation in fertile women

❑ Prepare your materials:
- textbook or (hormone) reference book
- Becton-Dickinson QTest Pregnancy Test kit (or similar); two required per experiment
- Becton-Dickinson QTest Ovulation Test kit (or similar)
- fresh urine samples:
  Pregnant (first trimester) female
  Nonpregnant female
  Ovulating female (see instructions)
- timer, watch, or clock
- paper lab wipes

❑ Read the directions and safety tips for this exercise **carefully** before starting any procedure.

A. Endocrine hormones

Fill in the hormone table presented in Lab Report 33. If you have already studied the endocrine hormones, try to fill it in without referring to a book or your notes. If this is your first exposure to endocrine hormones, you will need to use your textbook or a reference book. As you complete the table, you will find it a helpful study technique to shade in the rows of hormones that come from the same source with the same colored pen or pencil. A brief review of Lab Exercise 32 will be helpful.

> **SAFETY FIRST!** Activities B and C call for the handling of human urine specimens. Use the proper precautions for working with body fluids to protect yourself and others from the spread of disease. Dispose of your specimen as instructed by the laboratory supervisor.

B. Human chorionic gonadotropin

Human chorionic gonadotropin (HCG) is not found on the list of hormones in Lab Report 33 because it is not secreted from one of the major endocrine glands. Recent studies suggest that tiny amounts of HCG are found in tissues throughout the adult human body. However, it seems that the only time a physiologically significant amount is secreted is from the chorion during the early stages of development in the womb.

As you can see in Figure 33-1, *A*, **estrogen** and **progesterone** levels begin to increase at **ovulation** (release of a mature *ovum*). Both hormones promote conditions necessary for successful development of the offspring. For example, the **endometrium** (uterine lining) thickens and becomes more vascular. If fertilization and implantation does not occur, estrogen and progesterone levels fall, resulting in loss (sloughing) of the vascularized endometrial lining (*menstrual flow*). However, if pregnancy does begin, a decrease in estrogen and progesterone would result in the loss of the endometrium and the developing offspring. To prevent this loss, the offspring's tissue secretes large amounts of HCG. HCG stimulates the ovary's secretion of progesterone and estrogen, preventing the decrease in hormones that would otherwise occur. As the **placenta** develops, it begins to secrete its own progesterone and estrogen. HCG secretion decreases during the later phases of pregnancy. Because the placenta is secreting high levels of progesterone and estrogen by this time, stimulation of ovarian secretion is no longer required. See Figure 33-1, *B*.

High blood levels of many molecules are reflected in the urine. This is the case with the very high HCG levels during the first trimester of pregnancy. Because high levels of HCG are detectable in the urine almost immediately after fertilization of an ovum, HCG *urinalysis* is a common form of pregnancy testing.

One type of pregnancy test requires that a urine sample be processed, then injected into a female laboratory animal. If the urine stimulates ovulation or similar physiological changes, it is concluded that it contains the high levels of HCG typical of early pregnancy.

A more convenient HCG test uses *monoclonal antibodies,* which are protein molecules from cultured immune cells that react with HCG (and only with HCG or similar molecules). Anti-HCG antibodies react with HCG to produce a visible reaction. If a reaction occurs, HCG is present in the sample (and one concludes that the subject is pregnant). If no reaction occurs, HCG must be absent or low (and the subject not pregnant).

Over-the-counter (OTC) pregnancy test kits and many laboratory pregnancy screening tests are based on the antibody method. Demonstrate this test by performing antibody urinalysis with the Becton-Dickinson QTest Pregnancy Test kit. Any OTC test kit can be used, but this one has the advantage of having an *experimental control* component. Control tests help ensure that the results are valid and not the result of a side reaction or experimental error. If another test is used, follow the directions accompanying that test kit.

❏ 1 Obtain two fresh urine samples: one from a pregnant woman (in the first trimester) and one from a nonpregnant woman.

> **HINT** → Ideally, each sample is *fresh* urine from the first morning urination. Urine can be held for up to 12 hours, if kept refrigerated (not frozen). Warm the urine to room temperature before testing. If not the first morning urine, make sure the subject has not recently consumed a large volume of liquid. Always wear gloves, goggles, and other appropriate attire when handling human body fluids.

❏ 2 Arrange the test kit components as illustrated in Figure 33-2, *A*. Notice that the test strip has two white pads attached at one end (Figure 33-2, *B*). The distal pad is coated with anti-HCG antibodies and is called the RESULT PAD. The proximal pad, or ERROR CONTROL PAD, does not have anti-HCG antibodies and is the basis for the control test.

❏ 3 After removing the stoppers from all the vials, use the dropper to add urine to vial 1 *up to the line*. Mix by tapping the bottom of the tube until the solution is uniformly red, then allow it to stand for 1 minute. See Figure 33-3, *A*.

❏ 4 *Without touching the pads,* dip the test strip into vial 1 and let it stand there for *exactly 5 minutes*. See Figure 33-3, *B*.

❏ 5 As shown in Figure 33-3, *C,* hold the pads of the test strip under *cold* running water for at least 30 seconds. The pads should be white or very pale pink.

❏ 6 After shaking off the excess water, put the strip into vial 3 and let it stand for *exactly 10 minutes* (Figure 33-3, *D*).

❏ 7 Remove the test strip from vial 3 and blot it dry with lab wipes. Make sure it is completely dry. A blue color on the RESULT PAD indicates a

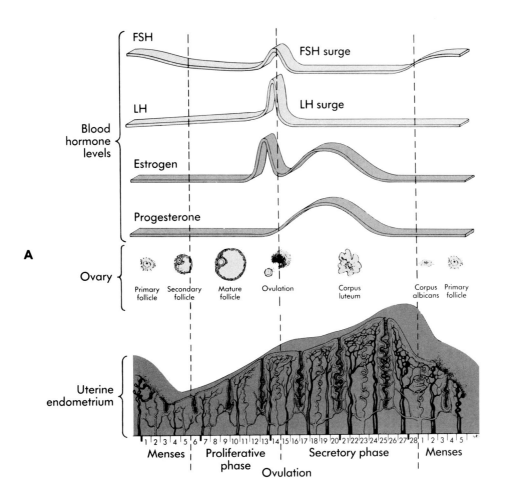
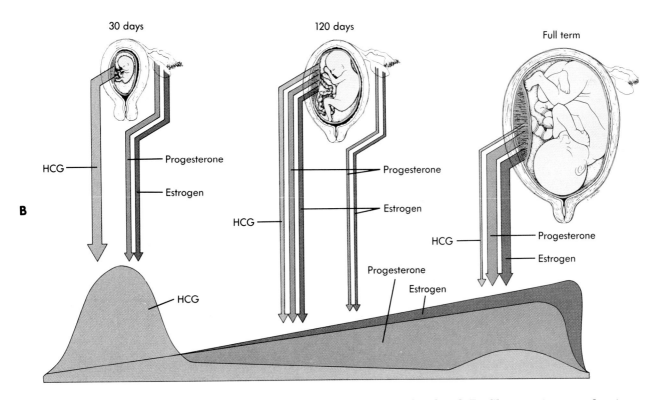

Figure 33-1 **A,** Female reproductive cycle, assuming ovum is not fertilized. **B,** Changes in reproductive hormones during pregnancy.

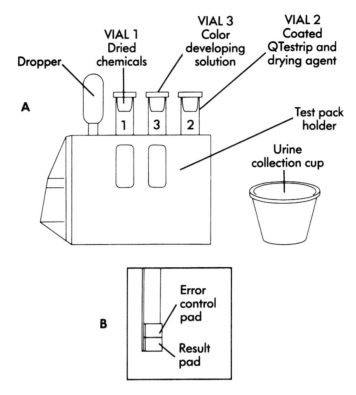

Figure 33-2 A, Pregnancy (or ovulation) test tkit. **B,** Test strip.

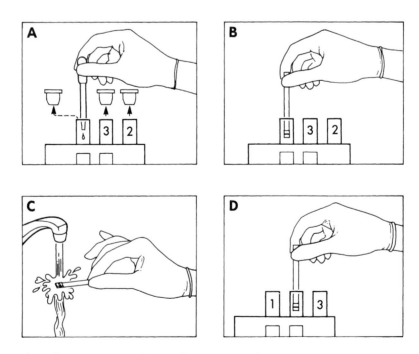

Figure 33-3 Procedure for pregnancy (or ovulation) test. See appropriate procedure for explanations.

positive antibody-HCG reaction (so HCG must be present in the sample). In a valid test, the ERROR CONTROL PAD is white (or very pale blue). If both pads are white (or both pale blue), the result is negative. Report your findings in Lab Report 33.

C. Luteinizing hormone

Luteinizing hormone (LH) is an anterior pituitary hormone that stimulates the secretion of sex hormones by the gonads. In the female, it stimulates secretion of estrogen and progesterone. As you can see in Figure 33-2, *A*, ovulation is immediately preceded by a surge in LH secretion to more than three times its baseline level. Because of that surge, the time of ovulation can be detected by means of a hormone urinalysis method similar to that used for pregnancy testing.

The ovulation test uses monoclonal anti-LH antibodies to test for high levels of LH in the urine. Because a series of tests on the same woman must be performed over a series of 5 days to demonstrate the LH surge, it is suggested that one student from the class be selected to perform the tests at home and report the results to the class. The ideal subject already charts her ovulation cycle, has regular cycles (varying 4 days or less), and is due to ovulate soon. The chart in the QTest Ovulation Test kit can help determine a good candidate, and the ideal time, for testing.

> **HINT** → The basic steps and equipment for the QTest Ovulation Test are the same as for the QTest Pregnancy Test but the *processing times vary.* Use the same figures as for the HCG test, but use the instructions that follow.

❏ 1 For each of 5 days tested, arrange the kit as shown in Figure 33-1, *A* and use a fresh sample collected at the *same time each day.*

❏ 2 Add urine to vial 1 *up to the line.* Mix by tapping the vial bottom until the solution is uniformly red. Let the vial stand for 5 minutes (Figure 33-3, *A*).

❏ 3 Put the test strip into vial 1 and let stand for *exactly 20 minutes,* as in Figure 33-3, *B*.

❏ 4 Rinse the pads under *cold* running water for at least 30 seconds. Shake off the excess water, then put the strip into vial 3 and let it stand for *exactly 10 minutes.* See Figure 33-3, *C* and *D*.

❏ 5 Remove the test strip and blot it completely dry. Read and record your results. The deeper the blue color on the RESULT PAD, the higher the LH level. An LH surge occurs on the day that the RESULT PAD is *clearly darker blue* than on previous days. Ovulation is expected in 24 to 44 hours. If the ERROR CONTROL PAD is darker than the RESULT PAD, the test is inconclusive.

NAME _____ DATE _____ SECTION _____

LAB REPORT 33

Hormones

Use your textbook or class notes to fill in this table.

Hormone	Gland (source)	Target	Action
Antidiuretic hormone			
Oxytocin			
Growth hormone			
Thyroid stimulating			
Adrenocorticotropic			
Melanocyte			
Luteinizing hormone			
Follicle stimulating			
Prolactin			
Thyroid hormones			
Calcitonin			
Parathyroid hormones			
Epinephrine			
Aldosterone			
Cortisol			

Continued

Hormone	Gland (source)	Target	Action
Melatonin			
Thymosin			
Insulin			
Glucagon			
Testosterone			
Estrogen			
Progesterone			
Prostaglandin(s)			

| Test | Sample | Color | | Result | Explanation |
		RESULT PAD	CONTROL PAD		
Pregnancy (HCG)	Pregnant			HCG ❑ present ❑ not present	
	Nonpregnant			HCG ❑ present ❑ not present	
Ovulation (LH)	Day 1			LH ❑ high ❑ low	
	Day 2			LH ❑ high ❑ low	
	Day 3			LH ❑ high ❑ low	
	Day 4			LH ❑ high ❑ low	
	Day 5			LH ❑ high ❑ low	

Discussion questions (answer these questions on a separate sheet of paper)

1. Explain how HCG secretion is regulated. Is it secreted by a pregnant woman or her offspring?

2. HCG depresses some reactions of the immune system. What adaptive advantage do you think this has?

3. How is LH secretion regulated? What effects does the changing blood levels of this hormone have?

4. Compare and contrast the reproductive roles of LH and HCG.

LAB EXERCISE 34

Blood

Blood is the fluid tissue that circulates within the cardiovascular system. Typical of connective tissues, it has a dominant extracellular matrix. In this case the matrix is a liquid called **plasma.** Suspended in the plasma are blood cells or **formed elements.** The characteristics of each person's blood vary with age, sex, metabolic condition, health condition, genetics, and other factors. Clinical evaluation of blood characteristics are important in assessing the condition of patients. This exercise provides demonstrations of just a few of the more common tests. Many blood tests are done by automated machines, at least in some areas, but it is important to understand the bases of these tests to interpret them correctly.

Before you begin

❏ Read the appropriate chapter in your textbook.

❏ Set your learning goals. When you finish this exercise, you should be able to
- name the components of human blood tissue and identify their functions
- demonstrate how common blood tests are performed
- interpret the meaning of important blood tests
- state the limitations of clinical blood testing

❏ Prepare your materials:
- blood sample (see SAFETY FIRST! box)
- sterile disposable lancets
- alcohol pads
- paper towels, lab wipes
- puncture-proof BIOHAZARD container
- disinfecting solution
- clean microscope slides
- bibulous paper
- distilled water (in dropper bottle)
- Wright stain (in dropper bottle)
- wax pencil
- hemocytometer
- Unipette WBC count kit
- Unipette RBC count kit
- heparinized capillary tubes
- microhematocrit centrifuge
- capillary tube sealing clay or plugs
- metric ruler (mm)
- Tallquist test paper and Hb scale (or hemoglobinometer kit)
- toothpicks
- blood typing sera: *anti-A, anti-B, anti-D*
- slide warming box
- microscope

❏ Read the directions and safety tips for this exercise **carefully** before starting any procedure.

> **SAFETY FIRST!** This exercise calls for the use of human blood samples. These tests should be performed as a demonstration by the instructor using packaged, contaminant-free blood. If students are to perform the tests, they should handle *only their own blood*. Any surface that comes into contact with blood *or MAY have been accidentally spattered* should be cleaned with a suitable disinfecting agent *immediately*. All used blood samples and disposable items, including wipes and towels, are to be discarded immediately in the appropriate BIOHAZARD container. Follow current laboratory practice by wearing a lab coat, gloves, and protective eyewear while in the lab.

A. Sampling blood

Blood sampling for clinical testing usually requires only tiny amounts of blood, often only a few drops. The usual method for collecting small blood samples is to draw it from a punctured finger. This activity tells you how blood can be drawn for clinical testing. Follow these steps yourself if you are collecting your own blood.

> **HINT** → If you are using your own blood for tests in this exercise, it may be best to draw all the blood needed for various tests at one time. Therefore, make sure that you have all your equipment standing by so that the blood can be collected and processed immediately. You may wish to collect blood in several heparinized capillary tubes.

SAFETY FIRST! If you suspect that you may have a communicable condition, such as hepatitis, or may be carrying the HIV or another virus, **do not draw your blood.** Even if you are working with your own blood, do not risk spreading infection in this way.

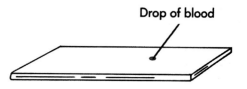

❑ 1 Clean your hands thoroughly with disinfecting hand soap (which should always be available at the lab sinks). Prepare the skin of the puncture site (see step 2) by dabbing it with a fresh alcohol pad. Do not use the pad again.

❑ 2 Using a sterile, disposable lancet, puncture the palmar surface of your nondominant hand's fourth finger. Use a quick, firm stroke.

❑ 3 Holding the arm down, wipe the first drop of blood away with a clean laboratory wipe. Immediately begin collecting drops of blood required for each test. *Do not squeeze the finger to draw more blood, because that will slow circulation through the finger.* If the blood stops flowing, you may have to puncture another finger. Begin at step 1 and *do not reuse the lancet or any other disposable item,* even if you are using the same finger. Dispose of all remnants of the blood collection in the labeled BIOHAZARD container.

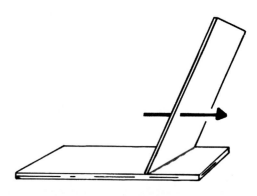

B. Blood smear

Investigate blood tissue by making a **blood smear.** As its name implies, it is merely some blood smeared on a slide, then prepared for microscopic examination.

❑ 1 Place a drop of blood (from a finger or capillary tube) on a clean microscope slide (about 2 cm from one end).

❑ 2 Hold the end of another clean slide at a 50-degree angle against the first slide and push the blood drop away from the middle of the first slide. Figure 34-1, *B* shows this move and how the blood drop spreads out to form a line of blood where the two slides meet. Now push the second slide toward the middle of the first slide. This makes an even smear across the first slide. Stop before reaching the end of the slide. Put the spreader slide directly into the BIOHAZARD container.

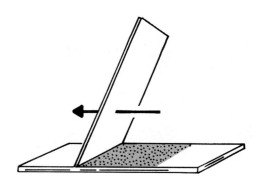

Figure 34-1 Preparing a blood smear. **A,** Drop blood near the end of the slide. **B,** Spread the blood in a line across the slide. **C,** *Pull* the blood across the slide to form a smear. **D,** Mark just outside the edges of the smear with a wax pencil (to contain the stain).

❑ 3 When the blood has dried, draw two lines on the sample slide with a wax pencil as shown in Figure 34-1, *D*. Mark on clean glass, not through the smear. Place a few drops of Wright stain on the smear, counting them as you do so. After 4 minutes, add an equal number of drops of distilled water to the smear. After 10 minutes, rinse the slide gently under tap water for 30 seconds, then blot it dry with bibulous paper.

❏ 4 Scan the slide under low power on your microscope until you find an area where the cells are thinly spread. Switch to high power and observe the blood cells on the slide. Do all the cells look the same? Are different cells (if there are any) distributed equally? What unique characteristics do these cells have (compared to other cells you have observed)? Describe the pattern or order of the cells in the tissue observed. Write your answers to these questions, and any other observations, in Lab Report 34.

HINT → If your microscope has an oil-immersion objective, you may want to use it to see the blood cells more clearly. After focusing on high-dry, rotate the nosepiece *half-way* to the high-oil objective. Put a drop of immersion oil on the slide, then rotate the high-oil objective into the oil drop. You may need to increase the lighting.

C. Differential white blood cell count

The larger, nucleated cells that you may have observed in the previous activity are **white blood cells,** or **leukocytes.** White blood cells are far less numerous than **red blood cells (erythrocytes)** and **platelets (thrombocytes).** White blood cells have numerous protective functions in the body, so their numbers sometimes change in response to changes in health.

Different leukocytes have different functions, so the changes may be reflected in changes in the relative proportions of different types of leukocytes. Clinically, information about relative proportions of leukocyte types is very useful. A test that gives this information is the **differential white blood cell count.**

❏ 1 Focus on a portion of the blood smear, as in the previous activity. Look around the field for any leukocytes.

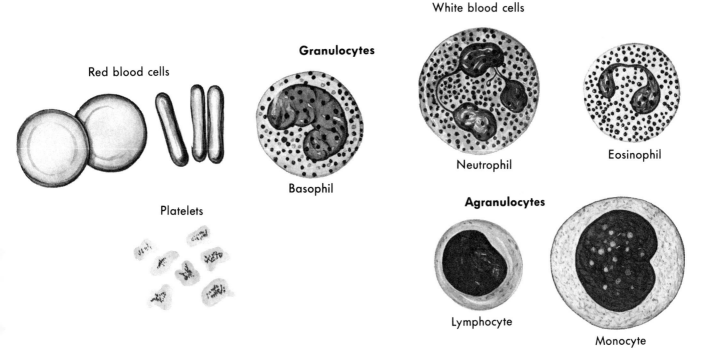

Figure 34-2 Representative human blood cells.

LANDMARK CHARACTERISTICS

White blood cells (**WBCs**) have nuclei and are distinctly larger than red blood cells (**RBCs**). **Agranulocytes** (*Agranular leukocytes*) have no granules (spots) visible in the cytoplasm, but **granulocytes** (*granular leukocytes*) do.

- **Monocytes** are very large agranular leukocytes with large, variably shaped nuclei.
- **Lymphocytes** are agranular WBCs that are almost as small as RBCs. Lymphocyte nuclei are sometimes so large that a lymphocyte appears to have no cytoplasm.
- **Neutrophils** are granular WBCs with small, pinkish granules and lobed nuclei that resemble several links of sausage.
- **Eosinophils** have red granules and two-lobed, dark nuclei.
- **Basophils** have fewer granules, which are a bluish tint and of variable size. Basophils have large, two-lobed or kidney-shaped nuclei.
- See LABORATORY REFERENCE Plates 71-75 and the figures in this exercise.

❑ 2 In Lab Report 34, mark off how many leukocytes you see of each type until you reach 100 total cells. Obviously, you'll have to move around to different fields to reach 100 cells. Move the slide in the manner shown in Figure 34-3 so that you do not count the same portion of the slide twice.

❑ 3 Determine the total number of WBCs of each type (the total for all types is 100). This number is also the percent (because 100 is the total, 1 cell is also 1% of the cells).

NORMAL BLOOD VALUES

Monocytes: 3% to 8% (↑ in chronic infections)
Lymphocytes: 20% to 25% (↑ in antibody reactions)
Neutrophils: 60% to 70% (↑ in acute infections)
Eosinophils: 2% to 4% (↑ in allergic reactions)
Basophils: 0.5% to 1%

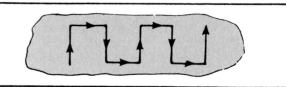

Figure 34-3 Move the slide in this manner to prevent counting the same cells more than once.

❑ 4 How do your results compare with the normal values given in the box?

D. White blood cell count

White blood cell (WBC) counts are clinical tests that estimate the number of leukocytes present in each cubic millimeter (mm^3) of blood. A known volume of a solution called the *diluent* is used to dilute a blood sample (also of a known volume) so that the cells are more spread out and therefore easier to count. The diluted sample is placed on a special glass microscope slide called a **hemocytometer.** The hemocytometer, shown in Figure 34-4, has two counting grids etched on it. When filled correctly, just the right volume is present under the coverslip and over the counting grid.

The instructions that follow are for the Unipette white blood cell counting kit. If you are using another method, follow the directions supplied by the manufacturer of the test kit or by your instructor. In any case, be sure to heed all safety advice.

> HINT → You will be using the hemocytometer for both the WBC count (this activity) and the RBC count (Activity E). You may want to perform them at the same time, filling one side of the hemocytometer with the sample for the WBC count and the other side of the hemocytometer with the sample for the RBC count.

❑ 1 After removing the plastic shield from the Unipette capillary tube collector (the narrow glass tube), touch the tip of the capillary tube to the blood sample until the capillary tube is completely full.

❑ 2 Puncture the top of the Unipette vial (containing WBC diluent) with the plastic shield from the capillary tube collector.

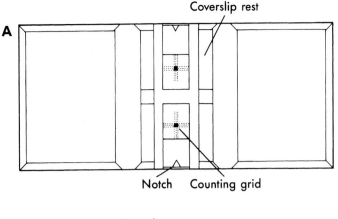
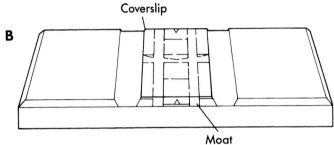

Figure 34-4 Hemocytometer slide. **A,** Viewed from top, without coverslip. **B,** Viewed from front, with coverslip.

☐ 3 Pinch the sides of the vial with your fingers, pushing some of the air out of the vial. With your other hand, put the tip of the capillary tube into the vial. As the cap starts to cover the neck of the vial, slowly release your grip on the vial. This will create suction that will draw the blood from the capillary tube and into the vial. You may need to squeeze some of the diluent back into the capillary tube to get all of the blood out of the tube. Be careful not to squeeze too hard, or you will lose some of your sample and thus ruin the test.

☐ 4 Gently shake the sample to mix it thoroughly, then allow it to stand for the amount of time specified in the Unipette kit.

☐ 5 After waiting the specified amount of time, remove the capillary tube from the vial. Turn the capillary tube around and place the cap end of the tube back on the vial so that the tube is now protruding outward from the neck of the vial. The vial and tube can now be used as a sort of dropper bottle to dispense the diluted blood sample. Practice using this setup by squeezing a few drops of the diluted sample into an approved waste container.

☐ 6 Place the coverslip on the hemocytometer slide. Now touch the tip of the capillary tube on the notch near the raised counting area on one side of the slide. The solution will be drawn under the coverslip. Continue until the entire raised counting area has been filled with solution (that is, until no air pockets are left).

☐ 7 Place the hemocytometer slide on your microscope (low power). Find the etched counting grid by scanning around in the area indicated in Figure 34-4. Once you have found it, adjust magnification, light, and focus so you can see both the cells and the lines of the grid. Center the microscopic field on one of the four areas for counting WBCs (indicated by "W" in Figure 34-5).

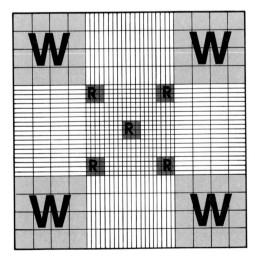

Figure 34-5 Detail of hemocytometer counting grid. Areas used for counting WBCs are marked "W" and the areas used for counting RBCs are marked "R."

❏ 8 Each area marked "W" in Figure 34-5 contains 16 squares. Count the number of cells present in all 16 blocks. Because all the RBCs have been destroyed by chemicals in the diluent, only WBCs will be present to count. To avoid counting cells more than once, count cells that are touching a line only if they are on the top or left edge of a block.

❏ 9 Add together the total number of WBCs in all four WBC counting areas. Multiply this total by 50 to arrive at the number of WBCs in each mm^3 of blood.

> **NORMAL BLOOD VALUES**
> Normal: 4,500-11,000 WBCs per mm^3
> ≠ in acute infections, trauma, some cancers
> Ø in some forms of anemia or during chemotherapy

E. Red blood cell count

Red blood cell (RBC) counts are tests that estimate the number of erythrocytes present in each cubic millimeter (mm^3) of blood. The RBC count is performed in virtually the same manner as the WBC count described in Activity D. However, in the RBC count a different diluent and dilution ratio are used.

The instructions that follow are for the Unipette red blood cell counting kit. If you are using another method, follow the directions supplied by the manufacturer of the test kit or by your instructor. In any case, be sure to heed all safety advice.

❏ 1 Follow steps 1 through 6 in Activity D, except this time using the Unipette kit for the *red blood cell count*.

❏ 2 Place the hemocytometer slide on your microscope and locate the etched counting grid as you did in Activity D, step 7. Center the microscopic field on one of the five areas for counting RBCs (indicated by "R" in Figure 34-5).

❏ 3 Each area marked "R" in Figure 34-5 contains 16 squares. Count the number of cells in all 16 blocks. If cells are touching a line, count them only if they are on the top or left edge of a block.

❏ 4 Add together the total number of RBCs in all five RBC counting areas on the grid. Multiply this total by 10,000 to arrive at the number of RBCs in each mm^3 of blood.

> **NORMAL BLOOD VALUES**
> Females: 4.2-5.4 million RBCs per mm^3
> Males: 4.5-6.2 million RBCs per mm^3
> ≠ in polycythemia or dehydration
> Ø in anemia (several forms), Addison's disease, systemic lupus erythematosus

F. Hematocrit

The blood is composed of a number of components. The plasma is the extracellular fluid matrix of blood tissue. Mostly water, it also contains ions (e.g., Na^+, Cl^-), **plasma proteins** (e.g., *antibodies, albumin*), and other dissolved substances (e.g., *urea*, O_2, glucose). The formed elements (blood cells or *blood solids*) include RBCs, WBCs, and platelets. You have already completed blood tests that look at numbers and proportions of WBCs and total numbers of RBCs. The blood test demonstrated in this activity explores the relative proportions of blood solids and blood plasma.

The **hematocrit** test is also called the **PCV** (packed cell volume) determination because it involves the packing of all the cells in a blood sample at one end of a tube. The sample is placed in a *centrifuge*, a machine that spins a tube of blood and allows centrifugal force to push the heavy blood cells to the bottom. Once the cells are packed, the proportion of cells to total volume can be calculated. The results are used to determine whether there is a shortage of RBCs (as in some *anemias*) or an excess of RBCs (*polycythemia*), or dehydration.

❏ 1 Fill a heparinized capillary tube about 75% full of blood by touching it to the blood drop on your punctured finger. See Figure 34-6, A. Heparinized tubes are red-tipped for easy identification.

❏ 2 Seal the unused end of the tube by holding it near the tip and pushing it into a tray of sealing clay at a 90-degree angle with a twisting motion. A plug of clay will then seal the tube. Remove the tube. Plastic plugs for this purpose are also available.

> **SAFETY FIRST!** Capillary tubes are easily broken, causing injury and contamination by the blood. Have the instructor demonstrate the proper technique for sealing them before attempting it yourself. Handle capillary tubes with gloved hands.

❑ 3 Place the tube in one of the numbered grooves of a microhematocrit centrifuge with the sealed end facing the perimeter of the centrifuge tray. Close the top of the centrifuge according to the manufacturer's instructions and spin the tubes for 4 minutes.

❑ 4 Remove your tube from the centrifuge and notice that the RBCs are now packed in one end of the tube. There may be a thin, lightly colored *buffy coat* topping the RBCs. This coat is actually the WBCs, which are slightly less dense than RBCs. The rest of the tube contains the yellowish, transparent plasma.

❑ 5 Lay the tube on a paper towel next to a metric ruler and measure the total length of the sample, from the clay-RBC border to the end of the plasma. Next, measure the RBC portion only. Report your results (to the nearest mm) in Lab Report 34.

❑ 6 Divide the length of the RBC section by the total sample length. Multiply your answer by 100 to arrive at the percent volume of packed red blood cells. This is the hematocrit value, reported with or without the percent symbol. Use the normal values given in the box on the following page to determine whether your results fall within the normal range.

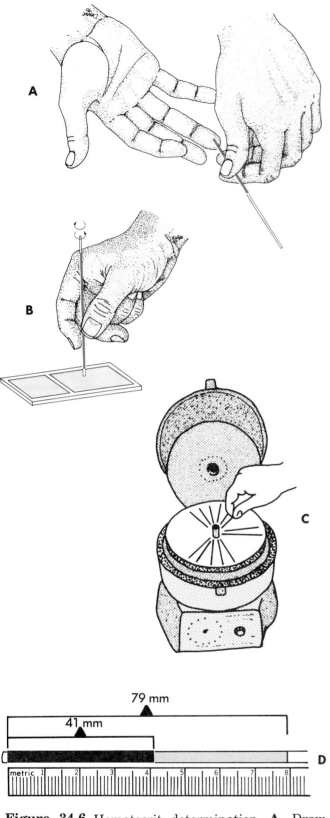

Figure 34-6 Hematocrit determination. **A,** Draw blood into a red-tipped capillary tube. **B,** Seal the unused end by twisting it in a tray of sealing clay. **C,** Place the tube, sealed end outward, in a centrifuge. **D,** Measure the total length and the packed cell length. In this example, $(41 \div 79) \times 100 = 52\%$.

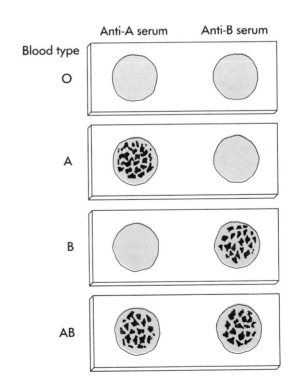

Figure 34-7 Possible results in the ABO test.

> **NORMAL BLOOD VALUES**
> Hematocrit: Male, 40% to 54%; Female, 38% to 47%
> ≠ in severe dehydration, shock, polycythemia
> Ø in anemia, leukemia, cirrhosis, hyperthyroidism

G. Hemoglobin determination

Hemoglobin is the pigment inside RBCs that has an affinity for O_2 (and CO_2), so it functions to transport blood gases. Although knowing the number or proportion of RBCs in the blood may hint at the total hemoglobin (Hb) content, a *hemoglobin determination* gives a relatively accurate figure. Most methods operate on the principle that the denser the color of the sample, the higher the content of Hb pigment. Of course, the Hb has to first be liberated from the RBCs and allowed to diffuse evenly throughout the plasma. The more accurate methods use *colorimeters* or *spectrophotometers* that measure the percentage of filtered light transmitted through the sample. Some clinics use a hand-held version of this type of instrument called a *hemoglobinometer*. A less accurate method, the *Tallquist method,* is used here to demonstrate hemoglobin determination. (If you have a hemoglobinometer and wish to use it, follow the directions that follow those for the Tallquist method.)

Tallquist method:

❏ 1 Place a drop of blood on the special paper in the Tallquist test kit. The RBCs break open in the paper.

❏ 2 Allow the drop to dry long enough to lose its glossy appearance but not long enough for it to become brown.

❏ 3 Compare the color of the blood spot to the colors in the Tallquist chart in your kit. Determine the Hb content by matching colors as closely as possible. Record your result in grams of Hb per 100 ml of blood (g/100 ml) in Lab Report 34. How does it compare with normal values?

Hemoglobinometer method:

❏ 1 Obtain a hemoglobinometer and locate the eyepiece, sample chamber, cover plate, and clip on the hemoglobinometer. Also obtain some hemolysis applicators.

❏ 2 Let a drop of sample blood fall on the open sample chamber.

❏ 3 Hemolyze (break open) the blood cells in the sample chamber by agitating gently with a hemolysis applicator. When hemolysis is complete (after about 45 seconds), the blood will have changed to a clear red solution.

❏ 4 Clip the cover on the sample chamber and slide the whole assembly into the slot on the side of the hemoglobinometer.

❏ 5 Press the illuminating button on the bottom of the hemoglobinometer as you look into the eyepiece. Move the sliding knob on the side of the hemoglobinometer until both sides of the split green field match in intensity.

❏ 6 Read the hemoglobin concentration directly from the scale and record your result in Lab Report 34.

> **NORMAL BLOOD VALUES**
> Hb: Male, 14 to 16.5 g/100ml; Female, 12 to 15 g/100ml
> ≠ in polycythemia, congestive heart failure, obstructive pulmonary disease, high altitudes
> Ø in anemia, hyperthyroidism, liver cirrhosis, severe hemorrhage

H. Blood typing

All cells have different proteins on the surface of their cell membranes that act as identification tags. The human immune system has cells and chemicals that can recognize proteins as *non-self* proteins. In this way, immune processes can try to destroy or inhibit proteins (and cells to which they may be attached) that are foreign to a person's body. The identifying proteins on cell surfaces are determined by heredity. If a foreign tissue is transplanted into someone's body, as in blood transfusions, the immune system will destroy new blood cells that don't have the *self* type of marker proteins.

To prevent this *tissue rejection* from happening, biologists have determined which blood cell marker proteins elicit life-threatening immune reactions. Biologists have also devised tests to determine the presence of these proteins and systems of naming them. This demonstration shows you how some of these **blood typing** tests work.

Before beginning the demonstration, you must learn some new terms. The term **antigen** refers to a molecule that elicits an immune response. In the case of blood transfusions, marker proteins on the donor's RBCs can be antigens (if they are non-self relative to the recipient). The term **antibody** refers to a type of plasma protein produced by the immune system. Antibodies react with non-self markers (antigens) and try to destroy or inhibit them. One type of antigen-antibody reaction is the **agglutination reaction,** in which antigens are "glued" together with antibodies like flies on flypaper. This clumping (agglutination) is the reaction commonly seen when incompatible blood types are transfused.

In the **ABO system** of blood typing, two blood antigens are important: the **A antigen** and the **B antigen.** People with *type A* blood have the A antigen but not the B antigen. *Type B* blood has the B antigen but not the A antigen. *Type AB* blood has both A and B antigens, whereas *type O* blood has neither A nor B antigens. To type blood in the ABO system, blood is mixed with *anti-A serum* and *anti-B serum.* Serum is plasma with the clotting factors removed. Anti-A serum contains antibodies that cause agglutination when the A antigen is present. Anti-B serum contains antibodies that react with the B antigen.

> **SAFETY FIRST!** Handle blood serum with the same caution that you would use when handling any blood sample.

❑ 1 Place a clean microscope slide on a paper towel and write "Anti-A" near the left side and "Anti-B" near the right side. Use a wax pencil to make two 1 cm circles on the slide, one on the right and one on the left.

❑ 2 Put a drop of fresh blood in each circle.

❑ 3 To the drop of blood in the left circle, add a drop of anti-A serum *without touching the dropper to the blood.* Similarly, add a drop of anti-B serum to the right blood drop. Using a different toothpick for each circle, mix the serum and the blood.

❑ 4 Check each mixture for a clumping (agglutination) reaction: the mixture changes from a uniform reddish color to distinct dark clumps in a transparent medium. If a reaction occurred in the left circle (anti-A), then the A antigen is present. If no reaction occurred, then the antigen is not present. According to the ABO system, what type is the blood (A, B, AB, or O)?

Another commonly used blood typing system is the **Rh system,** which deals with the **Rh antigen.** Sometimes called the *D antigen,* this blood cell marker does not elicit an immune reaction in someone who does not have it until the second exposure to the antigen because the immune system doesn't make **anti-D antibodies** until it has been exposed to the D antigen. Once exposed, however, an army of antibodies (and anti-D–producing cells) stands ready to react with the D antigen. A person with the antigen is said to be *Rh positive* and a person without this antigen is termed *Rh negative.* The Rh typing test is similar to the ABO test:

❑ 5 Draw a circle on a clean glass slide and place a drop of blood within it (as described earlier).

❑ 6 Add a drop of anti-D serum and place the slide on a slide warmer. This device warms the slide to about 45°C, and tilts it back and forth. Because the Rh reaction is less severe than reactions in the ABO system, such treatment is necessary to help the reaction so that it can be easily observed. Look for fine clumping. What Rh type is the sample?

NAME _____ DATE _____ SECTION _____

LAB REPORT 34

Blood

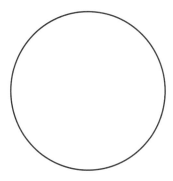

Specimen: *blood smear*
Total Magnification: _____

Observations: *blood smear*

DIFFERENTIAL WBC COUNT TALLY GRID					
Type:	**Monocyte**	**Lymphocyte**	**Neutrophil**	**Eosinophil**	**Basophil**
Tally marks:					
Totals:					
Percentages:					
Normal value:					
Interpretation:					

309

BLOOD TEST RESULTS

Subject:			Sex: ☐ male ☐ female

Test	Normal Value	Result	Interpretation
WBC count			
RBC count			
Hematocrit			
Hemoglobin METHOD:_____			
ABO type			
Rh type			

Fill-in

1. _____
2. _____
3. _____
4. _____
5. _____
6. _____
7. _____
8. _____
9. _____
10. _____
11. _____
12. _____
13. _____
14. _____
15. _____
16. _____
17. _____
18. _____
19. _____
20. _____

Fill-in (respond with the term that completes the statement correctly)

1. A blood cell protein that elicits an immune reaction may be called a(n) __?__.
2. The __?__ scale compares the color of blood-soaked paper to a standard, allowing one to determine hemoglobin content.
3. Lab materials soiled by a blood sample should be disinfected or disposed of in a puncture-proof __?__ container.
4. __?__ stain is used to make the cells in the blood smear more easily observable.
5. Blood is often collected in __?__ capillary tubes, rather than plain capillary tubes, to prevent clotting of the blood.
6. A(n) __?__ is a machine required for the hematocrit test.
7. The PCV test determines the ratio of __?__ volume to total blood volume.
8. __?__ is an oxygen-carrying pigment in the blood.
9. A person with a high differential count of __?__ WBCs is suspected of having an acute infection.
10. A person with a high differential count of __?__ WBCs is suspected of having a chronic infection.
11. Robert's hematocrit is 53%. Is this normal?
12. Maria's hematocrit also is 53%. Is this normal?
13. On a very hot day, Hiro has been outside doing heavy work and just collapsed. If you took a blood sample right now, would you find his hematocrit to be *higher/lower* than his usual hematocrit value?
14. The hemoglobinometer is *more/less* accurate than the Tallquist method.
15. Fred has just been informed that his Rh type is negative. Is this normal?
16. Fred has AB-blood. He needs a transfusion, and his wife (type AB+) is willing to donate her blood. Assuming Fred has never had any type of blood transfusion before, it is *likely/unlikely* that there will be an antigen-antibody reaction if the transfusion is done.
17. Kevin has B+ blood. His plasma is likely to contain anti-__?__ antibodies.
18. Bruce has been living in the mountains for years. The hemoglobin content of his blood is likely to have *risen/fallen* since he moved there.
19. Irma has cirrhosis of the liver. Her Hb value is likely to be *high/low*.
20. Hb values are usually reported in grams of Hb per __?__ of blood.

LAB EXERCISE 35

Structure of the Heart

The **heart** is a four-chambered, hollow organ composed primarily of cardiac muscle tissue. It contracts rhythmically, pumping blood into the **arteries**. After passing through tissues, blood returns to the heart by way of the **veins** and is pumped again. This exercise challenges you to explore the anatomy of the heart through the use of models and preserved specimens.

Before you begin
☐ Read the appropriate chapter in your textbook.

☐ Set your learning goals. When you finish this exercise, you should be able to
- describe the structure of the heart
- locate anatomical features of the heart in models and in preserved mammalian specimens
- explain the function of major heart structures

☐ Prepare your materials:
- dissectible models of the human heart
- preserved sheep heart
- dissection tools and trays
- wooden dowels (1 cm diameter × 12 cm), pencils, or dull probes

☐ Read the directions and safety tips for this exercise **carefully** before starting any procedure.

A. Human heart anatomy

Using dissectible models and the aid given in this exercise, find these features of the heart:

☐ 1 Identify these structures on the external aspect, ventral surface:
- **Interventricular sulcus**—This diagonal groove is located between the walls of the two lower heart chambers (**ventricles**). Along this groove lie the **anterior interventricular artery** and the **great cardiac vein.**
- **Auricles**—These are the flaplike outpouchings of the left and right **atria** (the upper heart chambers).
- **Atrioventricular sulci**—These are grooves between the walls of the atria above and the ventricles below. Locate the **small cardiac vein** and **right coronary artery** on the right and the **great cardiac vein** and **circumflex artery** on the left.
- **Aorta**—The largest artery of the body, it forms the **aortic arch** above the heart.
- **Pulmonary artery**—Somewhat smaller than the aorta, this vessel leaves the heart as a single **trunk** but soon branches to become the **left** and **right pulmonary arteries.**
- **Superior and inferior vena cava**—These two large veins communicate with the right atrium.
- **Apex**—The apex is the lower "point" of the heart.

☐ 2 Identify these features of the heart on the external aspect, dorsal surface:
- **Atria**—These are the upper left and right chambers. They have relatively thin walls.
- **Ventricles**—These are the lower left and right chambers. They have relatively thick walls.
- **Interventricular sulcus**—It is similar to that on the ventral surface. Locate the **middle cardiac vein** and the **posterior ventricular artery.**
- **Pulmonary veins**—These veins communicate with the atria.

☐ 3 Identify these features visible on the internal aspect:
- **Atria**—They are distinguished by their position and thin walls.
- **Ventricles**—They are thick-walled lower chambers. Note that one ventricle has thicker walls than the other. What functional adaptation does this represent?
- **Interventricular septum**—This heart wall separates the left and right ventricles from each other.

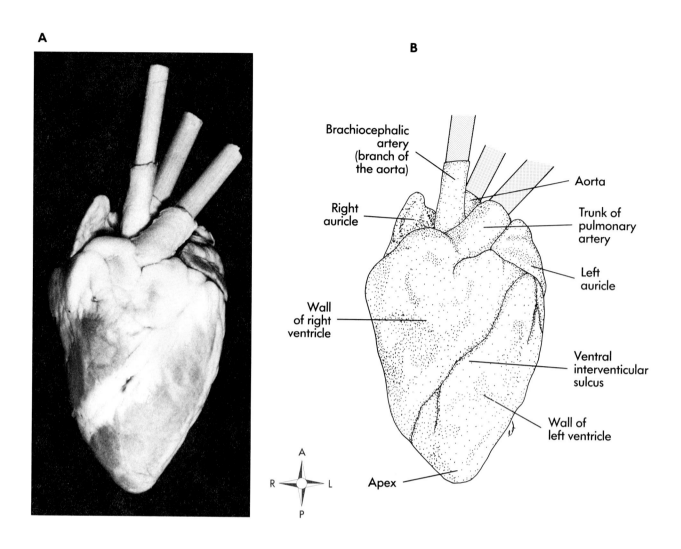
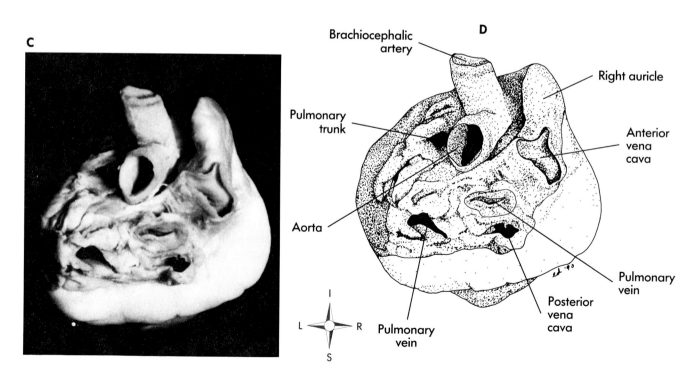

Figure 35-1 A, B, Ventral view of the sheep heart. **C, D,** Anterior view of the sheep heart, analogous to a superior view of the human heart.

- **Cuspid valves**—Also called **atrioventricular (AV) valves,** these valves ensure one-way flow of blood from the atria into the ventricles. The left AV valve, or **mitral (bicuspid)** valve, is composed of two cusps (flaps). The right AV valve, or **tricuspid valve,** has three cusps. Each cusp is attached to the wall of the ventricle below by means of fibrous **chordae tendineae** connected to fingerlike projections of the ventricular myocardium called **papillary muscles.**
- **Semilunar (SL) valves**—The right SL valve, or **pulmonary semilunar valve,** ensures one-way flow from the right ventricle into the pulmonary artery. The left SL valve, or **aortic semilunar valve,** is at the entrance of the aorta. Semilunar valves are each composed of thin-walled bags that hang from the walls of the vessel.
- **Myocardium**—This is the muscular layer of the heart wall.
- **Endocardium**—The thin endothelial lining of the heart chambers, it covers the beam-like **trabeculae** on the inner face of the myocardium.

B. Sheep heart dissection

The sheep heart is very similar in structure to the human heart. It is nearly the same size, so it makes an ideal study specimen.

> **SAFETY FIRST!** Observe the usual precautions when dissecting your specimen. Heed the safety advice accompanying the preservative and avoid cuts and punctures when using the dissecting tools. As always, dispose of your specimen as instructed.

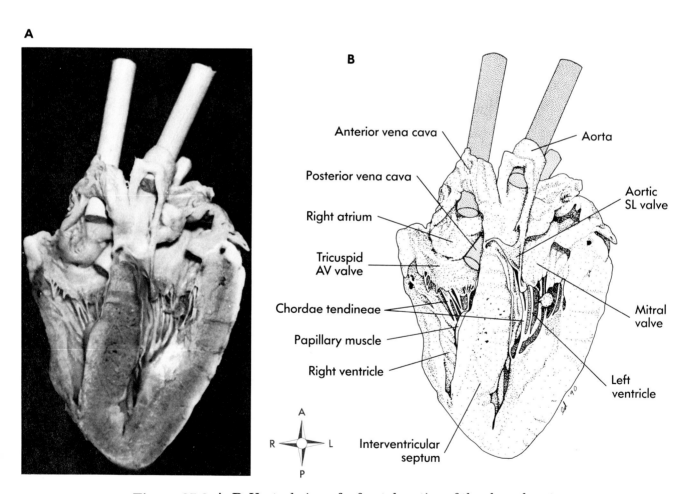

Figure 35-2 A, B, Ventral view of a frontal section of the sheep heart.

The Heart

RIGHT ATRIUM₁
LEFT ATRIUM₂
RIGHT VENTRICLE₃
LEFT VENTRICLE₄
INTERVENTRICULAR SULCUS₅
ANTERIOR INTER-VENTRICULAR ARTERY₆
GREAT CARDIAC VEIN₇
SMALL CARDIAC VEIN₈
RIGHT CORONARY ARTERY₉
CIRCUMFLEX ARTERY₁₀
LEFT CORONARY ARTERY₁₁
AORTA₁₂
PULMONARY ARTERY₁₃
SUPERIOR VENA CAVA₁₄
INFERIOR VENA CAVA₁₅
INTERVENTRICULAR SEPTUM₁₆
MYOCARDIUM₁₇
EPICARDIUM₁₈
MITRAL VALVE₁₉
TRICUSPID VALVE₂₀
CHORDAE TENDINEAE₂₂
PAPILLARY MUSCLE₂₃
AORTIC SEMILUNAR VALVE₂₄
PULMONARY SEMILUNAR VALVE₂₅

COLORING EXERCISE Using colored pens or pencils, shade in the figure and accompanying labels in contrasting colors of your choice as indicated by the red numerals.

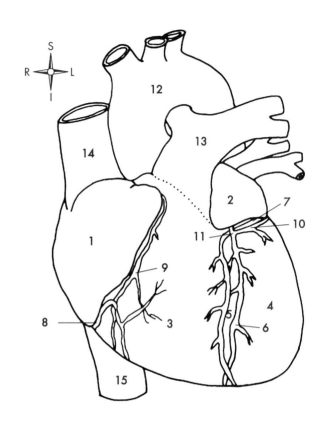
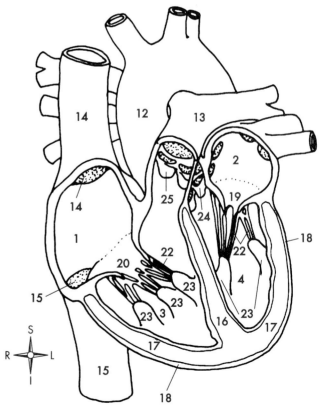

Figure 35-3

☐ 1 Orient yourself to the specimen. Using the photos as a guide, locate the dorsal and ventral surfaces. The shape of your specimen may have become distorted in shipping, so don't rely on shape as a guide. Recall that directions for the sheep heart are based on the fact that the sheep is a four-legged animal, so its *heart is oriented differently than in the human.*

☐ 2 Identify the structures of the external aspect of the sheep heart as you did for the human heart. Some adipose tissue may have to be removed so that you can see all the structures clearly. You may not be able to locate all of the *coronary vessels* because they are buried under fat. Many of the large heart vessels may have been cut very closely to the heart wall, so they appear as holes or short stubs. Use a wooden dowel or pencil to open the vessels for better viewing (as in the photo; Figure 35-1, *A*). Some identifications of external structures will be tentative until you open the heart and verify your observations.

☐ 3 Use a long knife or your scalpel and scissors to cut a frontal section in your specimen (as in the photo; Figure 35-2). Try to identify the internal heart features as you did with the human heart model.

HINT → LABORATORY REFERENCE Plates 70 and 71 show color photographs of a preserved sheep heart.

NAME _____ DATE _____ SECTION _____

LAB REPORT 35

Structure of the Heart

Use this table as a checklist for your study of the heart. Do not forget to fill in the function column.

Structure	Sheep	Human	Function(s)
Right atrium	☐	☐	
Left atrium	☐	☐	
Right ventricle	☐	☐	
Left ventricle	☐	☐	
Interventricular sulcus	☐	☐	
Anterior interventricular artery	☐	☐	
Great cardiac vein	☐	☐	
Small cardiac vein	☐	☐	
Right coronary artery	☐	☐	
Circumflex artery	☐	☐	
Left coronary artery	☐	☐	
Aorta	☐	☐	
Pulmonary artery	☐	☐	
Superior vena cava	☐	☐	
Inferior vena cava	☐	☐	
Interventricular septum	☐	☐	
Myocardium	☐	☐	
Epicardium	☐	☐	
Mitral valve	☐	☐	
Tricuspid valve	☐	☐	
Chordae tendineae	☐	☐	
Papillary muscle	☐	☐	
Aortic semilunar valve	☐	☐	
Pulmonary semilunar valve	☐	☐	

Put in order (arrange these structures in the order in which blood passes through them—assume the blood is about to leave the right atrium)

aorta
aortic semilunar valve
left ventricle
left atrium
lungs
mitral valve
pulmonary semilunar valve
pulmonary artery
right ventricle
superior/inferior vena cava
tissues of the body
tricuspid valve

Fill-in (complete each statement with the correct term)

1. The flaplike lateral wall of each atrium is called the __?__.
2. The __?__ valve is also known as the mitral valve or left AV valve.
3. The right AV valve is also known as the __?__ valve.
4. The aortic semilunar valve has __?__ pocketlike flaps of tissue.
5. The __?__ are fibrous structures that prevent the cuspid valves from prolapsing (bending backwards).
6. One-way flow of blood from the right ventricle is ensured by the presence of the __?__ valve.
7. Mitral valve prolapse, which is abnormal, may allow blood to enter the __?__ during contraction of the left ventricle.
8. The small cardiac vein and right coronary artery can be found along the right __?__ sulcus.
9. The great cardiac vein and anterior interventricular artery can be found along the anterior __?__ sulcus.
10. The __?__ is a muscular wall between the left and right ventricles.
11. The myocardium of the __?__ ventricle is thicker than the other.
12. The wall of the aorta is *thicker/thinner* than the wall of the superior vena cava.
13. The __?__ are beamlike processes of the inner face of the myocardium.
14. The "point" of the heart is called the __?__.
15. In the sheep heart, the right atrium is __?__ to the right ventricle.

LAB EXERCISE 36

Electrical Activity of the Heart

In the previous exercise, you learned some of the major anatomical features of the heart. This exercise challenges you to take a step beyond the basics and learn about the electrical nature of the heart. As part of your study, you will demonstrate one of the most often used clinical tools for assessing the health of the heart, **electrocardiography.**

This exercise calls for any of several electrocardiographs commonly available in educational laboratories (Activities A and B). A computer-based approach is presented as an alternate method in Activities C and D.

Before you begin

❑ Read the appropriate chapter in your textbook.

❑ Set your learning goals. When you finish this exercise, you should be able to
  ■ describe the conduction system of the heart
  ■ demonstrate electrocardiography
  ■ identify the features of a typical electrocardiogram and explain their significance

❑ Prepare your materials:
  ■ electrocardiograph apparatus or CARDIOCOMP system (with computer system)
  ■ alcohol swabs
  ■ ECG electrodes, electrode tape, and gel
  ■ table or cot
  ■ metric (mm) ruler

❑ Read the directions and safety tips for this exercise **carefully** before starting any procedure.

> **SAFETY FIRST!** The apparatus used in this exercise is powered by electricity, so appropriate cautions should be taken to avoid electrical hazards. Students with known heart problems may volunteer as subjects in the study of resting ECGs but should not participate in activities involving exercise or other stressful situations.

A. Electrocardiography

The **cardiac cycle** is the pattern of physiological events exhibited during each beat of the heart. One approach is to describe the cardiac cycle as an atrial, then ventricular, **systole** (contraction) and **diastole** (relaxation). In other words, the atria contract and relax, and the ventricles contract and relax, producing one pumping cycle of the heart. During the cardiac cycle, a variety of physiological characteristics of the heart and pumped blood measurably change. For example, the **blood pressure** within each chamber changes with the volume of the chamber and the strength of contraction of the myocardium. Certain *heart sounds* are heard as the valves snap shut and portions of the myocardium contract. The electrical properties of the heart change as action potentials travel through the myocardium. This electrical aspect of the cardiac cycle is of interest in this activity.

As you know, all muscle requires *excitation* of the cell membrane coupled with *contraction* involving filaments within the cell. The excitation, taking the form of action potentials, is an electrical event in which the polarity of electrical potential across the cell membranes changes. Because so much myocardial tissue depolarizes at more or less the same time, a great deal of current flows around the heart tissue, and the depolarization is easily detected from the outside of the body. Around the beginning of the twentieth century, a German scientist named Einthoven was one of the first to use a recording voltmeter to measure this electrical activity of the heart and to describe the changes found in a typical cardiac cycle. This technique is called *electrocardiography*.

The **electrocardiogram** (**ECG** or **EKG**) is a waveform graph produced by a recording voltmeter called an **electrocardiograph.** The graph is a line on a paper or monitor that rises and falls with changes in voltage (electric potential) between two points. Figure 36-1, *A* shows the voltage changes seen in one typical cardiac cycle. For convenience, each wave is given a name:

■ **P wave**—This wave represents depolarization of the atrial walls, which precedes atrial systole.
■ **QRS complex**—This set of waves results from the more or less concurrent ventricular

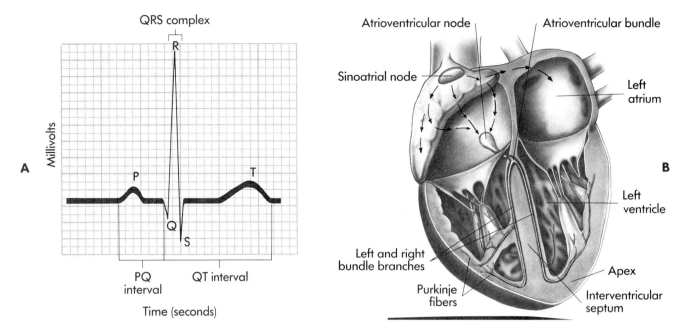

Figure 36-1 A, An idealized ECG pattern in one cardiac cycle. **B,** Anatomy of the conduction system of the heart.

depolarization (preceding ventricular systole) and atrial repolarization (signaling the onset of atrial diastole).

■ **T wave**—This wave represents ventricular repolarization, which precedes ventricular diastole.

The length of time during waves and between waves indicates the efficiency of conduction through the myocardium. Lengthened ECG portions are often evidence of *blocked conduction.* To understand the significance of conduction blockage, we must first outline the structure and function of the conduction system:

■ **Sinoatrial (SA) node**—This specialized section of cardiac muscle in the upper lateral wall of the right atrium generates its own action potentials at a faster rate than surrounding muscle, so it acts as a **pacemaker** for the rest of the heart. See Figure 36-1, *B*.

■ **Atrioventricular (AV) node**—A pacemaker in the lower right atrium subordinate to the SA node, it normally generates action potentials in response to potentials that travel to it through the atrial walls from the SA node.

■ **AV bundle (of His)**—This bundle of cardiac muscles in the interventricular septum is specialized to conduct action potentials rapidly. The *AV bundle branches* spread out as **Purkinje fibers** through the lateral ventricular walls.

Action potentials generated by the SA node travel across the atrial myocardium, triggering atrial systole. The AV node then picks up the signal and generates action potentials that travel rapidly through the AV bundle and its branches to stimulate the ventricular myocardium, resulting in ventricular systole.

The **PQ interval** (sometimes termed the *PR interval*) is the time required for a signal to pass from the SA node, through the atrial myocardium (depolarizing it), to the AV node and through its branches. The **QT interval** represents the time needed for complete depolarization and recovery of the ventricular myocardium.

Einthoven proposed three basic arrangements, or **leads,** of the ECG electrodes. Each lead is a set of one positive voltmeter electrode and one negative voltmeter electrode. These *standard leads,* or *appendicular leads,* are:

■ **Lead I**—One electrode on the left arm, one on the right arm
■ **Lead II**—One electrode on the right arm, one on the left leg
■ **Lead III**—One electrode on the left arm, one on the left leg

The electrode positions are shown in Figure 36-2.

The appendicular leads form what is called *Einthoven's triangle,* diagrammed in Figure 36-3. In the diagram, the heart is shown at the peak of the R wave, when the ventricles are partly depolarized. As you can see, an electric current flows from the positive area of the undepolarized region to the negative depolarized region. This difference in charge is detected by the ECG and graphed as the R wave. Because the direction of current flow is close to the angle of lead II, the R wave of lead II is larger than in lead III and lead I.

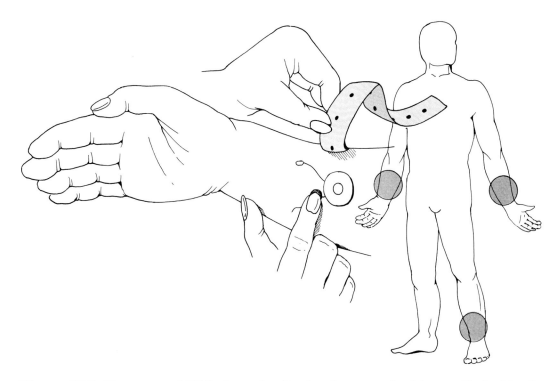

Figure 36-2 Placement of ECG electrodes for the three standard (appendicular) leads.

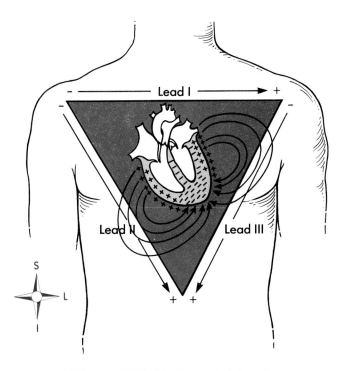

Figure 36-3 Einthoven's triangle.

Demonstrate the resting ECG:

☐ 1 Set up an electrocardiograph apparatus as directed by the instructions that accompany it. Set PAPER SPEED or TRACE VELOCITY to 25 mm/sec (2.5 cm/sec), the standard rate for ECG recording. Calibrate the SENSITIVITY or GAIN to 1 cm = 1 mV (see the instructions for your machine).

☐ 2 Have the (rested) subject lie on a cot or table or sit comfortably in a chair. Expose both ankles and wrists and clean a small area on each with alcohol pads. Put a *small* drop of electrode gel on each ECG electrode and place each on a different limb. Rub each electrode around gently to evenly distribute the gel, then anchor it with perforated rubber electrode tape pulled just tight enough to hold it in place.

☐ 3 Attach the limb wires to the appropriate electrode:
- LA to the left arm electrode
- RA to the right arm electrode
- LL to the left leg electrode
- RL to the right leg (for grounding the system)

☐ 4 Record lead I for about 20 cardiac cycles, then do likewise for leads II and III. Be sure to mark the part of the strip chart where you switched leads. If possible, record an ECG of every student in your group.

☐ 5 Using your lead II results, label a P wave, QRS complex, and T wave. If the P wave is absent and the heart rate is low, **SA node block** may be indicated. This could result from *ischemia* (reduced myocardial circulation) or *infarction* (myocardial damage).

☐ 6 Using lead II results, measure these values:
- Heart rate (in beats per minute)—Remember that the paper rate is 25 mm/sec (each mm = 0.04 sec). **Tachycardia** is the condition of a heart rate greater than 100 beats/min and may be caused by excitement (stress), high body temperature, or toxicity. **Bradycardia** is the condition of lowered heart rate (less than 60 beats/min). It is normal in conditioned athletes because they have a high stroke volume, or it may result from excessive vagal stimulation.
- PQ (PR) interval—If greater than 0.2 sec, a **first-degree heart block** could be present. This may result from inflammation of the AV bundle, which slows down conduction of a signal to the ventricles. If the PQ is 0.25 to 0.45 sec (and some P waves are not followed by QRS complexes), **second-degree heart block** may be present. This could result from AV node damage or excessive vagal stimulation. If the P wave seems independent of the QRS complex, with a P rate of about 100 per minute and a QRS of less than 40 per minute, **complete heart block** may be present.
- Your instructor may ask you to make other measurements and consult your textbook or a reference book for guidance in interpreting your results.

B. Deviations from the normal ECG

A variety of factors may result in ECG readings far different than those expected. For example, the conditions just described may produce variations in the resting heart rate. In this activity, you are challenged to demonstrate some factors that alter the standard ECG pattern.

☐ 1 Set up the ECG demonstration as described in activity A. Set the device to record lead II.

☐ 2 Once a normal ECG pattern is observed, have the subject breathe in and out deeply several times. What happens? What consequences might this have in a clinical setting?

☐ 3 Have the subject tense the muscles of the upper arm. What happens? What do you think causes this result? What implications does this observation have in clinical ECG applications?

☐ 4 Remove the ECG cables, leaving the electrodes in place. Have the subject exercise for 2 to 5 minutes, then quickly reconnect the ECG cables. With the subject again in resting position, record 15 or 20 cardiac cycles. Continue to sample 15 to 20 cycles every minute for the next 10 minutes. Record the heart rate for each sample. How do you explain your results? Does exercise alter the features of each ECG cycle? Why or why not?

☐ 5 What other factors might alter the ECG of the subject? As a group, discuss your hypotheses and, under the supervision of your instructor, perform experiments to test them (as time permits). Write a report of your experiment(s) and attach it to Lab Report 36. Be sure to include a statement of your hypothesis, an outline of your experimental design (methods), your results, and your interpretation of the results.

C. Computerized electrocardiography

The method given here uses Intelitool's CARDIOCOMP apparatus for recording ECGs using a personal computer system. The instructions given here are for the Apple-compatible version, but other versions are similar. The advantages of using this method, rather than that described earlier in this exercise, include ease of operation, ability to save data to a disk for later replay, ability to do on-screen measurements of ECG waves, and increasing familiarity with computer-based ECG systems, which are becoming more common in clinical settings. You must read and use Activities A and B in this exercise. This activity is intended only to given instruction regarding the use of an alternate technique.

> **SAFETY FIRST!** Observe the safety precautions given earlier in this exercise and in the CARDIOCOMP manual.

☐ 1 Set up the CARDIOCOMP system as instructed in the manual that accompanies the system. The electrode placement and connection of electrode wires is the same as in Activity A. As in that activity, the subject should be well rested and lying or sitting comfortably.

☐ 2 At the main menu select ELECTROCARDIOGRAM, then RUN EXAM STRIP (Apple) or EXPERIMENT MENU, then QUICK EXAM (DOS). The program will now ask these or similar questions:
- SIX OR TWELVE FRAMES? Select *six*.
- WITH INTERVAL ANALYSIS? Select *yes* if you want the program to measure ECG intervals, or *no* if not.
- SAVE EACH EXAM? Select *yes* if you want to replay the data at a later time, or *no* if you are only going to save the hardcopy printout.
- DO YOU WANT TO PRINT HARDCOPY? Select *yes* if you are given this option.

☐ 3 The AUTOMATIC CHANGE SETUP screen appears next. Press (S) to start recording data. When asked to select the data acquisition rate, select NORMAL SPEED. You should now see the data screen, and the program is beginning to collect ECG data. The data for six leads (the three standard leads and three **augmented leads**) will be automatically printed out in the standard strip chart format.

☐ 4 Study the resting ECG as described in Activity A and perform the experiments described in Activity B. If you have saved the data and wish to replay it at a later time, consult the CARDIOCOMP manual or follow the directions on the screen.

D. The augmented leads

The CARDIOCOMP program is designed to print out results of the standard (appendicular) leads, as well as three **augmented leads.** You may recall that Einthoven's triangle diagrammatically represents the direction of electric potential measurements using the three standard electrode arrangements (leads I, II, and III) (see Figure 36-4, *A*). Einthoven described a number of mathematical relationships among data produced by the three leads. For example, knowing data from any two leads is sufficient to mathematically reconstruct a data line for the third lead. Information from combinations of two leads can be manipulated to produce the three augmented, or *attenuated,* leads. These are not *bipolar* leads or electrode placements but are the result of combining two negative electrodes and measuring them against one positive electrode. Thus three more leads are produced:

- aV_R—The LA and LL are combined to form the negative electrode; RA is the positive electrode.
- aV_L—The RA and LL electrodes form the negative electrode; LA is the positive electrode.
- aV_F—The RA and LA electrodes form the negative pole and LL the positive pole.

Using Einthoven's triangle concept, the augmented leads are each shifted 90 degrees from the standard leads. See Figure 36-4, *B*.

Current clinical practice calls for recording twelve leads in all. So far, we have discussed the three standard leads and the three augmented leads. The remaining six leads are the **chest,** or **precordial,** leads. The chest leads use RA, LA, and LL combined to form the negative pole and each of six chest electrodes to form the positive poles. The chest leads are named V_1 through V_6.

What is the advantage of chest leads?

CARDIAC DYSRHYTHMIA

Various conditions such as inflammation of the endocardium (endocarditis) or myocardial infarction (heart attack) can damage the heart's conduction system and thereby disturb the normal rhythmical beating of the heart (Figure *A*). The term **dysrhythmia** refers to an abnormality of heart rhythm.

One kind of dysrhythmia is called a **heart block.** In *AV node block,* impulses are blocked from getting through to the ventricular myocardium, resulting in the ventricles contracting at a much slower rate than normal. On an ECG, there may be a large interval between the P wave and the R peak of the QRS complex (Figure *B*). *Complete heart block* occurs when the P waves do not match up at all with the QRS complexes—as in an ECG that shows two or more P waves for every QRS complex. A physician may treat heart block by implanting in the heart an *artificial pacemaker.*

Bradycardia is a slow heart rhythm—less than 50 beats per minute (Figure *C*). Slight bradycardia is normal during sleep and in conditioned athletes while they are awake (but at rest). Abnormal bradycardia can result from improper autonomic nervous control of the heart or from a damaged SA node. If the problem is severe, artificial pacemakers can be used to increase the heart rate by taking the place of the SA node.

Tachycardia is a very rapid heart rhythm—more than 100 beats per minute (Figure *D*). Tachycardia is normal during and after exercise and during the stress response. Abnormal tachycardia can result from improper autonomic control of the heart, blood loss or shock, the action of drugs and toxins, fever, and other factors.

Sinus dysrhythmia is a variation in heart rate during the breathing cycle. Typically, the rate increases during inspiration and decreases during expiration. The causes of sinus dysrhythmia are not clear. This phenomenon is common in young people and usually does not require treatment.

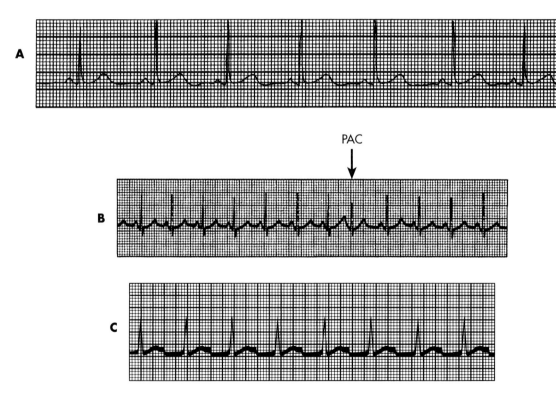

A, Normal ECG strip chart recording. **B,** AV node block. Very slow ventricular contraction (25 to 45 beats/min at rest); P waves widely separated from peaks of QRS complexes. **C,** Bradycardia. Slow heart rhythm (less than 60 beats/min); no disruption of normal rhythm pattern.

CARDIAC DYSRHYTHMIA—Cont'd

Premature contractions, or *extrasystoles,* are contractions that occur before the next expected contraction in a series of cardiac cycles. For example, *premature atrial contractions (PACs)* may occur shortly after the ventricles contract—seen as early P waves on the ECG (Figure *E*). Premature atrial contractions often occur with lack of sleep, too much caffeine or nicotine, alcoholism, or heart damage. Ventricular depolarizations that appear earlier than expected are called *premature ventricular contractions* (PVCs). PVCs appear on the ECG as early, wide QRS complexes without a preceding related P wave. PVCs are caused by a variety of circumstances that include stress, electrolyte imbalance, acidosis, hypoxemia, ventricular enlargement, or drug reactions. Occasional PVCs are not clinically significant in otherwise healthy individuals. However, in people with heart disease they may reduce cardiac output.

Frequent premature contractions can lead to **fibrillation,** a condition in which cardiac muscle fibers contract out of step with each other. This event can be seen in an ECG as the absence of regular P waves or abnormal QRS and T waves. In fibrillation, the affected heart chambers do not effectively pump blood. *Atrial fibrillation* occurs commonly in mitral stenosis, rheumatic heart disease, and infarction of the atrial myocardium (Figure *F*). This condition can be treated with drugs such as digoxin (a *digitalis* preparation) or by *defibrillation*—application of electrical shock to force cardiac muscle fibers to contract in unison. *Ventricular fibrillation* is an immediately life-threatening condition in which the lack of ventricular pumping suddenly stops the flow of blood to vital tissues (Figure *G*). Unless ventricular fibrillation is corrected immediately by defibrillation or some other method, death may occur within minutes.

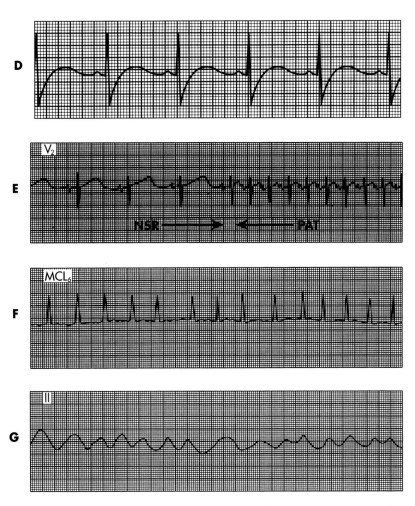

D, Tachycardia. Rapid heart rhythm (greater than 100 beats/min); no disruption of normal rhythm pattern. **E,** Premature atrial contraction (PAC). Unexpected, early P wave that is different from normal P waves; PR interval may be shorter or longer than normal; normal QRS complex; more than 6 PACs per minute may precede atrial fibrillation. **F,** Atrial fibrillation. Irregular, rapid atrial depolarizations; P wave rapid (greater than 300/min) with irregular QRS complexes (150 to 170 beats/min). **G,** Ventricular fibrillation. Complete disruption of normal heart rhythm.

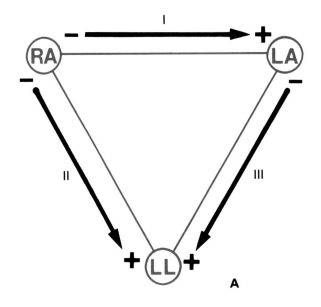
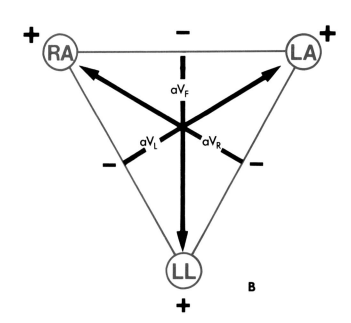

Figure 36-4 **A,** Einthoven's triangle. **B,** Einthoven's triangle superimposed with the three augmented leads.

NAME _____ DATE _____ SECTION _____

LAB REPORT 36

Electrical Activity of the Heart

Attach reports of any additional ECG experiments that you did.

Test condition	Heart rate	PX (PR) interval	Other observations	Interpretation
Resting				
Deep breathing				
Arm tensed				
Post-exercise: 0 min				
1 min				
2 min				
3 min				
4 min				
5 min				
6 min				
7 min				
8 min				
9 min				
10 min				

Use this grid to graph the heart rates observed after exercise. Be sure to label your graph completely and accurately.

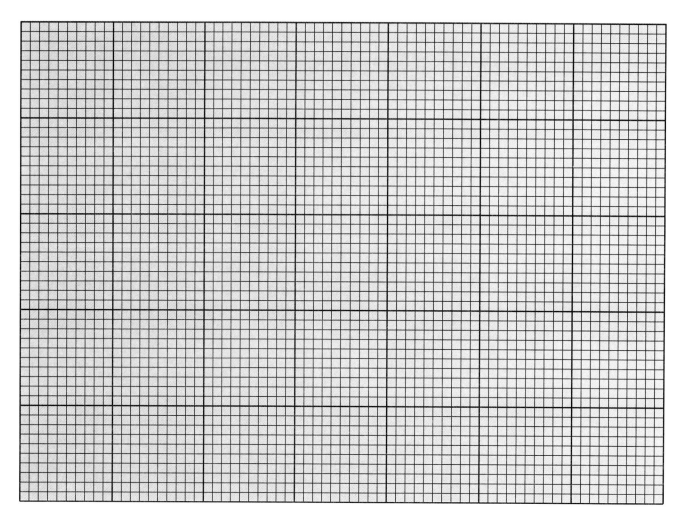

Put in order

_____ 1
_____ 2
_____ 3
_____ 4
_____ 5
_____ 6
_____ 7

Fill-in

_____ 1
_____ 2
_____ 3
_____ 4
_____ 5

Put in order (arrange these heart structures in the order through which electrical signals are conducted in a single cardiac cycle)

atrial myocardium
AV bundle branches
AV bundle
AV node
Purkinje fibers
SA node
ventricular myocardium

Fill-in (complete each statement with the correct term)

1. Relaxation of a heart chamber is called __?__.
2. Contraction of a heart chamber is called __?__.
3. The portion of the ECG that represents ventricular repolarization is the __?__.
4. Leads I, II, and III together are called the __?__ leads, or appendicular leads.
5. __?__ is the condition of elevated heart rate.

LAB EXERCISE 37

The Pulse and Blood Pressure

The heart pushes blood from the left ventricle into the aorta with great force, as you already know. The wall of the aorta and its branches stretch a little with the sudden burst of blood during each ventricular ejection. Burst after burst after burst of blood ejected from the heart produces bulge after bulge after bulge in the arterial wall. These **pressure waves,** known as the **pulse,** occur at more or less the same rhythm as the cardiac cycle. The pulse waves can be detected in clinical situations and used to determine many things about cardiovascular function. The rhythm and strength of the pulse in distal arteries can indicate something about how efficiently blood is being pumped into the circulation. The pressure of the waves can be measured indirectly and used to give information about overall arterial **blood pressure.** This exercise challenges you to find and measure some characteristics of blood pressure waves and use that skill to find some factors that affect cardiovascular function.

Before you begin

❏ Read the appropriate chapter in your textbook.

❏ Set your learning goals. When you finish this exercise, you should be able to
- explain the concept of the pressure wave, or pulse
- demonstrate detection of the pulse in a human subject
- use a sphygmomanometer to measure blood pressure
- list some factors that affect pulse and blood pressure

❏ Prepare your materials:
- sphygmomanometer
- stethoscope
- alcohol swabs
- watch or timer

❏ Read the directions and safety tips for this exercise **carefully** before starting any procedure.

A. The pulse

As blood is ejected from the left ventricle during each cardiac cycle, it is subjected to great pressure. Because fluids are unable to be compressed, this great pressure pushes the compliant walls of the arteries outward. Between ejections, the blood pressure is lower and the arterial walls bounce back to their original diameter. Because this occurs with each cardiac cycle, multiple waves of high pressure move into the arteries. These **pressure waves** are also called the **pulse.**

The rate and strength of the pulse can be measured easily with no special equipment. By simply pressing on the skin and pushing an artery against bone, one can feel the pressure waves. Counting the number of waves per minute gives the *pulse rate,* and the pulse strength (pressure) can be assessed by receptors in your fingertips.

Determine the pulse rate of a resting subject by placing your second and third fingers on the anterior surface of the wrist, over the point where the *radial artery* passes over the distal end of the radius. Do not use your thumb, or you may feel your own pulse waves. Using a watch, count the number of waves in 15 seconds. Multiply your result by 4 to determine the pulse rate in waves per minute. Try one or more of the other pulse points identified in Figure 37-1. Are the pulse waves of equal pressure at each pulse point? Why or why not?

B. Blood pressure

Biologists often measure the pressure of body fluids in units called **millimeters of mercury (mm Hg).** This unit is based on a measuring device called a *manometer,* which is an inverted tube of liquid mercury. As something presses against the mercury reservoir at the bottom of the tube, the column of mercury in the tube rises higher. The more pressure with which the mercury is pushed, the higher the column. Thus the millimeters the mercury rises (mm Hg) indicate the level of the pressure. A **sphygmomanometer** is a manometer with an air cuff attached to the reservoir; it is used to indirectly measure the changing blood pressures associated with the pulse.

❏ 1 Use a sphygmomanometer to measure the blood pressure of a resting subject. First, clean the ear pieces of a stethoscope with an alcohol swab and place them in your ears. Then, wrap the sphygmomanometer air cuff around the subject's left upper arm, just above the elbow.

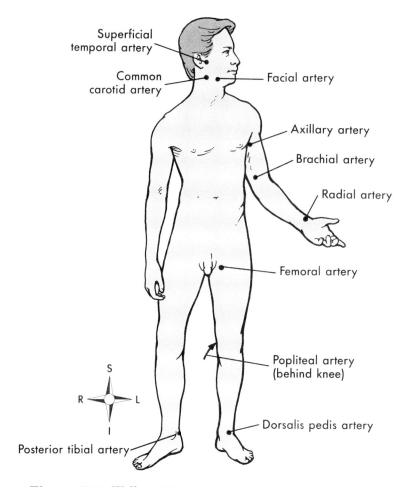

Figure 37-1 Well-positioned arteries for pulse detection.

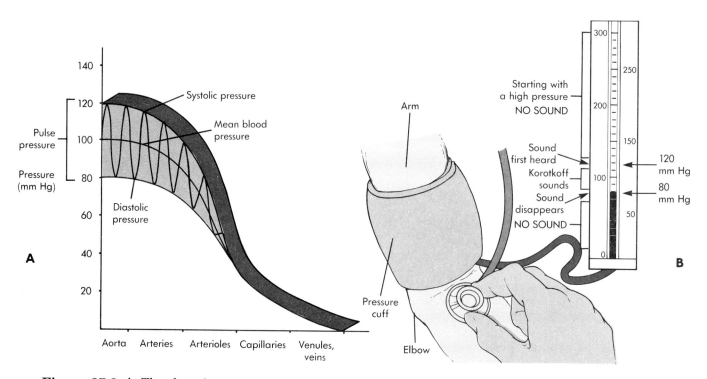

Figure 37-2 A, The changing pressures associated with the pulse wave. **B,** Using a sphygmomanometer.

❏ 2 Place the stethoscope sensor in the position shown in Figure 37-2, *B* and listen to the pulse in the *brachial artery*. Close the valve of the sphygmomanometer's air bulb and pump the cuff to a pressure of about 175 mm Hg or so.

❏ 3 Still listening to the artery and watching the mercury column, open the valve slightly and slowly release air. As you decrease the pressure, you will suddenly notice loud, tapping **Korotkoff sounds** in the artery. Note the pressure at which they begin. Continue to decrease pressure until the Korotkoff sounds disappear, also noting that point on the pressure scale. The point at which the sounds are first heard is the **systolic pressure,** the pressure of pulse wave's crest (see Figure 37-2, *A*). The sounds disappear when the **diastolic pressure,** the blood pressure between pulse waves, is reached. Blood pressure is usually expressed as systolic pressure over diastolic pressure. Average pressure is about "120 over 80," or *120 mm Hg/80 mm Hg*.

C. Variations in pulse and pressure

Determine the pulse rate and blood pressure in the situations described. Compare the results to each other and to the results for a resting subject. How do you account for the variations?

❏ 1 Test a variety of body positions: lying down (immediately, then after 5 minutes) and standing (immediately, then after 5 minutes).

❏ 2 Test a subject after exercise: immediately after 2 to 5 minutes of exercise, then once per minute for the next 10 minutes of recovery. Compare your results to those for the postexercise ECG (Lab Exercise 36).

NAME _____ DATE _____ SECTION _____

LAB REPORT 37

The Pulse and Blood Pressure

Test condition	Pulse rate	Pulse strength	Systolic blood pressure	Diastolic blood pressure	Pulse pressure (systolic minus diastolic)
Resting: *radial*					
Resting: _____					
Resting: _____					
Lying down:					
0 min					
5 min					
Standing:					
0 min					
5 min					
After exercise:					
0 min					
1 min					
2 min					
3 min					
4 min					
5 min					
6 min					
7 min					
8 min					
9 min					
10 min					

The next page is marked *Interpretation*. In the space provided, write your interpretation of the results reported in the preceding table. For example, tell why the strength of the resting pulse rate may have varied depending on where it was detected. Look for similarities and differences from row to row (how does pulse rate in different body positions compare?). Look for relationships from column to column (does pulse strength relate to pulse pressure?). Use the questions given in the exercise itself for further guidance.

Interpretation:

LAB EXERCISE 38

The Circulatory Pathway

Blood pumped from the right ventricle of the heart travels to and through the lungs before returning to the left atrium. This pathway is termed the **pulmonary circulation,** or the *pulmonary loop.* Blood pumped from the left ventricle travels to and through the other organs and tissues of the body. This pathway is termed the **systemic circulation** (*systemic loop*). These two pathways together with the heart form the basic structure of the cardiovascular system. In this exercise you are challenged to learn some of the major blood vessels that form the circulatory pathways of the human body.

Before you begin

❑ Read the appropriate chapter in your textbook.

❑ Set your learning goals. When you finish this exercise, you should be able to
- locate major veins and arteries in models and charts
- describe the circulatory pathways to and from major body regions
- describe the fetal circulatory plan and the changes in circulation that occur around the time of birth

❑ Prepare your materials:
- model of the human body (showing major vessels)
- charts of human circulatory routes (including fetal circulation)

❑ Read the directions and safety tips for this exercise **carefully** before starting any procedure.

A. Arteries

Arteries are blood vessels that conduct blood away from the heart and toward tissues. In the pulmonary circulation, **pulmonary arteries** conduct deoxygenated blood to the lungs. In the systemic circulation, the **aorta** and its branches conduct oxygenated blood toward the systemic tissues. Small arteries are usually called **arterioles.** Arterioles conduct blood into a network of even smaller vessels, or **capillaries.** On a chart, then in a model of the human body, locate the major human arteries listed. As you do so, note how they form routes to major regions of the body.

❑ 1 Find these portions of the aorta:
- **Ascending aorta**—First portion, before the aorta bends inferiorly
- **Aortic arch**—Bend of the aorta, just superior to the heart
- **Descending aorta**—The remainder of the aorta
- **Thoracic aorta**—Portion of the descending aorta that is within the thorax
- **Abdominal aorta**—Portion of the descending aorta that is within the abdomen

❑ 2 Locate these arteries of the head and neck:
- **Brachiocephalic**—First branch from the aorta
- **Right common carotid**—Medial branch of the brachiocephalic artery
- **Right subclavian**—Lateral branch of the brachiocephalic artery
- **Left common carotid, left subclavian**—Second and third branches, respectively, of the aorta
- **Internal carotid, external carotid**—Branches of the common carotid arteries
- **Vertebral**—Branches of the subclavians
- **Basilar**—Artery formed by the fusion of vertebral arteries
- **Circle of Willis**—Circular system of arteries around the brain's base, formed by branches of the basilar artery and the internal carotids

❑ 3 Identify these arteries of the upper limb:
- **Axillary artery**—Continuation of the subclavian artery inferior to the clavicle
- **Brachial**—Continuation of the axillary artery in the upper arm
- **Ulnar**—Medial branch of the brachial artery
- **Radial**—Lateral branch of the brachial artery

❑ 4 Identify these branches of the descending aorta:
- **Intercostal**—Branches from the aorta to the intercostal muscles
- **Phrenic**—Branches from the aorta to the diaphragm
- **Celiac**—Branch of the aorta to the stomach, pancreas, and liver

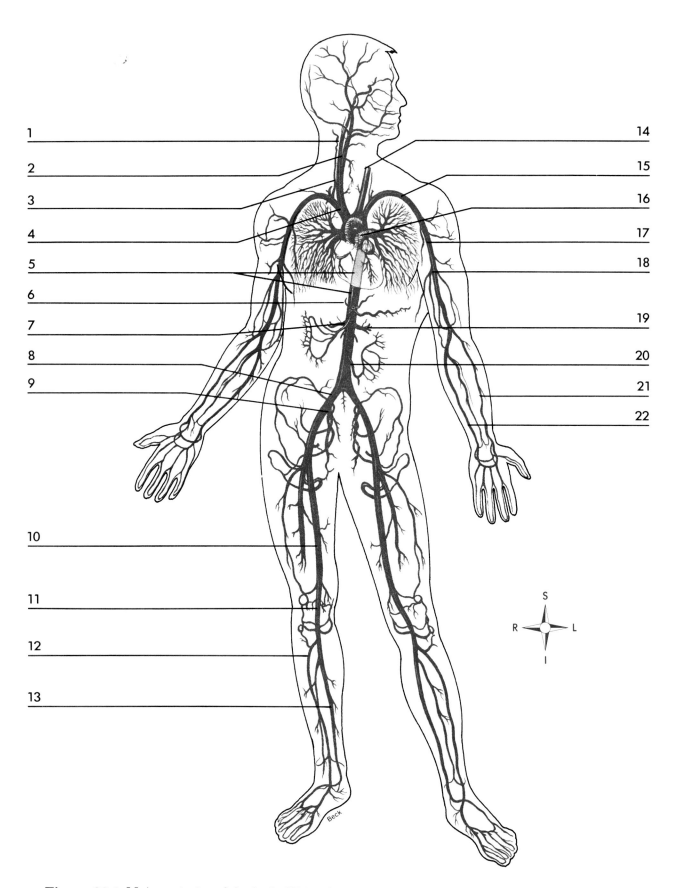

Figure 38-1 Major arteries of the body. Write the names of the arteries indicated by label lines.

336

- **Superior mesenteric**—Branch of the aorta to the small intestine and proximal large intestine
- **Inferior mesenteric**—Branch of the aorta to the distal regions of the large intestine
- **Renal**—Branches from the aorta to the kidneys
- **Suprarenal**—Branches from the aorta to the adrenal glands
- **Testicular, ovarian**—Branches from the aorta to the gonads
- **Common iliac**—Branch from the inferior end of the descending aorta toward a leg

❏ 5 Identify these arteries of the pelvis and lower limbs:
- **Internal iliac**—Medial branch of the common iliacs
- **External iliac**—Lateral branch of the common iliac
- **Femoral**—Continuation of the external iliac artery in the thigh
- **Popliteal**—Continuation of the femoral artery in the popliteal area (posterior knee)
- **Anterior tibial**—Anterior branch of the popliteal artery
- **Posterior tibial**—Posterior branch of the popliteal artery

B. Veins

Veins are blood vessels that conduct blood toward the heart. In the pulmonary circulation, the **pulmonary veins** return oxygenated blood from the lungs. In the systemic circulation, the **superior vena cava** returns deoxygenated blood from the head, neck, thorax, and arms. The **inferior vena cava** returns deoxygenated blood from the rest of the systemic loop. **Venules** are small veins. In charts, then in a model, locate the listed major veins. As with the arteries, note how they form routes from major body regions.

❏ 1 Identify these veins of the head and neck:
- **Brachiocephalic**—Medial branch into the superior vena cava
- **Subclavian**—Lateral branch into the brachiocephalic vein
- **Internal jugular**—Medial branch into the brachiocephalic vein
- **External jugular**—External vein of the neck that returns blood to the subclavian vein

❏ 2 Identify these veins of the upper limb and thorax:
- **Axillary**—Medial branch into the subclavian vein
- **Basilic vein**—Superficial vein that empties into the axillary vein
- **Brachial**—Upper arm vein that continues into the axillary region as the axillary vein
- **Cephalic**—Lateral, superficial branch into the subclavian vein
- **Azygos**—Unpaired branch into the posterior aspect of the superior vena cava
- **Hemiazygos, accessory hemiazygos**—Two sets of multiple veins that empty into the azygos
- **Intercostal**—Veins that empty into the azygos vein (right) and hemiazygos or accessory hemiazygos veins (left)

❏ 3 Identify these tributaries of the inferior vena cava:
- **Hepatic**—Vein from the liver to the inferior vena cava
- **Renal**—Vein from the kidney to the inferior vena cava
- **Testicular, ovarian**—Vein from the gonad to the inferior vena cava
- **Common iliac**—Two branches that fuse to become the inferior vena cava
- **Internal iliac**—Medial branch of the common iliac (in the pelvis)

❏ 4 A **portal circulation** is a set of vessels that begins and ends with capillary networks. In other words, blood is returned to a second set of capillaries before being returned to the heart. The **hepatic portal system** is an important part of the venous systemic circulation. It returns blood from digestive organs to the liver, rather than directly to the heart. Identify these hepatic portal vessels:
- **Hepatic portal**—From the veins of abdominal organs to the liver
- **Superior mesenteric**—From the small intestine to the hepatic portal vein
- **Inferior mesenteric**—From the large intestine, joining the splenic vein to the hepatic portal vein
- **Splenic**—From the spleen to the hepatic portal vein
- **Gastroepiploic**—From the stomach, joining the splenic vein to the hepatic portal vein

What physiological purpose can be served by the presence of the hepatic portal system?

❏ 5 Identify these veins of the lower limbs:
- **External iliac**—Lateral branch into the common iliac vein (in the pelvis)
- **Femoral**—Major lateral branch into the external iliac vein
- **Great saphenous**—Major medial, superficial branch into the external iliac vein
- **Popliteal**—Posterior branch into the femoral vein, on the posterior of the knee

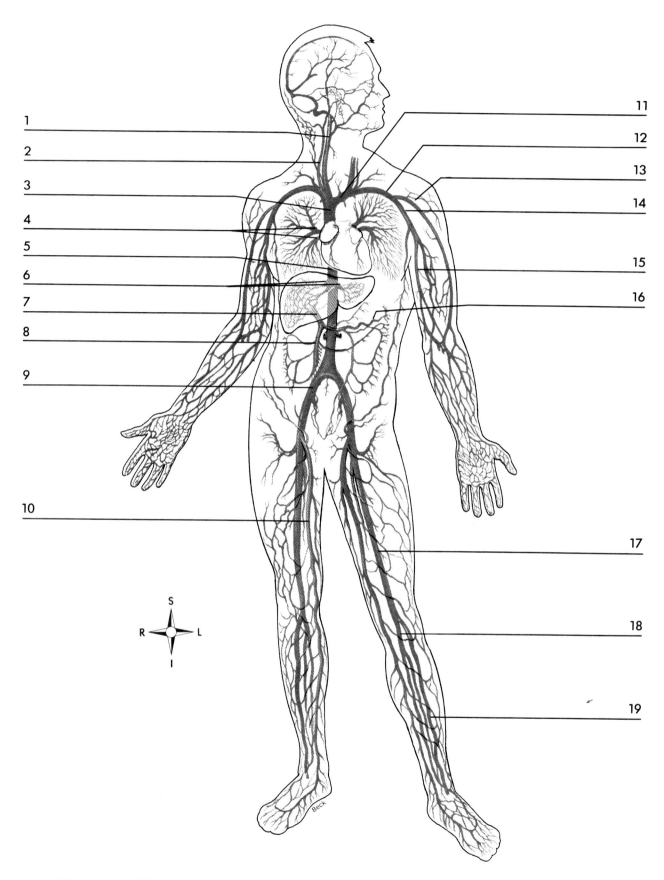

Figure 38-2 Major veins of the body. Write the names of the veins indicated by label lines.

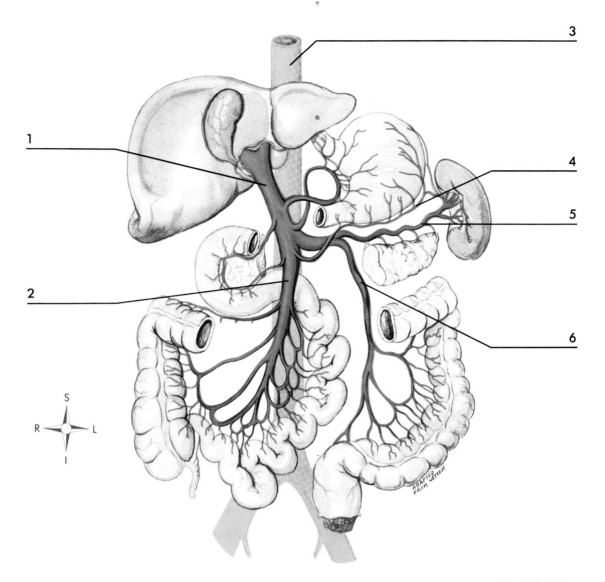

Figure 38-3 The hepatic portal system. Write the names of the vessels indicated by label lines.

- **Small saphenous**—Lateral, superficial branch into the popliteal vein, lateral to the tibia
- **Anterior tibial**—Branch into the popliteal vein, on the anterior aspect of the tibia
- **Posterior tibial**—Posterior to the tibia, draining into the popliteal vein

C. Fetal circulation

Some of the many dramatic anatomical and physiological changes that occur at parturition are the changes in the cardiovascular circulation. The fetal circulatory plan functions well in the circumstances found in utero but must change quickly if a newborn is to survive outside the uterus. Examine Figure 38-5, or a model or chart of fetal circulation, and note these special features:

❏ 1   The **placenta** has already been described. Located on the inside wall of the uterus, one function of the placenta is the exchange of nutrients and wastes between fetal and maternal blood. Most materials move by means of diffusion, crossing the membranes that separate the two blood systems.

❏ 2   The **umbilical vein** conducts oxygenated fetal blood from the placenta, through the umbilical cord, to and through the liver, and into the **ductus venosus.** The ductus venosus drains into the inferior vena cava, which returns the blood to the fetal heart (right atrium).

❏ 3   From the right atrium, oxygenated blood may flow through the **foramen ovale** into the left atrium. This effectively bypasses the entire pulmonary loop. Some right atrial blood moves into the right ventricle. Right ventricular blood exits the heart by way of the pulmonary artery. Near the branching of the pulmonary artery into left and right branches, the **ductus arteriosus** provides another bypass of the pulmonary loop. Blood tends to bypass circulation to the lungs here and at

ANGIOGRAPHY

Angiography is a radiographic (x-ray) imaging method used to visualize blood vessels. A *radiopaque* dye, a water-soluble material that absorbs x-rays so that it casts a heavy shadow on the film, is injected near the vessels of interest. When used to visualize the heart and nearby vessels, the resulting image is called an *angiocardiogram*. When used to visualize veins, the resulting image is called a *venogram*. When angiography is used to image arteries, the result is called an *arteriogram*.

Angiography is used to visualize the structure of blood vessels to determine cardiovascular health.

Some abnormalities that may be seen in an angiogram include:

- blood vessel injuries
- *aneurysms,* abnormal local distensions of a vessel wall
- blockage, perhaps by *atherosclerosis* (buildup of cholesterol and other matter on the inside of a vessel wall)

Figure 38-4 is an arteriogram in which radiopaque dye has been injected into the femoral artery. Try to identify each artery labeled in the figure.

1 _____

2 _____

3 _____

4 _____

5 _____

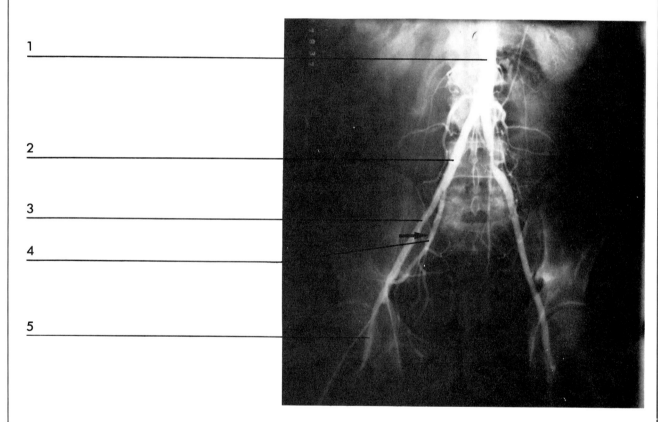

Figure 38-4 Arteriogram. Try to identify the arteries indicated by label lines.

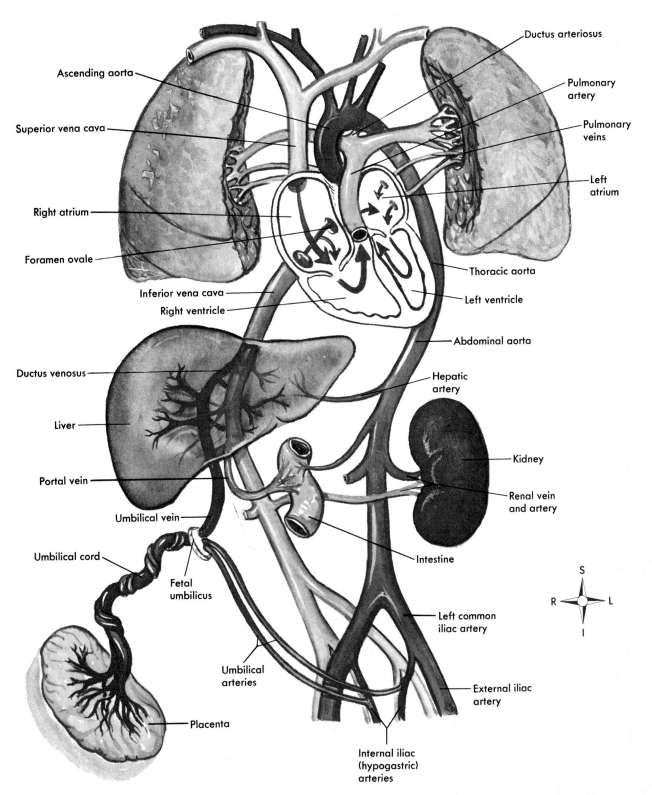

Figure 38-5 The fetal circulatory plan.

the right atrium because of the high resistance to blood flow in the developing lungs. Only a small proportion of blood reaches the lungs, just enough for its continued development.

❏ 4 Blood in the systemic circulation may enter the two **umbilical arteries** that branch from the internal iliac arteries. The umbilical arteries conduct partially deoxygenated blood through the umbilical cord to the placenta. Follow the possible circulatory paths through the fetal body several times, noting the differences between fetal and adult patterns of blood circulation.

❏ 5 Your model or chart may indicate the changes that occur at parturition to shift the fetal circulatory pattern to the adult pattern. The major changes are noted here:

- Flow through the umbilical arteries and vein stops when the umbilical cord is cut and tied, and the vessels degenerate. The umbilical vein and ductus venosus are replaced by ligaments.
- When the lungs expand during the respirations that begin after birth, pulmonary resistance to blood flow is reduced. Therefore, blood tends to flow through the pulmonary loop and tends not to flow through the foramen ovale or into the ductus arteriosus. Increased left atrial pressure forces a flap of the septum over the foramen ovale, which eventually seals the opening completely. The ductus arteriosus constricts and is later replaced by a fibrous ligament.

NAME _____ DATE _____ SECTION _____

LAB REPORT 38

The Circulatory Pathway

Figure 38-1

_____ 1
_____ 2
_____ 3
_____ 4
_____ 5
_____ 6
_____ 7
_____ 8
_____ 9
_____ 10
_____ 11
_____ 12
_____ 13
_____ 14
_____ 15
_____ 16
_____ 17
_____ 18
_____ 19
_____ 20
_____ 21
_____ 22

Put in order

　left ventricle　 1
_____ 2
_____ 3
_____ 4
_____ 5
_____ 6
_____ 7
_____ 8
_____ 9
_____ 10
_____ 11
_____ 12
_____ 13
_____ 14
_____ 15
_____ 16
_____ 17
_____ 18

Identify

_____ 1
_____ 2
_____ 3
_____ 4
_____ 5
_____ 6
_____ 7
_____ 8

Put in order (arrange these structures in the order in which blood passes through them in a circuit from the heart to the foot and back—the first one is done for you)

　　　abdominal aorta
　　　anterior tibial artery
　　　anterior tibial vein
　　　aortic arch
　　　ascending aorta
　　　common iliac vein
　　　common iliac artery
　　　external iliac artery
　　　external iliac vein
　　　femoral vein
　　　femoral artery
　　　foot
　　　inferior vena cava
　　　left ventricle
　　　popliteal artery
　　　popliteal vein
　　　right atrium
　　　thoracic aorta

Identify (tell what structure is described in each item)

1. Portion of the aorta in the abdomen
2. Continuation of the subclavian artery inferior to the clavicle
3. Artery that supplies the diaphragm
4. Vein that drains the liver
5. Vessel that conducts blood from digestive organs to the liver
6. Major medial tributary of the external iliac vein
7. Unpaired vein that drains into the posterior aspect of the superior vena cava
8. First major branch of the aorta

Figure 38-2

1. _____
2. _____
3. _____
4. _____
5. _____
6. _____
7. _____
8. _____
9. _____
10. _____
11. _____
12. _____
13. _____
14. _____
15. _____
16. _____
17. _____
18. _____
19. _____

Figure 38-3

1. _____
2. _____
3. _____
4. _____
5. _____
6. _____

Figure 38-4

1. _____
2. _____
3. _____
4. _____
5. _____

Concept Mapping

To learn the concept of how blood flows through various circulatory routes, draw a schematic diagram (of a design of your choice) that clearly shows the structures through which blood flows in the following routes:

1. Hepatic portal circulation (include routes all the way to and from the heart in your diagram)
2. Fetal circulation
3. Systemic circulation: to and from any organ of your choice

Vessel	Tissues supplied or drained
Brachiocephalic artery	
Left common carotid artery	
Left subclavian artery	
Celiac artery	
Phrenic arteries	
Superior mesenteric artery	
Inferior mesenteric artery	
Renal artery	
Suprarenal artery	
External iliac artery	
Internal iliac artery	
Hepatic vein	
External iliac vein	
Internal iliac vein	
Renal vein	
Testicular vein	
Ovarian vein	
Phrenic vein	
Superior mesenteric vein	
Inferior mesenteric vein	
Gastroepiploic vein	
Cystic vein	

LAB EXERCISE 39

The Lymphatic System

The **lymphatic system** has several primary functions: it participates in maintaining extracellular fluid balance, it absorbs fats and other substances from the digestive tract, and it functions in the **immune system** defense of the body. To do this, the lymphatic system is composed of a number of vessels and other structures. This exercise presents some of the basics of lymphatic anatomy and physiology.

Before you begin

❑ Read the appropriate chapter in your textbook.

❑ Set your learning goals. When you finish this exercise, you should be able to
- describe the major organs of the lymphatic system and find them in a chart
- identify the features of a lymph node
- briefly state the function of lymph nodes

❑ Prepare your materials:
- charts or models of the lymphatic system
- microscope
- prepared microslides:
  Lymphatic vessel l.s.
  Lymph node c.s.
  Palatine tonsil c.s.
  Spleen c.s.
  Thymus c.s.

❑ Read the directions and safety tips for this exercise **carefully** before starting any procedure.

A. Overview of the lymphatic system

Use a chart or model of the lymphatic system to find these gross features:

❑ 1 Locate some of the **lymphatic vessels.** These vessels are similar to veins in the structure of their walls and the presence of many valves. Lymph vessels collect **lymph** from interstitial spaces in the regions served and conduct it toward the blood circulation. Lymph, or *lymphatic fluid,* is an excess from interstitial areas. The **right lymphatic duct** is a large collecting vessel that receives lymph from the superior right quadrant of the body. The **thoracic duct** similarly conducts lymph received from the rest of the body. Each duct empties into a subclavian vein.

❑ 2 Lymph nodes are small, round organs located at irregular intervals in the network of lymph vessels. Like all lymph organs, they contain **lymphatic tissue** comprised mainly of lymphocytes (a category of white blood cell). They are distributed unevenly, with major aggregations of nodes in the neck (**cervical nodes**), armpit (**axillary nodes**), and groin (**inguinal nodes**).

❑ 3 The **tonsils** are lymph organs that surround the openings of the mouth, nose, and throat into the lower digestive and respiratory tracts. The **palatine tonsils** are on each lateral side of the opening of the mouth into the throat. The **pharyngeal tonsils** are at the top of the throat (*pharynx*) near the posterior of the nasal cavity. The **lingual tonsils** are on the posterior of the tongue's base.

❑ 4 The **spleen** and **thymus** should already be familiar to you. The spleen is located in the upper left abdominal cavity and the thymus is located in the anterior mediastinum.

> **HINT** → A microscopic study of thymic tissue is offered in Lab Exercise 32.

B. The lymph node

There are numerous lymph nodes throughout the lymphatic system, as Figure 39-1 shows. Lymph usually passes through one or more lymph nodes on its path toward the blood circulation. The materials in lymph may stimulate lymphocyte development (an immune response) or may be destroyed by *macrophages* in the node. Identify these structures associated with the lymph node in a chart or model:

❑ 1 **Afferent lymph vessels** conduct lymph into a lymph node, whereas **efferent lymph vessels** conduct lymph out of each node.

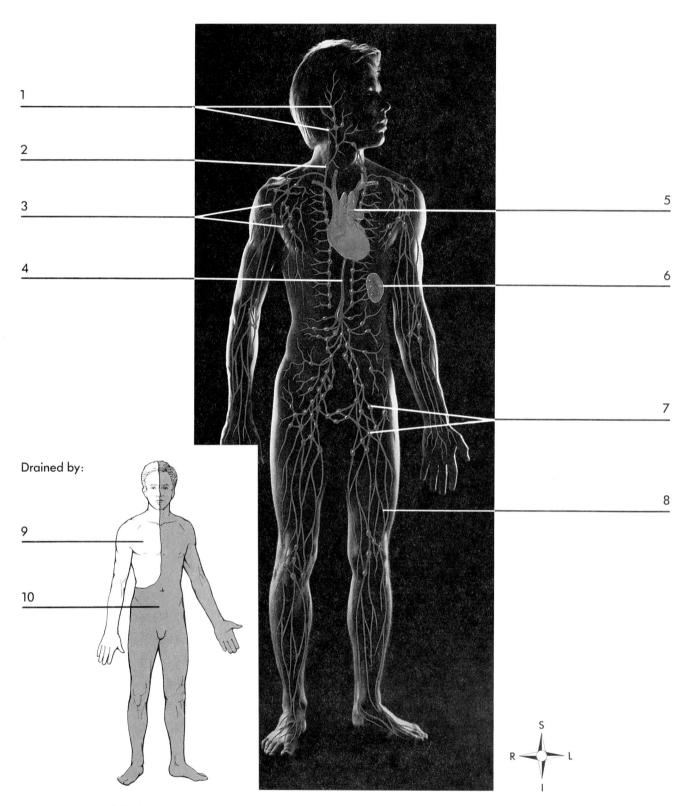

Figure 39-1 Label the lymphatic structures indicated. The inset shows the drainage regions served by the two main lymphatic ducts. Name them.

346

The Tonsils

PALATINE TONSILS₁
PHARYNGEAL TONSILS₂
LINGUAL TONSILS₃

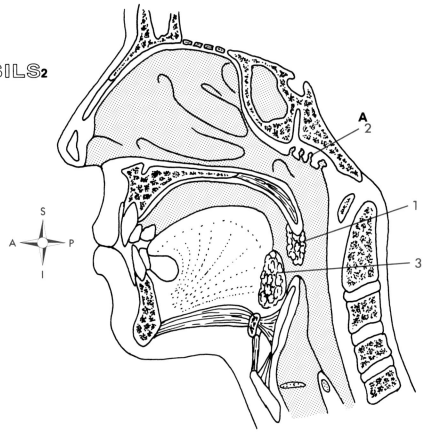

The Lymph Node

AFFERENT VESSEL₁
EFFERENT VESSEL₂
LYMPH NODULE₃
  GERMINAL CENTER₄
LYMPH SINUS₅
CAPSULE₆
ARTERY₇
VEIN₈

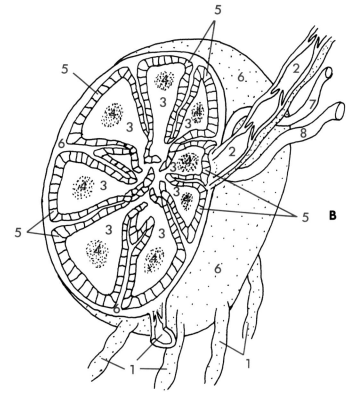

Figure 39-2

347

❏ 2 Within the node, lymph tissue often forms dense masses called **lymph nodules.** Each nodule contains a **germinal center** composed of rapidly dividing lymphocytes. Surrounding each node is a space through which lymph circulates; this space is called a **lymph sinus.** The sinuses also contain macrophages.

❏ 3 The fibrous **capsule** functions as the outer boundary and support structure for each lymph node. The inner surface of the capsule has inward extensions that separate the nodules from one another.

B. Microscopic studies of lymphatic structures

To complete our study of the lymphatic system, we will briefly explore some of the microscopic features of structures with which you are already familiar. Please refer to the appropriate plates in the LABORATORY REFERENCE, as well as the figures in this exercise and in your textbook, as you try to identify the structures described in this section.

> **SAFETY FIRST!** Do not forget the rules for safe use of the microscope.

❏ 1 **Lymphatic vessel**—As described in Section A of this exercise, lymphatic vessels are similar in structure to veins in the structure of their walls and the presence of regulatory valves that ensure one-way flow of fluid. Scan your specimen to locate at least one of these valves. See LABORATORY REFERENCE, Plate 66.

> **LANDMARK CHARACTERISTICS**
> Compare your specimen with that in LABORATORY REFERENCE Plates 76 to 78. You will notice relatively thin walls compared to blood vessels of the same size and more valves than normally seen in veins of the same size.

❏ 2 **Lymph node and tonsil**—See Section A, Step 2, for an overview of the structure of the lymph node and Section A, Step 3, for an overview of the structure of tonsils. Also see Figure 39-1 and LABORATORY REFERENCE Plates 67 and 68.

> **LANDMARK CHARACTERISTICS**
> A well-prepared cross section of a lymph node should show all the features shown in Figure 39-1. Lymph nodes are often rounded with a distinct capsule that projects inward to separate the tissue of the cortex into sinuses that contain lymph nodules. Each lymph nodule will be seen to contain a pinkish rounded germinal center, or *follicle,* surrounded by a distinct purple line. Toward the middle of the node, you may see the medulla—which is made up of cords of lymphocytes and plasma cells, with macrophages spanning the spaces between the cords. As LABORATORY REFERENCE Plate 68 shows, a tonsil is essentially a smaller version of a lymph node, that is, a standalone nodule with a single germinal center.

❏ 3 **Spleen**—The spleen, like the lymph node, has an outer capsule that divides it into subsections. Arteries that enter these compartments are surrounded by masses of developing lymphocytes, forming a whitish tissue called **white pulp.** Surrounding the white pulp is a network of reticular fibers through which blood tissue is flowing. This bloody meshwork is called **red pulp.** See Figure 39-3 and LABORATORY REFERENCE Plate 69.

> **LANDMARK CHARACTERISTICS**
> The capsule and its inward extensions (trabeculae) are normally pale, easy-to-find structures. Within the compartments of the spleen, try to find an area where there are one or more small arteries close together. Surrounding the arteries will be a darkly stained region of cells—this is the white pulp. The red pulp is less densely stained and surrounds the darker white pulp. You may notice some fine reticular fibers in the red pulp.

❏ 4 **Thymus**—The thymus has a dual role in that it functions as both a lymphatic organ and an endocrine gland. It serves as a site for the development of T-lymphocytes (immune cells) and secretes a hormone that influences lymphocyte development. The gland is surrounded by a *capsule* that extends inward, forming *trabeculae* that divide the thymus into lobules. Each lobule has a **cortex** largely filled with

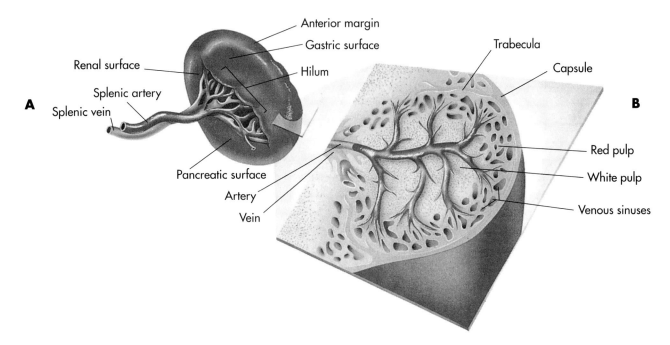

Figure 39-3 Spleen. **A,** Medial aspect of the spleen. Notice the concave surface that fits against the stomach within the abdominopelvic cavity. **B,** Section showing the internal organization of the spleen.

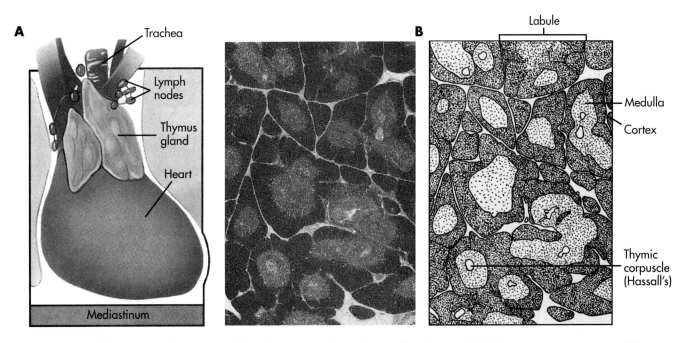

Figure 39-4 Thymus. **A,** Location of the thymus within the mediastinum. **B,** Microscopic structure of the thymus showing several lobules, each with a cortex and a medulla.

lymphocytes and a medulla with few lymphocytes. In the medulla, small spheres called thymic (Hassall's) corpuscles, whose function is unclear, can be found. See Figure 39-4 and LABORATORY REFERENCE Plate 67.

> **LANDMARK CHARACTERISTICS**
> The capsule and its inward extensions (trabeculae) are normally pale, easy-to-find structures. Near these walls, the cortex can be seen as a region of darkly stained lymphocytes. Farther away from the walls, the more lightly stained medullary region can be seen. Occasional thymic corpuscles (circles with a nonuniform interior) can sometimes be seen.

CLINICAL APPLICATIONS

Lymphocytes may be involved in two major types of responses to the presence of potentially threatening foreign substances. In **antibody-mediated immunity,** B *lymphocytes* may produce antibodies that react with specific antigens on an invading cell or molecule. In **cell-mediated immunity,** T *lymphocytes* secrete **lymphokines** that signal other immune responses and often destroy antigen-containing cells directly.

The antibody-antigen reactions associated with immune responses have been used by clinical biologists for years. Several examples are given. Think about each example and answer the questions posed.

1. Biologists often use antibodies to test for the presence of certain antigens in a particular substance. In Lab Exercises 24 and 25, you witnessed demonstrations of this technique. For each antigen listed, indicate the substance tested (e.g., blood, urine) and give a brief summary of the antibody-antigen reaction involved.

HUMAN CHORIONIC GONADOTROPIN:

LUTEINIZING HORMONE:

A and B ANTIGENS:

D ANTIGEN:

2. *Rheumatoid arthritis* is an inflammatory disease affecting joint tissues. This disease is known to have an *autoimmune* component. Autoimmunity is an immune response inappropriately directed toward normal self antigens. Many persons with rheumatoid arthritis have *rheumatoid factor (RF)* present in their blood. RF is an abnormal antibody. Not all victims of the disease have RF in their plasma, and RF is known to be present in conditions other than rheumatoid arthritis. Based on this information and your previous study, outline a simple test to detect the presence of RF in plasma. If your test is used to screen for rheumatoid arthritis, does a positive result (i.e., RF is present) mean that the person definitely has the disease? Why or why not?

LAB REPORT 39

The Lymphatic System

Figure 39-1

_____ 1
_____ 2
_____ 3
_____ 4
_____ 5
_____ 6
_____ 7
_____ 8

Ducts:

_____ 9
_____ 10

Multiple choice

_____ 1
_____ 2
_____ 3
_____ 4
_____ 5
_____ 6
_____ 7

Multiple choice (choose the best response—only one is correct)

1. Lymphatic tissue contains
    a. mainly lymphocytes
    b. mainly erythrocytes
    c. mainly monocytes

2. The large lymphatic organ found in the left hypochondriac region is the
    a. thymus
    b. palatine tonsil
    c. spleen
    d. lingual tonsil
    e. inguinal lymph node

3. The lymph from tissues of the lower limbs is conducted by the
    a. right lymphatic duct
    b. thoracic duct

4. Lymph nodes function to
    a. provide a site for lymphocyte reproduction
    b. anchor lymphatic vessels
    c. provide a site for phagocytosis of foreign cells or debris
    d. a and b are correct
    e. a and c are correct

5. Lymph vessels that conduct lymph into a lymph node are termed
    a. efferent lymphatic vessels
    b. efferent arterioles
    c. afferent arterioles
    d. afferent lymphatic vessels
    e. none of the above

6. The lymphatic system functions to
    a. maintain extracellular fluid balance
    b. collect excess fluid from the interstitial spaces
    c. defend the body against foreign cells and molecules
    d. a and c are correct
    e. a, b, and c are correct

7. Lymph drains directly into the blood at
    a. the right subclavian vein
    b. the left subclavian vein
    c. the subclavian arteries
    d. a and b are correct
    e. none of the above

Lymphatic structure	Location	Function(s)
Lymphatic capillaries		
Lymphatic vessels		
Right lymphatic duct		
Thoracic duct		
Lymph node		
Palatine tonsils		
Pharyngeal tonsils		
Lingual tonsils		
Spleen		
Thymus		

Sketch (draw a cross section of a lymph node and label its parts)

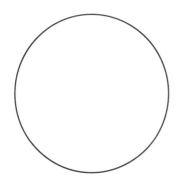

Specimen: *lympatic vessel l.s.*
Total Magnification: _____

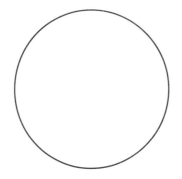

Specimen: *lymph node c.s.*
Total Magnification: _____

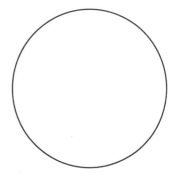

Specimen: *palatine tonsil c.s.*
Total Magnification: _____

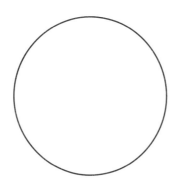

Specimen: *spleen c.s.*
Total Magnification: _____

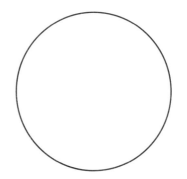

Specimen: *thymus c.s.*
Total Magnification: _____

LAB EXERCISE 40

Dissection: Cardiovascular and Lymphatic Systems

In this exercise, you will continue your study of cardiovascular and lymphatic anatomy by dissecting a whole preserved specimen.

Activity A provides directions for studying the cardiovascular and lymphatic anatomy of the cat. Activity B is an alternate activity, providing directions for studying the cardiovascular and lymphatic anatomy of the fetal pig.

Before you begin

❑ Read the appropriate chapter in your textbook.

❑ Set your learning goals. When you finish this exercise, you should be able to
- dissect the cardiovascular and lymphatic anatomy of a preserved cat or fetal pig
- identify the following in a dissected mammalian specimen:
  Heart and its major features
  Major arteries
  Major veins
  Lymph nodes

❑ Prepare your materials:
- preserved (injected) cat or fetal pig
- dissection tools and trays
- storage container (if specimen is to be reused)

❑ Read the directions and safety tips for this exercise **carefully** before starting any procedure.

> **SAFETY FIRST!** Observe the usual precautions when working with a preserved specimen. Heed the safety advice accompanying preservatives used with your specimen. Use protective gloves while handling your specimen. Avoid injury with dissection tools. Dispose of or store your specimen as instructed.

A. Cardiovascular and lymphatic anatomy of the cat

This activity challenges you to identify the major systemic blood vessels of the cat, features of the cat's heart, and lymphoid organs inside the cat's body. To find these structures, you must cut open the cat's thoracic and abdominopelvic cavities. Assuming you have already removed the cat's skin (Exercise 10), you can simply cut through the ventral aspect of the thoracic and abdominal musculature. Do this by inserting your scissors into the ventral abdominal wall (without damaging any internal organs), lifting up the scissors, and cutting along the midline toward the head and tail. You will need to cut through the costal cartilage that connects the ribs to the sternum. Transverse cuts from this initial cut will produce "doors" to the internal body cavities that can be pulled back to expose the viscera. You may have to cut the ribs to see the thoracic viscera. Carefully move the viscera out of the way—without cutting or removing them—to see the structures indicated in this activity.

> **HINT** → In double-injected specimens, *arteries* are filled with red latex and *veins* are filled with blue latex. This makes these vessels easy to identify. In specimens that are not injected, you can often distinguish veins and arteries by the thickness of their walls—veins often have thinner walls than arteries. Veins and arteries can also be distinguished by their *locations*—systemic veins lead toward a vena cava and systemic arteries branch from the aorta.

❑ 1 Identify the **aorta** extending from the **heart.** Distinguish these regions of the aorta:
- **Ascending aorta**
- **Aortic arch**
- **Thoracic portion of descending aorta**
- **Abdominal portion of descending aorta**

> **HINT** → Figures 40-1 through 40-4 illustrate many of the structures of the cat listed in this activity. Useful information can also be found in the LABORATORY REFERENCES Plates 83 and 84.

❏ 2 Identify these branches of the **ascending aorta:**
  - **Coronary arteries**

❏ 3 Identify these branches of the **aortic arch:**
  - **Brachiocephalic artery** (on right side only)
  - **Subclavian arteries**
  - **Axillary arteries**
  - **Brachial arteries**
  - **Vertebral arteries**
  - **Costocervical arteries**
  - **Common carotid arteries**

❏ 4 Identify these branches of the **thoracic aorta:**
  - **Intercostal arteries**
  - **Anterior phrenic arteries**

❏ 5 Identify these branches of the **abdominal aorta:**
  - **Posterior phrenic arteries**
  - **Celiac artery**
  - **Splenic artery**
  - **Anterior mesenteric artery**
  - **Suprarenal arteries**
  - **Renal arteries**
  - **Gonadal (testicular or ovarian) arteries**
  - **Lumbar arteries**
  - **Posterior mesenteric artery**
  - **External iliac arteries**
  - **Internal iliac arteries**
  - **Femoral arteries**

❏ 6 Identify the large **anterior vena cava,** which corresponds to the *superior vena cava* in the human. Identify these tributaries of the anterior vena cava in the cat:
  - **Azygous vein**
  - **Internal mammary vein**
  - **Brachiocephalic vein**
  - **Axillary vein**
  - **Subclavian vein**
  - **Subscapular vein**
  - **Transverse scapular vein**
  - **Brachial vein**
  - **Long thoracic vein**
  - **Internal jugular vein**
  - **External jugular vein**
  - **Transverse jugular vein**
  - **Posterior facial vein**
  - **Anterior facial vein**
  - **Submental vein**

❏ 7 Identify the large **posterior vena cava,** which corresponds to the *inferior vena cava* in the human. Identify these tributaries of the posterior vena cava in the cat:
  - **Adrenolumbar vein**
  - **Renal vein**
  - **Common iliac vein**
  - **External iliac vein**
  - **Internal iliac vein**
  - **Femoral vein**
  - **Greater saphenous vein**

❏ 8 Examine the **heart** of the cat. Identify as many features as you can, using the information in Lab Exercise 35 as a guide.

❏ 9 Locate the **spleen,** a lymphoid organ near the stomach. Try to find one or more **lymph nodes** or **lymphatic vessels.**

ANATOMICAL ATLAS OF THE CAT (CARDIOVASCULAR SYSTEM)

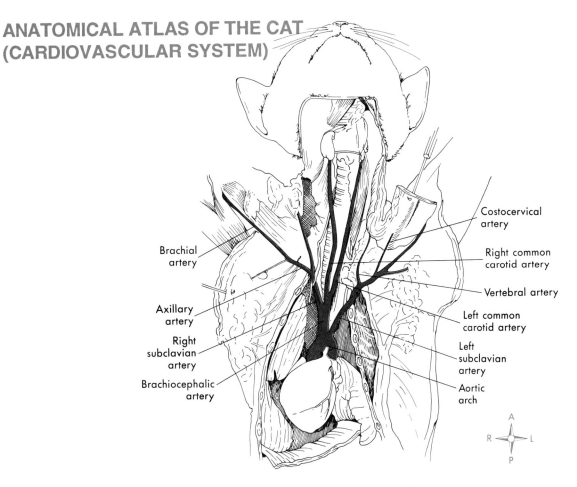

Figure 40-1　Arteries of the cat. Thorax, ventral view.

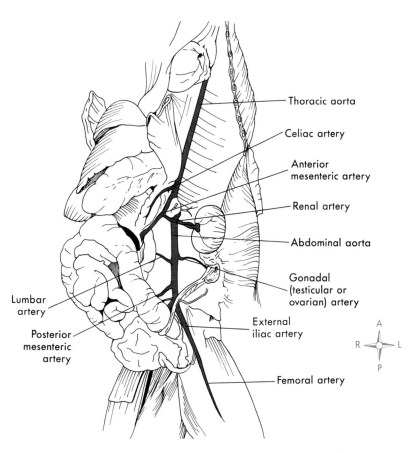

Figure 40-2　Arteries of the cat. Abdomen, ventral view.

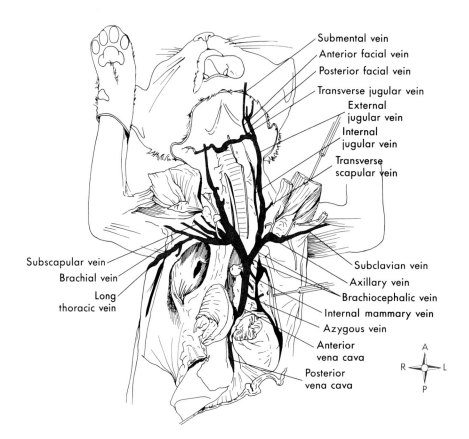

Figure 40-3 Veins of the cat. Thorax, ventral view.

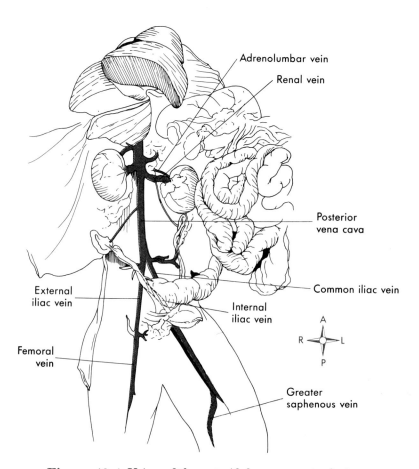

Figure 40-4 Veins of the cat. Abdomen, ventral view.

B. Cardiovascular and lymphatic anatomy of the fetal pig

This activity challenges you to identify the major systemic blood vessels of the fetal pig, features of the heart, and lymphoid organs inside the fetal pig's body. To find these structures, you must cut open your specimen's thoracic and abdominopelvic cavities. Assuming you have already removed the skin (Exercise 10), you can simply cut through the ventral aspect of the thoracic and abdominal musculature. Do this by inserting your scissors into the ventral abdominal wall (without damaging any internal organs), lifting up the scissors, and cutting along the midline toward the head and tail. You will need to cut through the costal cartilage that connects the ribs to the sternum. Transverse cuts from this initial cut will produce "doors" to the internal body cavities that can be pulled back to expose the viscera. You may have to cut the ribs to see the thoracic viscera. Carefully move the viscera out of the way—without cutting or removing them—to see the structures indicated in this activity.

> **HINT** → In double-injected specimens, *arteries* are filled with red latex and *veins* are filled with blue latex. This makes these vessels easy to identify. In specimens that are not injected, you can often distinguish veins and arteries by the thickness of their walls—veins often have thinner walls than arteries. Veins and arteries can also be distinguished by their *locations*—systemic veins lead toward a vena cava and systemic arteries branch from the aorta.

❏ 1 Identify the **aorta** extending from the **heart.** Distinguish these regions of the aorta:
- **Ascending aorta**
- **Aortic arch**
- **Thoracic portion of descending aorta**
- **Abdominal portion of descending aorta**

> **HINT** → Figures 40-5 through 40-7 illustrate many of the structures of the cat listed in this activity. Useful information can also be found in the Laboratory Reference Plates 83 and 84.

❏ 2 Identify these branches of the **ascending aorta:**
- **Coronary arteries**

❏ 3 Identify these branches of the **aortic arch:**
- **Brachiocephalic artery** (on right side only)
- **Subclavian arteries**
- **Axillary arteries**
- **Brachial arteries**
- **Common carotid arteries**

❏ 4 Identify these branches of the **thoracic aorta:**
- **Intercostal arteries**

❏ 5 Identify these branches of the **abdominal aorta:**
- **Celiac artery**
- **Splenic artery**
- **Anterior mesenteric artery**
- **Renal arteries**
- **Gonadal (testicular or ovarian) arteries**
- **Posterior mesenteric artery**
- **Umbilical arteries**
- **External iliac arteries**
- **Internal iliac arteries**
- **Femoral arteries**

❏ 6 Identify the large **anterior vena cava,** which corresponds to the *superior vena cava* in the human. Identify these tributaries of the anterior vena cava in the fetal pig:
- **Internal mammary vein**
- **Brachiocephalic vein**
- **Axillary vein**
- **Subclavian vein**
- **Internal jugular vein**
- **External jugular vein**

❏ 7 Identify the large **posterior vena cava,** which corresponds to the *inferior vena cava* in the human. Identify these tributaries of the posterior vena cava in the fetal pig:
- **Renal vein**
- **External iliac vein**
- **Internal iliac vein**
- **Femoral vein**

❏ 8 Examine the **heart** of the fetal pig. Identify as many features as you can, using the information in Exercise 35 as a guide.

❏ 9 Locate the **spleen,** a lymphoid organ near the stomach. Try to find one or more **lymph nodes** or **lymphatic vessels.**

ANATOMICAL ATLAS OF THE PIG (CARDIOVASCULAR SYSTEM)

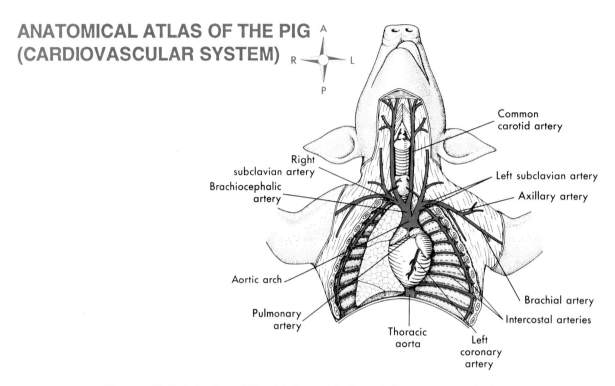

Figure 40-5 Arteries of the fetal pig. Neck and thorax, ventral view.

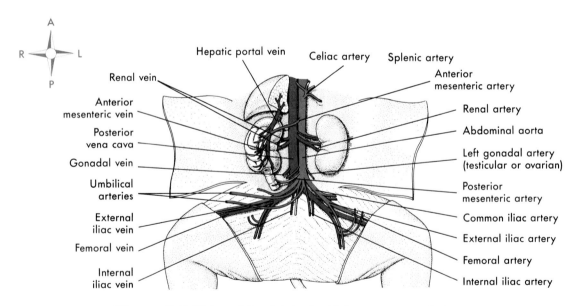

Figure 40-6 Veins of the fetal pig. Abdomen, ventral view.

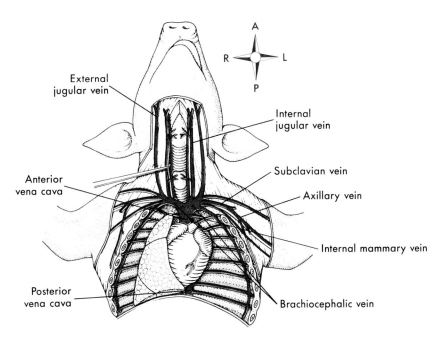

Figure 40-7 Veins of the fetal pig. Neck and thorax, ventral view.

| NAME | DATE | SECTION |

LAB REPORT 40

Dissection: Cardiovascular and Lymphatic Systems

A. Cat dissection checklist

■ **Arteries**
- aorta
- ascending aorta
  - coronary arteries
- aortic arch
  - brachiocephalic artery (on right side only)
  - subclavian arteries
  - axillary arteries
  - brachial arteries
  - vertebral arteries
  - costocervical arteries
  - common carotid arteries
- thoracic portion of descending aorta
  - intercostal arteries
  - anterior phrenic arteries
- abdominal portion of descending aorta
  - posterior phrenic arteries
  - celiac artery
  - splenic artery
  - anterior mesenteric artery
  - suprarenal arteries
  - renal arteries
- gonadal (testicular or ovarian) arteries
- lumbar arteries
- posterior mesenteric artery
- external iliac arteries
- internal iliac arteries
- femoral arteries

■ **Veins**
- anterior vena cava
  - azygous vein
  - internal mammary vein
  - brachiocephalic vein
  - axillary vein
  - subclavian vein
  - subscapular vein
  - transverse scapular vein
  - brachial vein
  - long thoracic vein
  - internal jugular vein
  - external jugular vein
  - transverse jugular vein
  - posterior facial vein
  - anterior facial vein
  - submental vein
- posterior vena cava
  - adrenolumbar vein
  - renal vein
  - common iliac vein
  - external iliac vein
  - internal iliac vein
  - femoral vein
  - greater saphenous vein

■ **Other organs**
- heart
- spleen
- lymph nodes
- lymphatic vessels

B. Fetal pig dissection checklist

■ **Arteries**

❏ aorta

❏ ascending aorta

  ❏ coronary arteries

❏ aortic arch

  ❏ brachiocephalic artery (on right side only)

  ❏ subclavian arteries

  ❏ axillary arteries

  ❏ brachial arteries

  ❏ common carotid arteries

❏ thoracic portion of descending aorta

  ❏ intercostal arteries

❏ abdominal portion of descending aorta

  ❏ celiac artery

  ❏ splenic artery

  ❏ anterior mesenteric artery

  ❏ renal arteries

  ❏ gonadal (testicular or ovarian) arteries

  ❏ posterior mesenteric artery

  ❏ umbilical arteries

  ❏ external iliac arteries

  ❏ internal iliac arteries

  ❏ femoral arteries

■ **Veins**

❏ anterior vena cava

  ❏ internal mammary vein

  ❏ brachiocephalic vein

  ❏ axillary vein

  ❏ subclavian vein

  ❏ internal jugular vein

  ❏ external jugular vein

❏ posterior vena cava

  ❏ renal vein

  ❏ external iliac vein

  ❏ internal iliac vein

  ❏ femoral vein

■ **Other organs**

❏ heart

❏ spleen

❏ lymph nodes

❏ lymphatic vessels

LAB EXERCISE 41

Respiratory Structures

This exercise is the first of two that deal with the **respiratory system.** In the activities presented here you will explore the essential structure of the *respiratory tract.* Lab Exercise 42, on the other hand, will deal with some aspects of respiratory function.

The respiratory tract consists of the **nose** and **nasal cavity, pharynx, larynx, trachea, bronchi,** and **lungs.** During **ventilation,** a process that includes *respiratory cycles* of **inspiration** then **expiration,** air moves in and out of the respiratory tract. During inspiration air moves through the tract to the lungs, and during expiration it moves out. While in the lungs, air exchanges oxygen and carbon dioxide in a process called **gas exchange.** Keep these important functions of the system in mind as you explore the gross and microscopic structures presented in this exercise.

Before you begin

❑ Read the appropriate chapter in your textbook.

❑ Set your learning goals. When you finish this exercise, you should be able to
- describe the major organs of the respiratory system
- locate respiratory structures on a chart or model
- demonstrate the structure of respiratory organs in a preserved specimen
- trace the movement of air through the respiratory tract
- locate structural features of the trachea in a microscopic specimen
- locate structural features of lung tissue in a microscopic specimen

❑ Prepare your materials:
- chart or model of the human respiratory system
- dissection tools and large tray
- preserved specimen: *sheep pluck*
- microscope
- prepared microslide: *mammalian trachea c.s.*
- prepared microslide: *human lung c.s.*

❑ Read the directions and safety tips for this exercise **carefully** before starting any procedure.

A. Gross anatomy

Using a chart or model of the respiratory tract (perhaps a dissectible human head and torso), locate the structural features of the respiratory tract listed.

❑ 1 **Nose** is a term often reserved for the external cartilage and bone forming the anterior wall of the **nasal cavity** but may refer to the entire cavity. Identify these nasal features:
- **Nasal septum**—Wall dividing the nasal cavity into right and left portions
- **Nares**—*External nares* or the *nostrils* and the *internal nares* or openings from the posterior nasal cavity into the pharynx (singular of nares is *naris*)
- **Hard palate**—Floor of the nasal cavity
- **Conchae**—Three curved ridges extending from each lateral wall of the cavity: the *superior nasal conchae, middle nasal conchae,* and *inferior nasal conchae* (singular of conchae is *concha*)
- **Paranasal sinuses**—Air-filled, mucus-lined spaces in the skull that communicate with the nasal cavity: *maxillary, frontal, ethmoid,* and *sphenoid sinuses*
- **Nasolacrimal duct**—Tube that conducts tears from the conjunctiva over the eyes and empties into the nasal cavity

❑ 2 Identify these features of the **pharynx,** or throat:
- **Nasopharynx**—The upper of three regions of the pharynx, this one is posterior to the nasal cavity. The **uvula** is an extension of the floor of the nasopharynx, or **soft palate,** and serves as the inferior boundary landmark of the nasopharynx. The *pharyngeal tonsils* are on the posterior wall, and the lateral walls have the inferior opening of the *auditory (Eustachian) tube* from the middle ear.
- **Oropharynx**—It is the middle portion of the pharynx, posterior to the oral cavity. The *palatine* and *lingual tonsils* are found here.
- **Laryngopharynx**—The most inferior portion of the pharynx, it is posterior to the larynx.

❑ 3 Identify these features of the voice box, or **larynx:**
- **Laryngeal cartilages**—Nine pieces of cartilage (6 pairs and 3 single pieces) that fit

together to form the wall of the larynx: **thyroid, cricoid, epiglottis, cuneiforms** (pair), **corniculates** (pair), and **arytenoids** (pair)
- **Vestibular folds** (false vocal cords)—Superior of two pairs of ligaments stretched across the lateral portions of the larynx cavity
- **Vocal folds** (true vocal cords)—Inferior of two pairs of laryngeal ligaments

❑ 4 The **trachea,** or windpipe, conducts air between the larynx above and the bronchi below. It has C-shaped cartilage rings embedded in its anterior and lateral walls for support. At its inferior end, the trachea divides into the left and right **primary bronchi,** each of which enters a lung at a **hilum.**

❑ 5 The **lungs** are divided, on a gross level, into **lobes,** which can be subdivided into **lobules.** The left lung has a *superior lobe* and *inferior lobe.* The right lung has a *superior, middle,* and *inferior lobe.* The lungs are found in the *pleural cavities* in the left and right portions of the thorax. A serous membrane called the **pleura** covers each lung (*visceral pleura*), then folds back to line the pleural cavity (*parietal pleura*). The thin space between the visceral and parietal layers is filled with serous **pleural fluid.**

❑ 6 Once in a lung, each primary bronchus branches to form a **bronchial tree.** The successive levels of bronchial branching include:
- **Secondary bronchi**—Each serves a separate lobe.
- **Tertiary bronchi**—Each is a branch of a secondary bronchus; each tertiary bronchus serves a lobule.
- **Bronchioles**—Several levels of small branches of the airway exist within each lobule. Near the end of the branching, **terminal bronchioles** give rise to **respiratory bronchioles,** which in turn subdivide to form **alveolar ducts** that end as groups of air sacs called **alveoli.** The smaller bronchioles are not represented in many models but can be found in some charts and figures.

❑ 7 The **diaphragm** is a sheet of skeletal muscle just inferior to both lungs. The diaphragm contracts during inspiration to reduce pressure inside the thorax, drawing air into the respiratory tract.

B. Dissection

This activity calls for individual dissection or a demonstrated dissection of a *sheep pluck.* The pluck is a portion of the respiratory tract and other organs removed from an animal when it is slaughtered. Notice that the heart and diaphragm are still intact. This is a good opportunity to reinforce your understanding of their anatomical relationship to one another.

> **SAFETY FIRST!** Remember to observe the standard precautions for dissecting a preserved specimen.

❑ 1 Identify the larynx, trachea, primary bronchi, lungs, heart, and diaphragm.

❑ 2 Examine the lungs. Can you distinguish the lobes? How many lobes are in each lung? Is it the same number as in the human?

❑ 3 Use a scissors to cut along the dorsal wall of the respiratory tract to open the larynx and trachea. What features can you identify? Continue your cut through a primary bronchus and into the bronchial tree. How many levels of branching can you see? Can you name the different branches?

C. Microscopic anatomy

> **SAFETY FIRST!** Use the microscope and prepared slides with caution. Do not forget the electrical hazards and possibility of cuts from broken slides.

❑ 1 Examine a prepared cross section of a mammalian trachea. Identify these features:
- **Lining**—It is composed of *pseudostratified columnar epithelium.* Locate *cilia* and *goblet cells* if you can.
- **Outer wall**—It is composed of *smooth muscle* and assorted connective tissues. Locate portions of the C-shaped cartilage supports in the wall of the trachea. What type of cartilage is it?

❑ 2 Examine a prepared cross section of human lung tissue. Identify this feature:
- **Alveolus**—It is an air sac that communicates with a respiratory bronchiole. Try to locate the *alveolar epithelium,* the simple squamous epithelium forming the wall. Is any of the connective tissue between alveoli visible in your specimen?

❑ 3 Try to locate *pulmonary capillaries* that surround the alveoli. In a cross section, they appear as circles of simple squamous epithelium (and sometimes have blood cells within them).

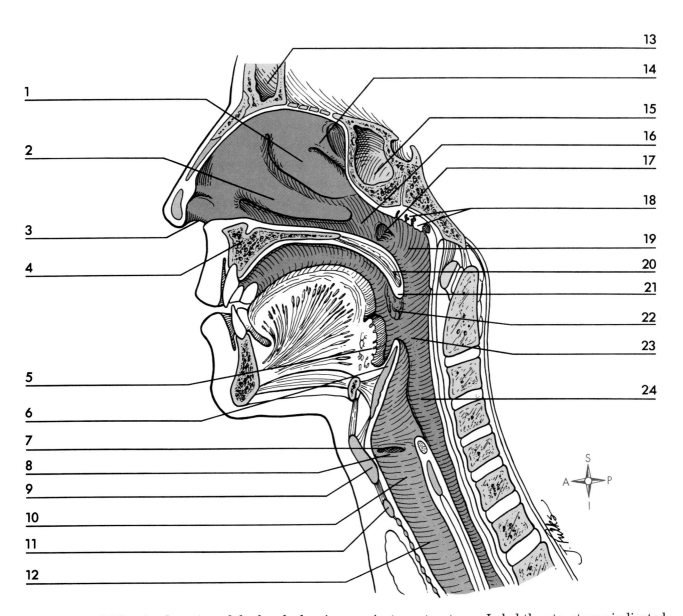

Figure 41-1 Midsagittal section of the head, showing respiratory structures. Label the structures indicated.

Respiratory Anatomy

COLORING EXERCISE Using colored pens or pencils, shade in the figure and accompanying labels in contrasting colors of your choice as indicated by the red numerals.

NASAL CAVITY₁
NASOPHARYNX₂
OROPHARYNX₃
LARYNGOPHARYNX₄
LARYNX₅
TRACHEA₆
PRIMARY BRONCHI₇
SECONDARY BRONCHI₈
TERTIARY BRONCHI₉
BRONCHIOLE₁₀
TERMINAL BRONCHIOLE₁₁

RESPIRATORY BRONCHIOLE₁₂
ALVEOLAR DUCT₁₃
ALVEOLUS₁₄

LOBES (Shade lightly)
RIGHT SUPERIOR LOBE₁₅
RIGHT MIDDLE LOBE₁₆
RIGHT INFERIOR LOBE₁₇
LEFT SUPERIOR LOBE₁₈
LEFT INFERIOR LOBE₁₉

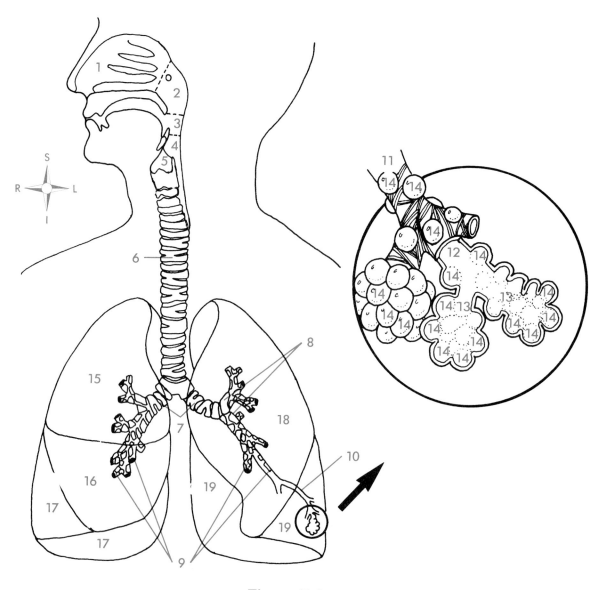

Figure 41-2

NAME _____ DATE _____ SECTION _____

LAB REPORT 41

Respiratory Structures

Use this table as a checklist for your study of the human model and the sheep pluck. Use your textbook for help with the section that asks for functions.

Structure	Human	Pluck	Function(s)
Nasal cavity	☐	☐	
Nasal septum	☐	☐	
Nares: internal, external	☐	☐	
Hard palate	☐	☐	
Soft palate	☐	☐	
Conchae: superior, middle, inferior	☐	☐	
Paranasal sinuses: maxillary, frontal, ethmoid, sphenoid	☐	☐	
Nasopharynx	☐	☐	
Oropharynx	☐	☐	
Laryngopharynx	☐	☐	
Laryngeal cartilages	☐	☐	
Vestibular folds	☐	☐	
Vocal folds	☐	☐	
Trachea	☐	☐	
Primary bronchi	☐	☐	
Secondary bronchi	☐	☐	
Tertiary bronchi	☐	☐	
Lungs: lobes, lobules	☐	☐	
Pleurae: visceral, parietal	☐	☐	
Bronchioles	☐	☐	
Terminal bronchioles	☐	☐	
Respiratory bronchioles	☐	☐	
Alveolar ducts	☐	☐	
Alveoli	☐	☐	

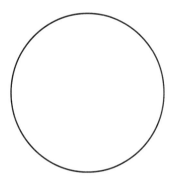

Specimen: *trachea c.s.*

Total Magnification: _____

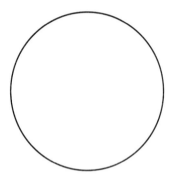

Specimen: *human lung c.s.*

Total Magnification: _____

Sketch and label the sheep pluck specimen:

Figure 41-1

_____ 1
_____ 2
_____ 3
_____ 4
_____ 5
_____ 6
_____ 7
_____ 8
_____ 9
_____ 10
_____ 11
_____ 12
_____ 13
_____ 14
_____ 15
_____ 16
_____ 17
_____ 18
_____ 19
_____ 20
_____ 21
_____ 22
_____ 23
_____ 24

Put in order

_____ 1
_____ 2
_____ 3
_____ 4
_____ 5
_____ 6
_____ 7
_____ 8
_____ 9

Put in order (arrange these structures in the order through which air passes during *inspiration*)

larynx
nasal cavity
pharynx
primary bronchi
respiratory bronchioles
secondary bronchi
terminal bronchioles
tertiary bronchi
trachea

LAB EXERCISE 42

Dissection: Respiratory System

In this exercise, you will continue your study of respiratory anatomy by dissecting a whole preserved specimen.

Activity A provides directions for studying the respiratory anatomy of the cat. Activity B is an alternate activity, providing directions for studying the respiratory anatomy of the fetal pig. Activity C is also an alternate activity that asks you to dissect a sheep pluck, or lower respiratory tract and heart of a sheep.

Before you begin

❑ Read the appropriate chapter in your textbook.

❑ Set your learning goals. When you finish this exercise, you should be able to
- dissect the respiratory anatomy of a preserved cat, fetal pig, or sheep
- identify the following in a dissected mammalian specimen:
Larynx
Trachea and bronchial tree
Lungs
Diaphragm

❑ Prepare your materials:
- preserved cat, fetal pig, or sheep pluck
- dissection tools and trays
- storage container (if specimen is to be reused)
- dissection microscope (optional)

❑ Read the directions and safety tips for this exercise **carefully** before starting any procedure.

> **SAFETY FIRST!** Observe the usual precautions when working with a preserved specimen. Heed the safety advice accompanying preservatives used with your specimen. Use protective gloves while handling your specimen. Avoid injury with dissection tools. Dispose of or store your specimen as instructed.

A. Respiratory anatomy of the cat

This activity challenges you to identify the major respiratory structures of the cat. Assuming you have already removed the cat's skin (Exercise 10) and opened the thoracic cavity (Exercise 40), you simply have to cut through the muscles of the neck and head to expose internal features of the anterior airway (equivalent to the upper respiratory tract in humans). Be careful as you do this to avoid damaging the specimen to the point that you really cannot tell where the structures are normally located in relation to one another. Remember to carefully move the viscera out of the way—without cutting or removing them—to see the structures indicated in this activity.

❑ 1 Identify the **external nares,** the openings of the **nasal cavity.** Notice that a **nasal septum** divides the nasal cavity into two air passages. If your instructor directs you to do so, cut into the nose and mouth of the cat to expose the nasal cavity and **pharynx.** In the nasal cavity, try to identify the nasal **turbinates** (*conchae*) projecting from the lateral walls. Can you identify the three major regions of the pharynx: *nasopharynx, oropharynx,* and *laryngopharynx*?

❑ 2 Locate and identify the **larynx,** or voice box, deep in the neck. The **epiglottis, thyroid cartilage,** and **cricoid cartilage** of the larynx are easily visible on the ventral aspect.

❑ 3 Identify the **trachea** extending posteriorly from the larynx. Notice the cartilage rings that support this structure.

> **HINT** → Figure 42-1 illustrates many of the structures of the cat listed in this activity. Useful information can also be found in the LABORATORY REFERENCE Plate 83.

❑ 4 Identify the right and left **primary bronchi.** Notice that each extends to a different **lung,** where it splits to form several **secondary bronchi.**

ANATOMICAL ATLAS OF THE CAT (RESPIRATORY STRUCTURES)

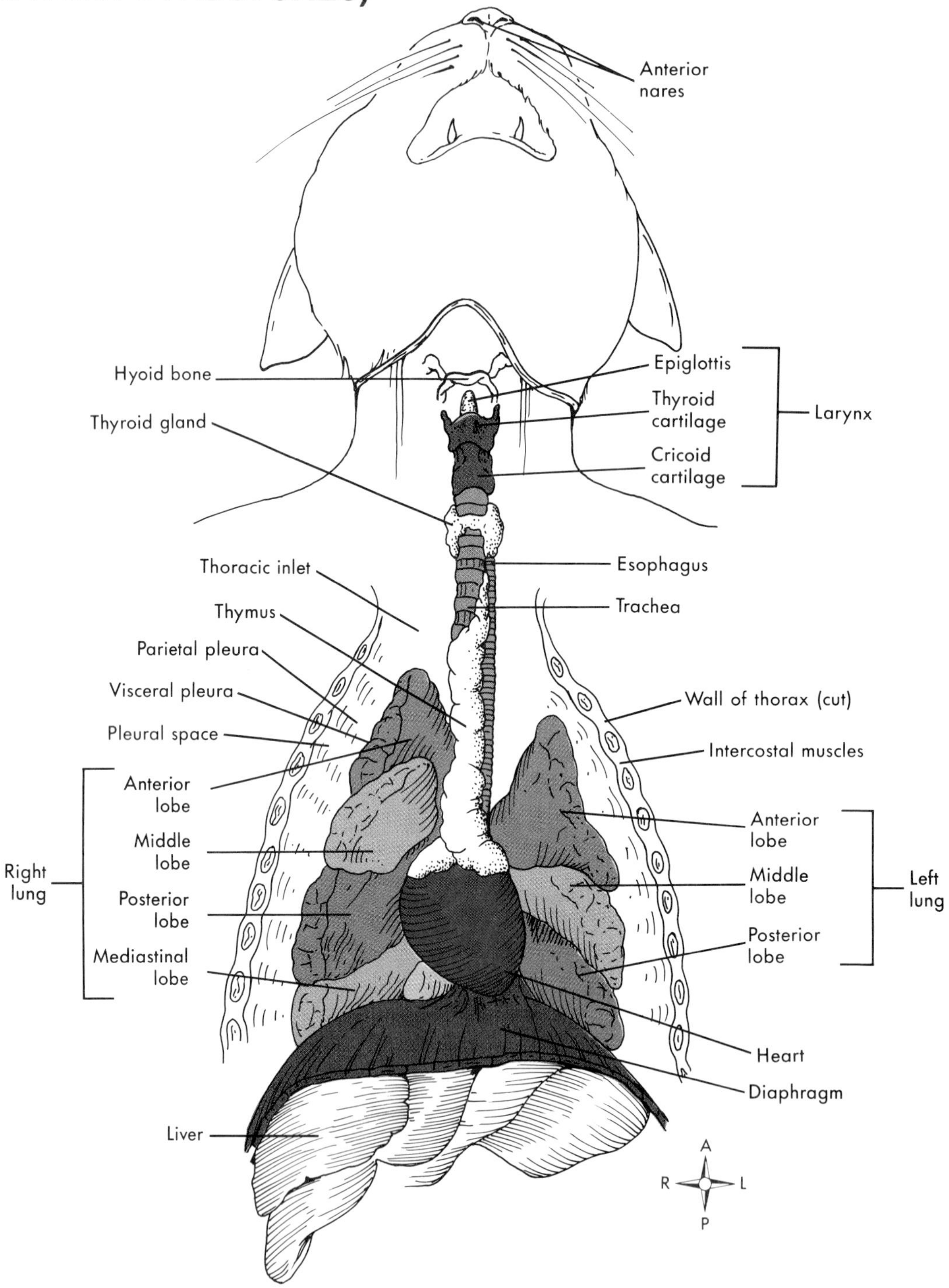

Figure 42-1 Respiratory anatomy of the cat. Ventral view.

❑ 5 Each secondary bronchus serves a distinct **lobe** of a lung. Identify the following lobes in *each* of your specimen's lungs:
- **Anterior lobe** (left and right)
- **Middle lobe** (left and right)
- **Posterior lobe** (left and right)
- **Mediastinal lobe** (right only)

❑ 6 If your instructor directs you to do so, slice into the tissue of one of the lobes and remove a small section of tissue. Examine the internal structure of this tissue with a dissection microscope or hand lens to reveal air passages that constitute distal elements of the **bronchial tree** and gas-exchange tissues. You may make a deeper cut into the lobe to reveal other elements of the bronchial airways.

❑ 7 Locate the following structures of the wall of the thoracic cavity:
- **Intercostal muscles**
- **Thoracic inlet**
- **Parietal pleura**
- **Visceral pleura**
- **Pleural space**
- **Diaphragm**
- **Phrenic nerve** (innervates diaphragm)

B. Respiratory anatomy of the fetal pig

This activity challenges you to identify the major respiratory structures of the fetal pig. Assuming you have already removed the animal's skin (Exercise 10) and opened the thoracic cavity (Exercise 40), you simply have to cut through the muscles of the neck and head to expose internal features of the anterior airway (equivalent to the upper respiratory tract in humans). Be careful as you do this to avoid damaging the specimen to the point that you really cannot tell where the structures are normally located in relation to one another. Remember to carefully move the viscera out of the way—without cutting or removing them—to see the structures indicated in this activity.

❑ 1 Identify the **external nares,** the openings of the **nasal cavity.** Notice that a **nasal septum** divides the nasal cavity into two air passages. If your instructor directs you to do so, cut into the nose and mouth of the fetal pig to expose the nasal cavity and **pharynx.** In the nasal cavity, try to identify the nasal **turbinates** (*conchae*) projecting from the lateral walls. Can you identify the three major regions of the pharynx: *nasopharynx, oropharynx,* and *laryngopharynx?*

❑ 2 Locate and identify the **larynx,** or voice box, deep in the neck.

❑ 3 Identify the **trachea** extending posteriorly from the larynx. Notice the cartilage rings that support this structure.

> HINT → Figure 42-2 illustrates many of the structures of the fetal pig listed in this activity. Useful information can also be found in the LABORATORY REFERENCE Plates 85 and 86.

❑ 4 Idenfity the right and left **primary bronchi.** Notice that each extends to a different **lung,** where it splits to form several **secondary bronchi.**

❑ 5 Each secondary bronchus serves a distinct **lobe** of a lung. Identify the following lobes in *each* of your specimen's lungs:
- **Anterior lobe** (left and right)
- **Middle lobe** (right only)
- **Posterior lobe** (left and right)

❑ 6 If your instructor directs you to do so, slice into the tissue of one of the lobes and remove a small section of tissue. Examine the internal structure of this tissue with a dissection microscope or hand lens to reveal air passages that constitute distal elements of the **bronchial tree** and gas-exchange tissues. You may make a deeper cut into the lobe to reveal other elements of the bronchial airways.

❑ 7 Locate the following structures of the wall of the thoracic cavity:
- **Intercostal muscles**
- **Thoracic inlet**
- **Parietal pleura**
- **Visceral pleura**
- **Pleural space**
- **Diaphragm**

C. Respiratory anatomy of the sheep

This activity challenges you to identify the major respiratory structures in a *sheep pluck* specimen. This specimen includes a portion of the respiratory system, the heart, and other nearby structures removed from the carcass of a sheep. The organs of the sheep pluck are similar in size to the adult human (depending on the particular sheep and the particular human, of course).

❑ 1 Locate and identify the **larynx,** or voice box, at the anterior end of the sheep pluck.

❑ 2 Identify the **trachea** extending posteriorly from the larynx. Notice the cartilage rings that support this structure.

ANATOMICAL ATLAS OF THE PIG (RESPIRATORY STRUCTURES)

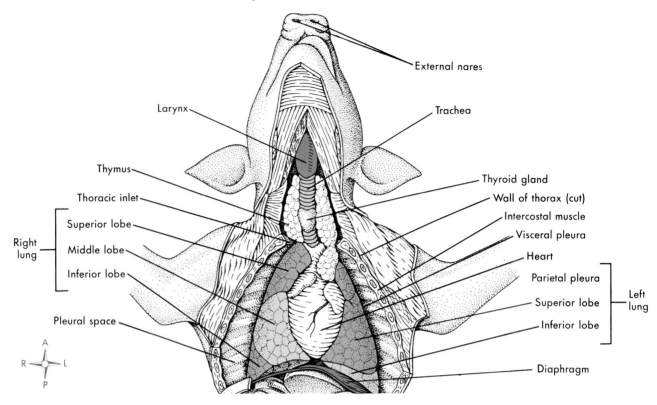

Figure 42-2 Respiratory anatomy of the fetal pig. Ventral view.

HINT → Figures found throughout Exercises 41 and 42 are useful in identifying many of the structures of the sheep pluck listed in this activity.

☐ 3 Identify the right and left **primary bronchi.** Notice that each extends to a different **lung,** where it splits to form several **secondary bronchi.**

☐ 4 Each secondary bronchus serves a distinct **lobe** of a lung. Identify the lobes in *each* of your specimen's lungs. How many are there in each lung? How does this compare to the human?

☐ 5 If your instructor directs you to do so, slice into the tissue of one of the lobes and remove a small section of tissue. Examine the internal structure of this tissue with a dissection microscope or hand lens to reveal air passages that constitute distal elements of the **bronchial tree** and gas-exchange tissues. You may make a deeper cut into the lobe to reveal other elements of the bronchial airways.

NAME _____ DATE _____ SECTION _____

LAB REPORT 42

Dissection: Respiratory System

A. Cat dissection checklist
- external nares
- nasal cavity
- nasal septum
- turbinates (conchae)
- pharynx
    - nasopharynx
    - oropharynx
    - laryngopharynx
- larynx
    - epiglottis
    - thyroid cartilage
    - cricoid cartilage
- trachea
- primary bronchi
- lung
- secondary bronchi
- lobes of the lungs
    - anterior lobe (left and right)
    - middle lobe (left and right)
    - posterior lobe (left and right)
    - mediastinal lobe (right only)
- wall of the thoracic cavity
    - intercostal muscles
- thoracic inlet
- parietal pleura
- visceral pleura
- pleural space
- diaphragm
- phrenic nerve

B. Fetal pig dissection checklist
- external nares
- nasal cavity
- nasal septum
- turbinates (conchae)
- pharynx
    - nasopharynx
    - oropharynx
    - laryngopharynx
- larynx
- trachea
- primary bronchi
- lung
- secondary bronchi
- lobes of the lungs
    - anterior lobe (left and right)
    - middle lobe (right only)
    - posterior lobe (left and right)
- wall of the thoracic cavity
- intercostal muscles
- thoracic inlet
- parietal pleura
- visceral pleura
- pleural space
- diaphragm

C. Sheep pluck dissection checklist
- larynx
    - epiglottis
    - thyroid cartilage
    - cricoid cartilage
    - other features visible?
- trachea
- primary bronchi
- lung
- secondary bronchi
- lobes of the lungs
    - how many (left lung)? _____
    - how many (right lung)? _____

LAB EXERCISE 43

Pulmonary Volumes and Capacities

Spirometry is the use of instrumentation to determine the basic **pulmonary volumes** and **pulmonary capacities.** Pulmonary volumes, or *lung volumes,* are the amounts of air that are moved in and out of the respiratory tract during different phases of the respiratory cycle. Pulmonary capacities are combinations of pulmonary volumes, as you will see. Spirometry is important in many applications: research, clinical situations, and exercise training.

In this exercise, you will use a **spirometer** to measure pulmonary volumes and capacities. Activities D and E present a computer-based spirometry technique that can be used as an alternate method.

Before you begin

❑ Read the appropriate chapter in your textbook.

❑ Set your learning goals. When you finish this exercise, you should be able to
- define the volumes and capacities of the respiratory tract and explain their functional significance
- demonstrate the use of a spirometer to determine pulmonary volumes and capacities
- understand the effects of exercise on pulmonary volumes and capacities

❑ Prepare your materials:
- wet or dry (hand-held) spirometer
- one-way valve and mouthpiece (for wet spirometer) or disposable mouthpieces (dry)
- series of bowls of disinfectant
- noseclip
- BIOHAZARD bag
- stopwatch or timer
- computer system (if computerized version is used)

❑ Read the directions and safety tips for this exercise **carefully** before starting any procedure.

A. Pulmonary volumes and capacities

Inspiration, or **inhalation,** is one of two major phases of the respiratory cycle. **Expiration,** or **exhalation,** is the other. The movement of air into and out of the respiratory tract during these phases is termed **ventilation** and is critical to the normal functioning of the entire body. As mentioned in the introduction, a spirometer is an instrument that measures the volume of air moved during the different phases of the respiratory cycle. This activity calls for the use of a *wet spirometer,* which is a vessel inverted in a container of water. A breathing tube communicates with the inverted vessel, which rises in the water as it fills with air. The rise of the inverted vessel is proportional to the volume exhaled into the instrument. A recording spirometer, as shown in Figure 43-2, produces a wavy **spirogram.** You will probably use a spirometer that shows the rise on a dial or linear scale. You may instead use a hand-held *dry spirometer,* which is a gauge that measures air volume as you exhale into it.

Before using the spirometer, you should practice breathing normally while wearing a noseclip (or pinching your nostrils shut). The noseclip is needed to prevent the natural tendency to breath through your nose, which invalidates the spirometry results.

> **SAFETY FIRST!** A one-way valve and mouthpiece used with a wet spirometer prevents inhalation of potentially contaminated air from inside the spirometer's air passages. However, be certain that the valves and mouthpieces are disinfected thoroughly by soaking them in a series of alcohol baths before reusing them. If a handheld spirometer is used, be sure to use a *new* disposable mouthpiece *and* disinfect the spirometer itself between uses.

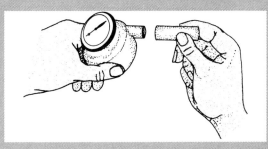

Figure 43-1 Insert a disposable mouthpiece over the stem of a handheld spirometer. Dispose of used mouthpieces in a BIOHAZARD container.

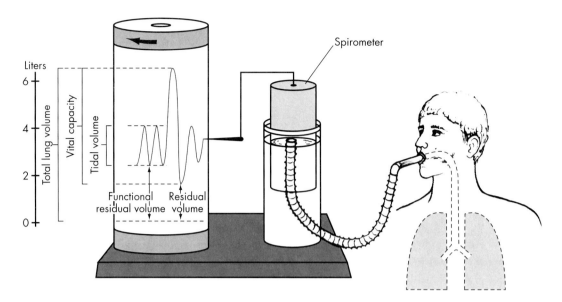

Figure 43-2 A recording wet spirometer. A recording spirometer yields a wave-form *spirogram*, but you will probably use a nonrecording spirometer that shows changing pulmonary volume as movements on a dial or scale.

❏ 1 **Tidal volume (TV)** is the volume of air moved into or out of the respiratory tract during normal breathing. Measure your TV as follows.
If using a one-way valve:
■ Place your mouth on the tube
■ Inhale normally
■ Exhale normally
■ Read the result and reset the apparatus to zero
If not using a one-way valve:
■ Inhale normally
■ Place your mouth on the tube
■ Exhale normally
■ Read the result and reset the apparatus to zero

> **HINT** → Pulmonary volumes are reported in either milliliters (ml) or liters (L). Remember that milliliters are a thousandth the size of liters. Therefore, a typical TV might be reported as 500 ml or 0.5 L. Be sure to indicate which units you are using when reporting your results.
>
> ■ Perform each test three times and average your results.
> ■ Typical volumes and capacities are given in the box on the facing page.

❏ 2 **Expiratory reserve volume (ERV)** is the volume of air that can be expired forcefully after a normal expiration. Measure the ERV as follows.
If using a one-way valve:
■ Place your mouth on the tube
■ Inhale normally
■ Exhale forcefully (to the maximum)
■ Read the result and subtract your TV—the difference is ERV; reset the apparatus to zero
If not using a one-way valve:
■ Inhale and exhale normally
■ At the end of the normal expiration, place your mouth on the tube and exhale forcefully
■ Read the result as ERV and reset the apparatus to zero

❏ 3 **Vital capacity (VC)** is the volume of air that can be expired after a forceful inspiration. Determine your vital capacity as follows.
If using a one-way valve:
■ Place your mouth on the tube
■ Inhale forcefully (to the maximum)
■ Exhale forcefully (to the maximum)
■ Read the VC value and reset the apparatus to zero
If not using a one-way valve:
■ Inhale forcefully (to the maximum)
■ Place your mouth on the tube
■ Exhale forcefully (to the maximum)
■ Read the VC value and reset the apparatus to zero

PULMONARY VOLUMES AND CAPACITIES

Volume	Definition	Typical value
Tidal volume (TV)	Volume moved into or out of the respiratory tract during a normal respiratory cycle	0.5 L
Inspiratory reserve volume (IRV)	Maximum volume that can be moved into the respiratory tract after a normal inspiration	3.0 L
Expiratory reserve volume (ERV)	Maximum volume that can be moved out of the respiratory tract after a normal expiration	1.1 L
Residual volume (RV)	Volume remaining in the respiratory tract after maximum expiration	1.2 L

Capacity	Formula	Typical value
Vital capacity (VC)	TV + IRV + ERV	4.6 L
Inspiratory capacity (IC)	TV + IRV	3.5 L
Functional residual capacity (FRC)	ERV + RV	2.3 L
Total lung capacity (TLC)	TV + IRV + ERV + RV	5.8 L

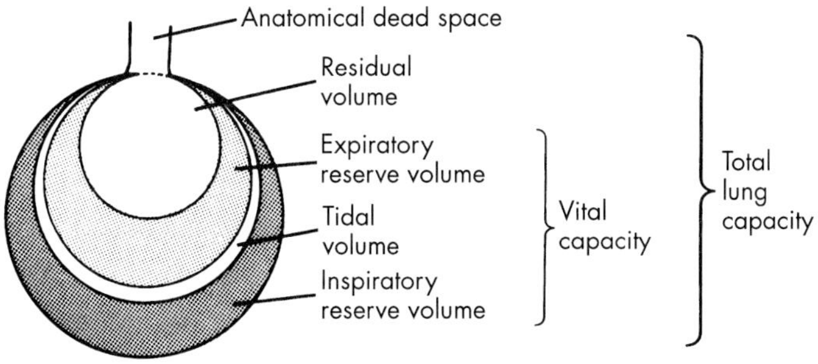

Figure 43-3 Diagram of pulmonary volumes and capacities.

❏ 4 **Inspiratory reserve volume (IRV)** is the volume of air that can be inspired forcefully after a normal inspiration. Because your spirometer is not really set up to do this, determine the IRV by subtracting TV and ERV from the vital capacity:

$$IRV = VC - (TV + ERV)$$

❏ 5 Analyze and interpret your results. Compare TV, ERV, VC, and IRV to the "textbook" values given in the box. How do you account for any differences?

B. Forced expiratory vital capacity

The **forced expiratory vital capacity (FVC)** test is useful in determining the nature of pulmonary conditions. In the FVC test, the subject maximally inhales then maximally exhales as rapidly as possible. The event is timed and a reading taken at exactly 1 second after the beginning of expiration. The 1-second volume is called the **forced expiratory volume** at 1 second, or **FEV_1**. The total amount expired, however long it takes, is the FVC (see Figure 43-4). If the FEV_1 is much less than 80% of the FVC, then airway obstruction is suspected. For example, airway constriction in asthma produces a low FEV_1/FVC percentage. The increased resistance to airflow in constricted airways prevents air from being expired rapidly.

The spirometry apparatus used here is not the best for measuring FEV_1 values. However, a rough estimate can be made by following the procedure described.

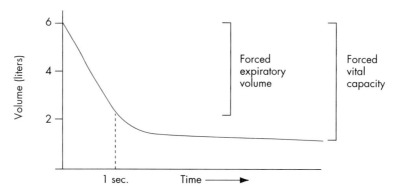

Figure 43-4 Spirogram of the forced expiratory vital capacity (FVC) test. In normal subjects, about 80% of the total forced vital capacity is exhaled during the first second of forced rapid expiration.

☐ 1 Designate a SUBJECT, a TIMER, and a RECORDER in your lab team. As the SUBJECT prepares to exhale into the spirometer tube, the TIMER stands ready to call "time" at exactly 1 second. The RECORDER will mark the spirometer scale at (or take note of) the volume indicated when "time" was called.

☐ 2 *If a one-way valve is used:*
 ▪ Have the SUBJECT place his or her mouth on the tube
 ▪ exhale as forcefully and rapidly as possible until the maximum expiration is reached
If a one-way valve is not used:
 ▪ Have the SUBJECT inhale maximally
 ▪ Have the SUBJECT place his or her mouth on the tube
 ▪ Have the SUBJECT exhale as forcefully and rapidly as possible until maximum expiration is reached

☐ 3 The TIMER calls "time" at 1 second after the beginning of exhalation, and the RECORDER marks or makes note of the spirometer volume at exactly that point. This volume is FEV_1. Record the volume at the end of forceful expiration. This volume is FVC.

HINT → This method is very inaccurate, but a reasonable approximation of FEV_1 can be obtained if you practice the procedure a few times. More accurate results can be obtained with recording spirometers or spirometers with FEV timing devices.

☐ 4 Determine the FEV_1/FVC percentage by using this equation:

$$FEV_1/FVC \% = (FEV_1 \div FVC) \times 100$$

Is the result 80% or above? Try the test on a subject with a known obstructive pulmonary condition. What are your results?

C. The effects of exercise

Try repeating Activity A, the determination of pulmonary volumes and capacities, with a person after exercise. After determining resting values, have the subject exercise heavily for 2 to 5 minutes. Immediately after exercise has stopped, measure the pulmonary volumes again. How have they changed? How do you account for the changes? Try measuring the volume after about 20 minutes of rest. Explain your results.

D. Using SPIROCOMP

The previous activities present some basic determinations of pulmonary volumes and capacities using traditional equipment. Activities D and E present a method that uses a traditional wet spirometer with computer-assisted data acquisition. Computer-assisted techniques have become the norm in clinical and research settings. These activities are based on Intelitool's SPIROCOMP apparatus for use with Apple-compatible and DOS-compatible personal computers. The SPIROCOMP program has the feature of collecting and saving group results so that they can be compared later. Your lab section may want to design an experiment to test a hypothesis. You can then use the program's built-in group data collection to gather your results.

> **SAFETY FIRST!** Heed the safety advice given previously in this exercise and in the SPIROCOMP manual.

The instructions given here parallel the tests outlined in the previous activities.

❏ 1 Use the arrow keys to select RUN SPIROCOMP or EXPERIMENT MENU at the main menu. If the spirometer has not been calibrated or if the chain has slipped and recalibration is required, the CALIBRATE SPIROCOMP routine will appear. When instructed by the program, lift the spirometer chain off the pulley by pulling up on the *weighted* side of the chain. Turn the pulley knob until the calibration number on the screen reads "0 (±5)." Replace the chain, making sure the pointer is at 0 liters. Press any key. Inflate to 5 liters *exactly* and press any key.

❏ 2 Answer the question about group records as you wish by pressing the key indicated on the screen. Next you will be asked information about the subject: initials (for identification), gender (M or F), age (in years), and height (in cm). Enter each value as asked, pressing (ENTER) after each entry. The data acquisition screen appears after you are finished.

❏ 3 Perform the ERV test: Put the mouthpiece in the subject's mouth (do not forget the noseclip). Press (E) while the subject inhales. The subject should follow these instructions, which appear on the screen:
- "Breathe normal [tidal] cycles" until told to stop.
- "Stop after [next] normal exhale"—wait for the next message.
- "Exhale fully [maximally]"—you will hear a bell sound.

The expiratory reserve volume (ERV) is automatically calculated for you.

> **HINT** → The SPIROCOMP manual contains additional information regarding control keys. Any test may be repeated, but only the last test is saved in the memory. Pressing the space bar will abort a test and the system will try to use the data already collected. Select (P) to print out data after a test, or series of tests, is complete. *Be sure to deflate the spirometer after each test.*

❏ 4 Perform the TV test: Press (T) while inhaling and "breathe normal [tidal] cycles" as directed. The system automatically determines average tidal volume (TV) after several cycles. If TV is high enough to fill the spirometer before the end of three cycles, press the space bar during inhalation of the second cycle.

❏ 5 Perform the VC test: Press (V) and follow the instructions telling the subject to inhale maximally, then exhale maximally. (Apple users press (V) as maximal exhalation begins.) Continue exhalation until a VC (vital capacity) graph begins to appear on the screen. This maneuver takes some practice.

> **HINT** → Notice that the program automatically calculates VC, IRV, FEV_1, and other values. The next section of this exercise explains the meaning of the graph's elements. Press (P) at any data screen to obtain a printout of the graph.

❏ 6 When you are ready to test the next subject press (N) and proceed as before. Continue until all subjects have been tested.

❏ 7 After the last test, press the (Esc) key to return to the main menu. Select DATA REVIEW (GRAPH) or REVIEW/ANALYZE MENU and GRAPH AVERAGES. A graph of averaged data will appear. Obtain a hardcopy printout by pressing (P). Press (Esc) to return to the main menu or REVIEW/ANALYZE MENU and select DATA REVIEW (SPREADSHEET) or VIEW SPREADSHEET. Apple users select GROUP AVERAGE or another report, then press (P) to obtain a printed copy.

E. The SPIROCOMP graph

Figure 43-5 approximates that seen in the SPIROCOMP data screen. It is a bar chart representing these values (from left):
- **ERV**—Expiratory reserve volume (value in liters is printed below this and other bars)
- **TV**—Tidal volume (averaged from three cycles)
- **cIRV**—Computed inspiratory reserve volume; recall that IRV is calculated by the equation:

$$cIRV = VC - (TV + ERV)$$

- **cVC**—computed vital capacity; notice how the bars are stacked, to represent:

$$cVC = ERV + TV + cIRV$$

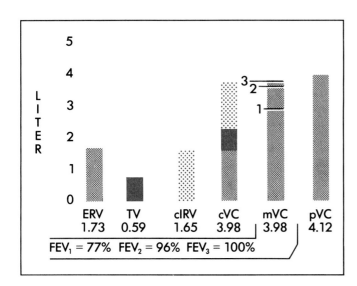

Figure 43-5 Data graph similar to that in the SPIROCOMP screen.

mVC—Measured vital capacity; this is the result from the VC test; the forced expiratory volumes (FEVs) at 1 second (FEV_1), 2 seconds (FEV_2), and 3 seconds (FEV_3) are given—actually, they are not volumes but are FEV_x/FVC percentage is expected to be about 80%.

pVC—Predicted vital capacity; this value is based on research data from subjects of the same size, age, and gender. (Use this value rather than the predicted vital capacity given in Lab Report 43.)

NAME_____ DATE_____ SECTION_____

LAB REPORT 43

Pulmonary Volumes and Capacities

Resting:

Volume or capacity	Expected value[1]	Actual value[2]	Difference	Interpretation
Tidal volume (TV)				
(Average)	0.5 L (500 ml)			
Expiratory reserve volume (ERV)				
(Average)	1.1 L (1100 ml)			
Vital capacity (VC)				
(Average)	4.6 L (4600 ml)			
Inspiratory reserve volume (IRV)	3.0 L (3000 ml)			

1. Predicted values are general approximations. Factors such as body surface area, sex, age, and athletic conditioning affect predicted values. If a "predicted vital capacity chart" is available, use the figure from that rather than the figure given here.
2. Be sure to indicate units for each reported value (milliliters or liters).

Trial	FEV_1	FVC	FEV_1/FVC %	Interpretation
1				
2				
3				
4				
Best value				

After exercise (0 min):

Volume or capacity	Expected value[1]	Actual value	Difference	Interpretation
TV				
ERV				
VC				
IRV				

1. Use the results from the first table on the previous page as your expected results. Be sure to use the correct units of volume when reporting your values.

After exercise (20 min):

Volume or capacity	Expected value[1]	Actual value	Difference	Interpretation
TV				
ERV				
VC				
IRV				

1. Use the results from the first table on the previous page as your expected results. Be sure to use the correct units of volume when reporting your values.

Discuss your results (comparing resting values to values after exercise):

LAB EXERCISE 44

Digestive Structures

The **digestive tract** is a series of hollow organs through which food passes: **mouth, pharynx, esophagus, stomach, small intestine,** and **large intestine.** Each portion is specialized for one or more aspects of the three major functions of the **digestive system:** *secretion, digestion,* and *absorption.* Accessory organs such as the **salivary glands, liver, gallbladder,** and **pancreas** have ducts that lead into the digestive tract and thus support digestive function.

This exercise challenges you to explore the structure and some of the essential functions of the digestive system.

Before you begin

❑ Read the appropriate chapter in your textbook.

❑ Set your learning goals. When you finish this exercise, you should be able to
- describe the structure of digestive organs and locate them in models and charts
- identify the principal function of each major digestive organ
- describe the basic histology of the gastrointestinal wall

❑ Prepare your materials:
- model of the human torso (dissectible) or chart
- human skull (with teeth intact)
- model or specimen of a tooth section
- microscope
- prepared microslides:
  Stomach wall c.s.
  Small intestine wall c.s.
  Salivary gland c.s.

❑ Read the directions and safety tips for this exercise **carefully** before starting any procedure.

A. Wall of the digestive tract

The wall of the digestive tract forms a hollow tube continuous with the external environment. Food passes along the tube in a process usually termed **motility.** The wall of the tract is composed of four layers (tunics). They are

- **Mucosa**—Inner lining, consisting of mucous epithelium, connective tissue (*lamina propria*), and a thin layer of smooth muscle (*muscularis mucosae*).
- **Submucosa**—Thick layer of connective tissue just superficial to the mucosa; the submucosa includes blood vessels, a network of nerve fibers called the *submucosal plexus,* and *submucosal glands.*
- **Muscularis**—Layers of smooth muscle surrounding the submucosa—usually an inner *circular layer* and an outer *longitudinal layer* (the stomach also has a middle *oblique layer*); the muscularis has a network of nerve fibers called the *myenteric plexus.*
- **Serosa (adventitia)**—Most superficial layer of the digestive wall, composed entirely of connective tissue (adventitia) or of connective tissue fused to serous epithelial tissue (serosa); the serous membrane extends away from the GI wall in folds called *mesenteries;* digestive glands with ducts that lead through the GI wall and into the *lumen* (hollow part) of the GI tract may lay outside the serosa or adventitia.

❑ 1 Locate the digestive tract in a dissectible model of the human torso. Some portions may have the wall cut in a cross section. If so, try to identify each of the four layers previously described. How does each layer differ from organ to organ?

> **SAFETY FIRST!** Observe the usual precautions when using the microscope and prepared slides.

❑ 2 Obtain a prepared slide of a sectioned stomach wall and examine it with your microscope. Identify the four layers of the wall. Examine the mucosa. Notice that it is composed of simple columnar epithelium that forms deep **gastric pits.** They are lined with **surface mucous cells,** which you may recognize as goblet cells. Follow a pit's lumen to the base, or **gastric gland.** You may see more goblet cells called **mucous neck cells.** Some of the very darkly stained cells are **chief cells.** The

383

very pale cells are **parietal cells.** What is the function of each of these cells?

❑ 3 Obtain a prepared slide of a small intestine section. Identify the four layers of the wall. In a true cross section, you cannot tell that the simple columnar epithelium forms large **circular folds,** but you may have noticed them in the model. Each circular fold has fingerlike projections called **villi** that are visible in your specimen.

❑ 4 Obtain a prepared slide of a salivary gland section. Identify the mucus-producing glandular cells by their characteristic light appearance that is produced by the transparent nature of the mucus that fills them. Try to find the cuboidal duct cells that form rings in a cross-sectioned specimen.

> **HINT** → Refer to LABORATORY REFERENCE Plates 87 through 89 for color micrographs of the specimens observed in this activity.

B. The upper digestive tract

Use a dissectible model or other aid to identify the digestive structures indicated.

❑ 1 The **mouth,** or **oral cavity,** is the beginning of the one-way human digestive tract. Locate these features:
- **Lips**
- **Buccinator muscle**—Muscle that forms most of the cheek walls
- **Tongue**—Mass of muscle covered with mucous epithelium containing *taste buds;* the **frenulum** is a fold of tissue anchoring its underside to the floor of the mouth
- **Hard palate**—Anterior, bony portion of the mouth's roof
- **Soft palate**—Posterior, soft portion of the mouth's roof (it ends in the conelike **uvula**)
- **Palatine tonsils**—Lymphatic structures on the posterior, lateral aspect of the mouth cavity

❑ 2 The 32 **teeth** of the normal adult are special features of the mouth cavity. Identify these tooth types in a skull or model (each type has a representative on each side, left and right, of each jaw, upper and lower):
- **Incisors**—*Central* and *lateral* cutting teeth
- **Canine**—Tearing tooth
- **First and second premolars**—Tearing and cutting teeth, each with two points or *cusps*
- **First, second, and third molars**—Grinding teeth with three flattened cusps

Find each of these features on a sectioned tooth or tooth model:
- **Crown**—Superior portion of the tooth
- **Neck**—Narrow portion, below the crown
- **Root**—Portion of the tooth within the **alveoli,** or spaces, of the mandible and maxillae
- **Pulp cavity**—Central space of the tooth filled with **pulp,** which is connective tissue with nerves and vessels
- **Dentin**—Hard, calcified tissue that forms most of the tooth shell
- **Enamel**—Hard, nonliving, mineralized substance that covers the dentin of the crown
- **Gingiva**—Mucous epithelial covering of the *alveolar ridges* of the jaws
- **Periodontal membrane**—Fibrous lining of the alveolar walls (**periodontal ligaments** project from this membrane to the root of the tooth, keeping it in place)

❑ 3 The **salivary glands** are exocrine glands whose ducts empty into the mouth cavity. **Saliva** produced by the glands is a watery solution containing mucus, mineral ions, and digestive enzymes. Identify these pairs of salivary glands in a model:
- **Parotid glands**—The largest pair, just anterior to each ear, each secretes a serous (watery) solution that flows through a **parotid duct** leading to the cheek wall across from the second upper molar.
- **Submandibular glands**—Medial to the angle of the mandible, these glands produce mostly serous fluid and some mucous fluid. Their ducts empty onto the mouth floor, just lateral to the frenulum.
- **Sublingual glands**—Smallest pair of salivary glands in the floor of the mouth, they produce mostly mucous fluid, which flows into the mouth through 10 or more small ducts in the mouth's floor.

❑ 4 The **pharynx,** or throat, is divided into three regions:
- **Nasopharynx**
- **Oropharynx**
- **Laryngopharynx**

❑ 5 The **esophagus** is a long, muscular tube connecting the pharynx to the stomach. Upper and lower **esophageal sphincters** may be represented in the model you are using.

❑ 6 The **stomach** is an enlargement of the digestive tract just inferior to the diaphragm, at the end of the esophagus. Locate these stomach features in a model or chart:
- **Cardiac opening**—Opening from the esophagus into the stomach region called the **cardiac region.**

Wall of the GI Tract

COLORING EXERCISE Using colored pens or pencils, shade in the figure and accompanying labels in contrasting colors of your choice as indicated by the red numerals.

SEROSA
VISCERAL PERITONEUM₁
CONNECTIVE TISSUE OF SEROSA₂
MUSCULARIS
LONGITUDINAL LAYER₃
CIRCULAR LAYER₄
SUBMUCOSA₅
MUCOSA
LAMINA PROPRIA₆
MUCOUS EPITHELIUM₇

LYMPH NODULE₈
GLAND IN SUBMUCOSA₉
MUCOSAL GLAND₁₀
SUBMUCOSAL PLEXUS₁₁
MYENTERIC PLEXUS₁₂
NERVE₁₃
BLOOD VESSEL₁₄
MESENTERY₁₅
GLAND OUTSIDE OF TRACT₁₆

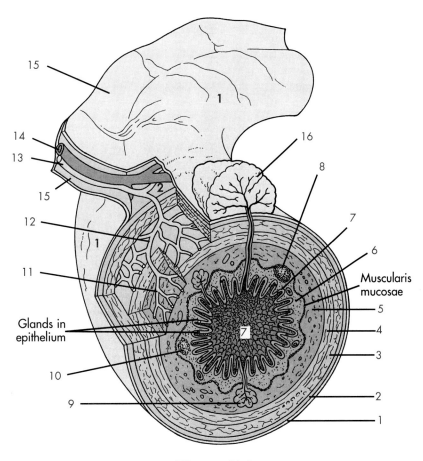

Figure 44-1

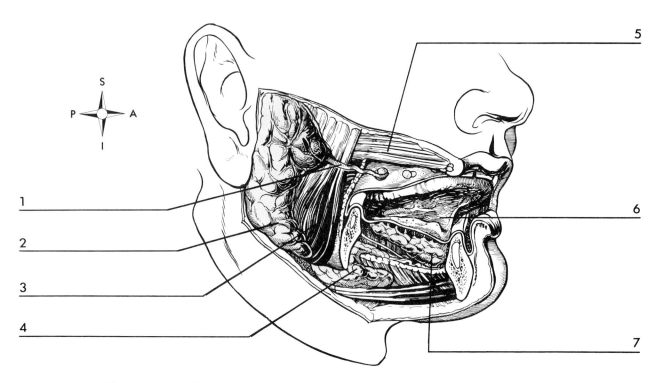

Figure 44-2 Identify the features of the mouth indicated by the label lines.

- **Fundus**—Region of the stomach superior to the cardiac opening.
- **Body**—Main portion of the stomach, forming a **greater curvature** and a **lesser curvature.**
- **Pyloric region**—Portion of the stomach near the **pyloric opening** into the small intestine and the **pyloric sphincter** that surrounds the opening.
- **Rugae**—Large folds of the mucosa.

C. The lower digestive tract

❑ 1 The **small intestine** is a long, narrow tube that folds to fill a large portion of the abdominal cavity. Find these features of the small intestine:
- **Duodenum**—Short (±25 cm), C-shaped beginning of the small intestine.
- **Jejunum**—Long (±2.5 m) middle section of the small intestine.
- **Ileum**—Very long (±3.5 m) end section of the small intestine.
- **Common bile duct**—Duct that empties into the duodenum.
- **Ileocecal junction**—Junction of the ileum with the large intestine, featuring an **ileocecal sphincter** and a one-way **ileocecal valve.**

❑ 2 The **liver,** a large gland in the superior right abdominal cavity, produces **bile** that is secreted into the duodenum. Two **hepatic ducts** conduct bile from the liver and fuse to form a **common hepatic duct.** The common hepatic duct joins the two-way **cystic duct** from the **gallbladder** (just inferior to the liver) to form the **common bile duct.** The common bile duct empties into the duodenum.

❑ 3 The **pancreas** is a gland cradled in the C of the duodenum that secretes digestive fluids. Exocrine pancreatic secretions flow through the **pancreatic duct** that joins the common bile duct.

❑ 4 The **large** intestine includes these features:
- **Cecum**—A blind sac extending inferiorly from the ileocecal junction; attached to the posterior wall of the cecum is a narrow, worm-like blind sac called the **appendix.**
- **Colon**—A long (±1.6 m) portion of the large intestine that is subdivided into **ascending colon, transverse colon, descending colon,** and **sigmoid colon;** the longitudinal layer of the colon's muscularis forms three bands of muscle called the **teniae coli;** the wall of the colon forms **haustra,** or pouches; the exterior aspect of the colon wall has fatty attachments called **epiploic appendages.**
- **Rectum**—Short, straight, muscular tube at the end of the sigmoid colon.
- **Anal canal**—Short (±2.5 cm), muscular canal at the end of the rectum that exits to the outside of the body by way of the **anus;** it features a smooth muscle **internal anal**

sphincter at its superior end and a skeletal muscle **external anal sphincter** at its inferior end.

☐ 5 The organs and inner wall of the abdominal cavity are covered with a serous membrane called the **peritoneum.** The portion that covers the organs is called the **visceral peritoneum,** and the portion that lines the cavity walls is called the **parietal peritoneum.** Extensions of the peritoneum, composed of a thin sheet of connective tissue sandwiched between two layers of serous membrane, support the visceral organs. These extensions are called **mesenteries.** Find these mesenteries in your model:

- **Lesser omentum**—It is located between the lesser curvature of the stomach and the liver (and diaphragm).
- **Greater omentum**—It connects the stomach's greater curvature to the transverse colon and posterior abdominal wall. It has a long, fatty, double fold that forms an apronlike pocket called the **omental bursa** over the front of the viscera.

The Mouth and Teeth

LIPS₁
BUCCINATOR
   MUSCLE₂
TONGUE₃
HARD PALATE₄
SOFT PALATE₅
UVULA₆
PALATINE TONSILS₇
OROPHARYNX₈

TEETH
CENTRAL INCISOR₉
LATERAL INCISOR₁₀
CANINE₁₁
FIRST PREMOLAR₁₂
SECOND
   PREMOLAR₁₃
FIRST MOLAR₁₄
SECOND MOLAR₁₅
THIRD MOLAR₁₆

TOOTH FEATURES
CROWN₁₇
NECK₁₈
ROOT₁₉
PULP CAVITY₂₀
PULP₂₁
DENTIN₂₂
ENAMEL₂₃
GINGIVA₂₄
PERIODONTAL
   MEMBRANE₂₅
PERIODONTAL
   LIGAMENTS₂₆
ALVEOLAR RIDGE₂₇

COLORING EXERCISE Using colored pens or pencils, shade in the figure and accompanying labels in contrasting colors of your choice as indicated by the red numerals.

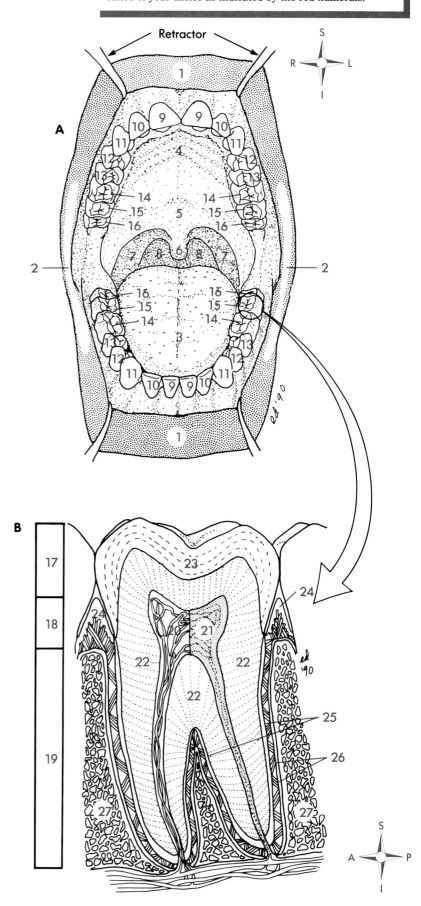

Figure 44-3

The Digestive System

ORAL CAVITY 1
PAROTID GLAND 2
SUBMANDIBULAR GLAND 3
SUBLINGUAL GLAND 4
PHARYNX 5
ESOPHAGUS 6
CARDIAC SPHINCTER 7
FUNDUS 8
CARDIAC REGION 9
BODY OF STOMACH 10
PYLORIC REGION 11
PYLORIC SPHINCTER 12
DUODENUM 13
JEJUNUM 14
ILEUM 15
COMMON BILE DUCT 16
ILEOCECAL VALVE 17
LIVER 18
COMMON HEPATIC DUCT 19
CYSTIC DUCT 20
GALLBLADDER 21
PANCREAS 22
PANCREATIC DUCT 23
CECUM 24
APPENDIX 25
ASCENDING COLON 26
TRANSVERSE COLON 27
DESCENDING COLON 28
SIGMOID COLON 29
RECTUM 30
ANAL CANAL 31

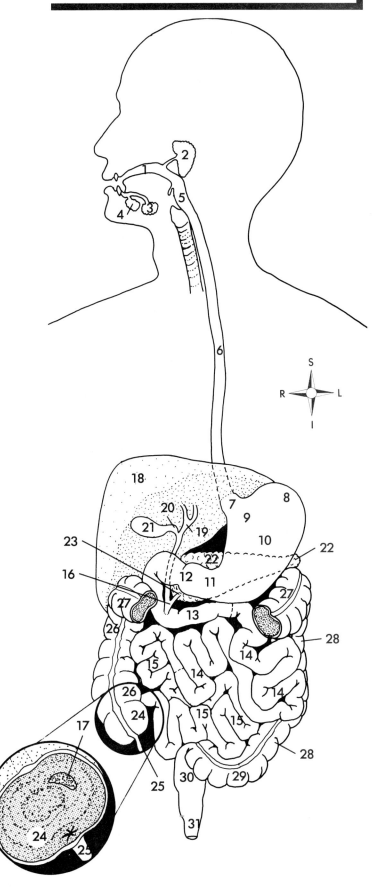

COLORING EXERCISE Using colored pens or pencils, shade in the figure and accompanying labels in contrasting colors of your choice as indicated by the red numerals.

Figure 44-4 Note: Many organs are shown separated from one another. In life, many digestive organs overlap one another when viewed from this perspective.

FLUOROSCOPY

Fluoroscopy is a type of radiographic imaging that uses a fluorescent screen rather than film. X-rays pass through the patient and are absorbed by the screen's coating. The energy is reemitted by the coating as light energy. Dense tissues cast a dark shadow on the screen. Therefore, continuous radiation allows a continuous, moving image in a moving patient.

Fluoroscopy with the contrast medium **barium sulfate** is used to visualize portions of the digestive tract. A *barium swallow,* or **upper GI series,** images the stomach and portions of the esophagus and small intestine. A *barium enema* (**lower GI series**) images the rectum and colon. A **hernia** of the digestive tract, where a portion of the tract is pushed out of a weak spot in the abdominopelvic wall, can be detected this way. **Ulcers,** holes in the digestive wall, can also be assessed in this manner.

Because fluoroscopy allows observation of the body in motion, it can be used to see the muscular movements of the digestive tract as the barium passes through it. A person can push on the abdomen from the outside to see how digestive organs withstand pressure similar to compression of abdominal muscles during heavy lifting.

The figures that follow are regular x-ray photographs taken of an upper and lower GI series. A fluoroscope image would be reversed (black on white, not the white on black seen here). Try to identify the labeled parts.

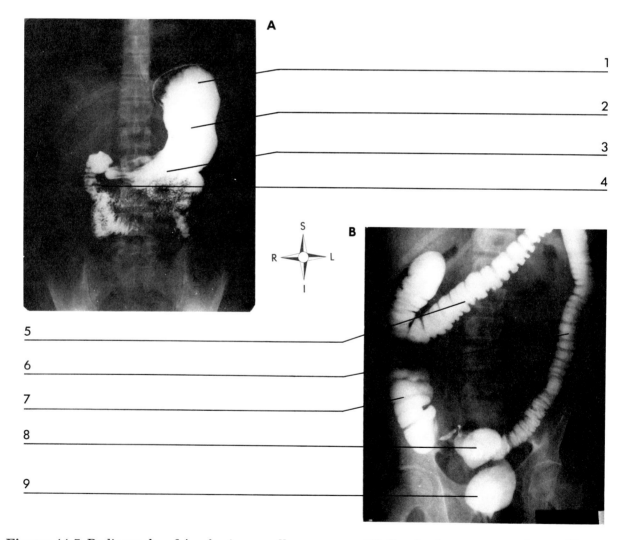

Figure 44-5 Radiographs of **A,** a barium swallow, or upper GI, **B,** a barium enema, or lower GI. Note: Fluoroscope images appear black on white, the reverse of the white-on-black images shown here.

NAME _____ DATE _____ SECTION _____

LAB REPORT 44

Digestive Structures

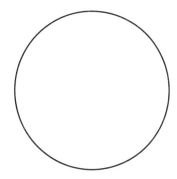

Specimen: *stomach wall c.s.*

Total Magnification: _____

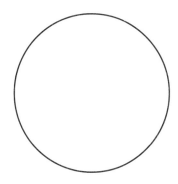

Specimen: *small intestine wall c.s.*

Total Magnification: _____

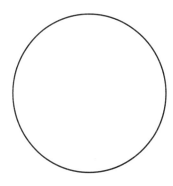

Specimen: *salivary gland c.s.*

Total Magnification: _____

Structure	Model	Function(s)
Oral cavity	☐	
Buccinator muscle	☐	
Tongue	☐	
Palate: hard, soft	☐	
Uvula	☐	
Palatine tonsils	☐	
Incisors: central, lateral	☐	
Canine teeth	☐	
Premolars: first, second	☐	
Molars: first, second, third	☐	
Parotid glands	☐	
Submandibular glands	☐	
Sublingual glands	☐	
Pharynx	☐	
Esophagus	☐	
Cardiac opening, region	☐	
Fundus	☐	

Structure	Model	Function(s)
Body of stomach	☐	
Stomach curvatures: greater, lesser	☐	
Pyloric opening, region	☐	
Rugae	☐	
Duodenum	☐	
Jejunum	☐	
Ileum	☐	
Circular folds	☐	
Common bile duct	☐	
Ileocecal junction: valve, sphincter	☐	
Common hepatic duct	☐	
Cystic duct	☐	
Pancreas	☐	
Pancreatic duct	☐	
Cecum	☐	
Appendix	☐	
Ascending colon	☐	
Transverse colon	☐	
Descending colon	☐	
Rectum	☐	
Anal canal, anus	☐	
Lesser omentum	☐	
Greater omentum	☐	

Figure 44-2

_____ 1
_____ 2
_____ 3
_____ 4
_____ 5
_____ 6
_____ 7

Figure 44-5

_____ 1
_____ 2
_____ 3
_____ 4
_____ 5
_____ 6
_____ 7
_____ 8
_____ 9

Fill-in

_____ 1
_____ 2
_____ 3
_____ 4
_____ 5
_____ 6
_____ 7

Fill-in (complete each statement with the correct term)

1. The stomach's mucosa forms large folds called __?__.

2. The two ducts that exit the liver and join to form the common hepatic duct are called __?__ ducts.

3. The colon is divided into __?__ sections.

4. __?__ are fatty extensions of the colon's outer wall.

5. A substance called __?__ covers the dentin of the tooth's crown.

6. The __?__ is the portion of the pharynx posterior to the mouth cavity.

7. The __?__ sphincter prevents stomach contents from flowing back into the esophagus.

LAB EXERCISE 45

Dissection: Digestive System

In this exercise, you will continue your study of digestive anatomy by dissecting a whole preserved specimen.

Activity A provides directions for studying the digestive anatomy of the cat. Activity B is an alternate activity, providing directions for studying the digestive anatomy of the fetal pig.

Before you begin

❏ Read the appropriate chapter in your textbook.

❏ Set your learning goals. When you finish this exercise, you should be able to
- dissect the digestive anatomy of a preserved cat or fetal pig
- identify the primary organs of the digestive tract in a dissected mammalian specimen
- identify the accessory organs of digestion in a dissected mammalian specimen

❏ Prepare your materials:
- preserved cat or fetal pig
- dissection tools and trays
- storage container (if specimen is to be reused)

❏ Read the directions and safety tips for this exercise **carefully** before starting any procedure.

> **SAFETY FIRST!** Observe the usual precautions when working with a preserved specimen. Heed the safety advice accompanying preservatives used with your specimen. Use protective gloves while handling your specimen. Avoid injury with dissection tools. Dispose of or store your specimen as instructed.

A. Digestive anatomy of the cat

This activity challenges you to identify the major digestive structures of the cat. Assuming you have already removed the cat's skin (Exercise 10) and opened the abdominopelvic cavity (Exercise 40), you simply have to move the viscera out of the way—without cutting or removing them—to see the structures indicated in this activity.

> **HINT** → Figure 45-1 illustrates many of the structures of the cat listed in this activity. Useful information can also be found in the LABORATORY REFERENCE Plate 75.

❏ 1 Identify the **peritoneum**, which is the serous membrane that lines the wall of the abdominopelvic cavity (*parietal peritoneum*) and covers the organs within the cavity (*visceral peritoneum*). Identify these extensions or folds of the peritoneum:
- **Greater omentum**
- **Mesentery**

> **HINT** → The *greater omentum* is a large, fatty fold of the peritoneum that lies like an apron over the ventral surfaces of the abdominopelvic organs. You must lift or cut this "apron" out of the way to see most of the digestive organs described in the following steps.

❏ 2 Identify the **liver** with its *left lobes* and *right lobes*. Locate the **gallbladder** at the posterior surface of the right lobes of the liver. Trace the **cystic duct** from the gall bladder and the **hepatic ducts** from the lobes of the liver. Note where they join to form the **common bile duct**.

❏ 3 Locate the curved **stomach**. Identify these regions of the stomach:
- **Fundus**
- **Body**
- **Pylorus**

Cut the stomach open and identify the mucosal folds called **rugae**.

ANATOMICAL ATLAS OF THE CAT
(DIGESTIVE STRUCTURES)

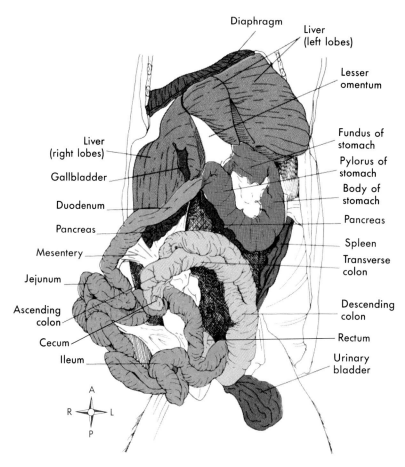

Figure 45-1 Digestive anatomy of the cat. Abdominopelvic cavity, ventral view.

❏ 4 Identify the **pancreas.** It is comprised of two lobes of glandular tissue, one that is embedded in the *mesentery* along the duodenum and one that extends along the edge of the stomach.

❏ 5 The **small intestine** extends posteriorly from the pylorus of the stomach. Identify these portions of the small intestine:
- **Duodenum**
- **Jejunum**
- **Ileum**

❏ 6 The **large intestine** attaches to the distal end of the ileum of the small intestine. Locate these portions of the large intestine:
- **Cecum**
- **Ascending colon**
- **Transverse colon**
- **Descending colon**
- **Rectum**
- **Anus**

B. Digestive anatomy of the fetal pig

This activity challenges you to identify the major digestive structures of the fetal pig. Assuming you have already removed the animal's skin (Exercise 10) and opened the abdominopelvic cavity (Exercise 40), you simply have to move the viscera out of the way—without cutting or removing them—to see the structures indicated in this activity.

> **HINT** → Figure 45-2 illustrates many of the structures of the fetal pig listed in this activity. Useful information can also be found in the LABORATORY REFERENCE Plates 76, 85, and 86.

❏ 1 Identify the **peritoneum,** which is the serous membrane that lines the wall of the abdominopelvic cavity (*parietal peritoneum*) and covers the organs within the cavity (*visceral peritoneum*). Identify these extensions or folds of the peritoneum:
- **Greater omentum**
- **Mesentery**

> **HINT** → The *greater omentum* is a large, fatty fold of the peritoneum that lies like an apron over the ventral surfaces of the abdominopelvic organs. You must lift or cut this "apron" out of the way to see most of the digestive organs described in the following steps.

❏ 2 Identify the **liver** with its *left lobes* and *right lobes*. Locate the **gallbladder** at the posterior surface of the right lobes of the liver. Trace the **cystic duct** from the gallbladder and the **hepatic ducts** from the lobes of the liver. Note where they join to form the **common bile duct.**

❏ 3 Locate the curved **stomach.** Identify these regions of the stomach:
- **Fundus**
- **Body**
- **Pylorus**

Cut the stomach open and identify the mucosal folds called **rugae.**

❏ 4 Identify the **pancreas,** a mass of glandular tissue that lies within the curve of the duodenum of the small intestine.

❏ 5 The **small intestine** extends posteriorly from the pylorus of the stomach. Identify these portions of the small intestine:
- **Duodenum**
- **Jejunum**
- **Ileum**

❏ 6 The **large intestine** attaches to the distal end of the ileum of the small intestine. There are two main portions of the large intestine. The proximal portion is a coiled loop called the **spiral colon.** The distal portion is the **rectum,** which terminates at the **anus.**

ANATOMICAL ATLAS OF THE FETAL PIG
(DIGESTIVE STRUCTURES)

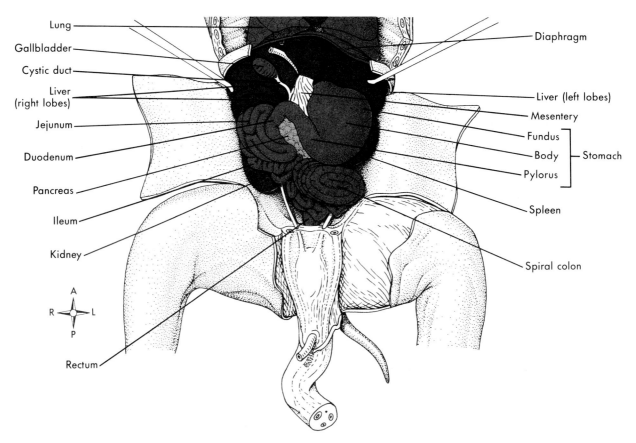

Figure 45-2 Digestive anatomy of the fetal pig. Abdominopelvic cavity, ventral view.

NAME _____ DATE _____ SECTION _____

LAB REPORT 45

Dissection: Digestive System

A. Cat dissection checklist
- ❏ parietal peritoneum
- ❏ visceral peritoneum
- ❏ greater omentum
- ❏ mesentery
- ❏ mesocolon
- ❏ liver
  - ❏ left lobes
  - ❏ right lobes
- ❏ gallbladder
- ❏ cystic duct
- ❏ hepatic ducts
- ❏ common bile duct
- ❏ stomach
  - ❏ fundus
  - ❏ body
  - ❏ pylorus
  - ❏ rugae
- ❏ pancreas
- ❏ small intestine
  - ❏ duodenum
  - ❏ jejunum
  - ❏ ileum
- ❏ large intestine
  - ❏ cecum
  - ❏ ascending colon
  - ❏ transverse colon
  - ❏ descending colon
  - ❏ rectum
  - ❏ anus

B. Fetal pig dissection checklist
- ❏ parietal peritoneum
- ❏ visceral peritoneum
- ❏ greater omentum
- ❏ mesentery
- ❏ liver
  - ❏ left lobes
  - ❏ right lobes
- ❏ gallbladder
- ❏ cystic duct
- ❏ hepatic ducts
- ❏ common bile duct
- ❏ stomach
  - ❏ fundus
  - ❏ body
  - ❏ pylorus
  - ❏ rugae
- ❏ pancreas
- ❏ small intestine
  - ❏ duodenum
  - ❏ jejunum
  - ❏ ileum
- ❏ large intestine
  - ❏ spiral colon
  - ❏ rectum
  - ❏ anus

LAB EXERCISE 46

Enzymes and Digestion

Digestion refers to the breaking down of complex nutrients into simpler ones. Digestion involves mechanical processes that simply break the nutrients into smaller chunks and chemical processes that split large molecules into smaller molecules. In this exercise, you will look at examples of both types of digestive processes. As an example of a mechanical process, you will use **bile salts** secreted by the liver to break apart lipid droplets. You will use pancreatic **enzymes** to chemically break apart complex carbohydrate and lipid molecules. Enzymes are proteins that catalyze biochemical reactions.

Before you begin

❑ Read the appropriate chapter in your textbook.

❑ Set your learning goals. When you finish this exercise, you should be able to
- describe the action of pancreatic enzymes on carbohydrates and lipids
- explain the effects of temperature on enzyme activity
- perform tests to determine the presence of starch and sugars
- demonstrate the action of bile on lipids

❑ Prepare your materials:
- distilled water
- pancreatin powder (and spatula)
- 1% starch solution
- 1% maltose solution
- Benedict's reagent
- ice water bath (with test tube rack)
- warm water bath (37°C, with tube rack)
- hot plate
- Lugol's reagent
- disposable 1 ml measuring droppers
- test tubes (approx. 10 ml), tube caps
- test tube racks and tongs
- wax pencils
- beakers (500 ml)
- bile salt powder (and spatula)
- vegetable oil (dyed with Sudan B)
- litmus cream

❑ Read the directions and safety tips for this exercise **carefully** before starting any procedure.

A. Observing carbohydrate digestion

As you know, carbohydrates are molecules composed of carbon, hydrogen, and oxygen arranged in chemical units called **saccharides. Polysaccharides** are complex carbohydrates, such as *starch* and *glycogen,* composed of many saccharide units. **Disaccharides** such as *sucrose, lactose,* and *maltose* are types of complex sugar composed of two saccharide units. **Monosaccharides** such as the simple sugars *glucose (dextrose), galactose,* and *fructose* are single saccharide units. Because the digestive tract absorbs carbohydrates only in their simplest form, monosaccharides, digestive processes must break polysaccharides and disaccharides into smaller units. Figure 46-1 presents a simplified scheme of how carbohydrates are digested in the human. **Amylases,** enzymes that break some of the bonds holding polysaccharides together, are found in the saliva and in pancreatic secretions. The disaccharides formed after the amylases have had their effect are then acted on by **disaccharidase** enzymes to yield monosaccharides that can be absorbed into the blood.

In this exercise, you will use an extract of pancreatic secretions called *pancreatin*. Pancreatin contains pancreatic amylase, as well as pancreatic **lipase** and pancreatic **proteolytic** (protein-digesting) enzymes.

> **SAFETY FIRST!** Ask your lab instructor to demonstrate the proper procedure for handling chemicals and glassware and for dealing with spills and other accidents. Wear protective clothing and eyewear. Never point the open end of a test tube toward a person. Be careful to avoid contaminating chemicals by using a dropper or spatula for more than one type of substance.

❑ 1 First, practice the tests used to analyze solutions for the presence of starch and its sugar products:
- **Lugol's test**—Lugol's test uses *Lugol's reagent* (an iodine solution) to test for the presence of starch. If Lugol's iodine turns

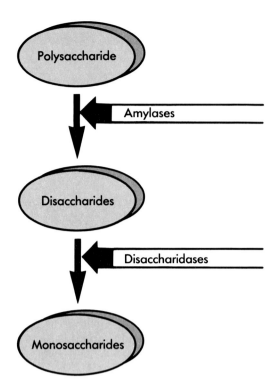

Figure 46-1 A simple scheme of carbohydrate digestion.

from its normal orange color to a deep blue-black, starch is present. Mark a test tube **1a,** and put in 2 ml (two full droppers) of starch solution. Add a few drops of Lugol's reagent. Notice the color change. Anytime you see this result for Lugol's test, you have a positive result (meaning starch *is* present). Mark another tube **1b** and put in 2 ml distilled water and a few drops of Lugol's reagent. The solution should be a shade of orange, which is a negative result (meaning starch *is not* present). Save the tubes and their contents for later reference.

> **SAFETY FIRST!** Be careful of electrical and heat hazards when doing Benedict's test. Review safety procedures with the instructor before beginning this test.

- **Benedict's test**—Benedict's test uses Benedict's reagent to determine whether a solution contains certain types of sugar. Mark a test tube **2a** and add 2 ml of maltose solution. To that add 2 ml of Benedict's reagent. In a tube marked **2b** put 2 ml of distilled water and 2 ml of Benedict's reagent. Place each tube in a beaker of boiling water on a hot plate for 3 minutes. At the end of 3 minutes, remove the tubes and place them in your rack. Tube **2a** should show a positive result, or a change from blue to another color, meaning sugar is present. Tube **2b** is an example of a negative result, with the contents remaining blue. Save these tubes for later reference.

❏ 2 Mark eight test tubes: **3a, 3b, 4a, 4b, 5a, 5b, 6a,** and **6b**. To each add 2 ml of starch. Next, simultaneously add a small amount of pancreatin (just enough powder to cover the end of the spatula) to each tube. Immediately:
- Place tubes **3a** and **3b** in an ice water bath (0° to 1°C)
- Place tubes **4a** and **4b** in a rack at room temperature (20° to 25°C)
- Place tubes **5a** and **5b** in a warm water bath set at body temperature (37°C)
- Place tubes **6a** and **6b** in a boiling water bath (100°C)

Allow each tube to stand in its respective bath for 30 minutes. Remove tubes **6a** and **6b** from the boiling water earlier if they decrease to 0.5 ml or less.

❏ 3 At the end of 30 minutes, place all test tubes in the rack. Test all the **a** tubes for starch by adding a few drops of Lugol's reagent to each. Test all the **b** tubes for sugar by adding 2 ml of Benedict's reagent to each and boiling them for 3 minutes. Record and interpret your results in Lab Report 46. Did starch digestion occur in any of the tubes? Which ones? How do you know? Did temperature affect digestion? How? Can you explain this result?

B. Observing lipid digestion

As described in Figure 46-3, lipids undergo two phases of digestion. The first is a physical process called **emulsification.** Emulsification breaks up large fat droplets into small fat droplets, increasing the fat's surface area and allowing lipase molecules access to the lipid molecules. **Bile salts** secreted by the liver act as *emulsifying agents* in the human digestive tract. The second phase of lipid digestion is a chemical process involving *lipase*. Lipase catalyzes the breakdown of complex lipid molecules into smaller units such as **fatty acid, glycerol, monoglyceride, cholesterol,** and **phospholipids.** Pancreatin, an extract of pancreatic secretions, contains the digestive enzyme pancreatic lipase.

❏ 1 Observe the emulsifying action of bile by performing this demonstration:
- Mark a test tube **7a** and add 2 ml of distilled water and 2 ml of vegetable oil.
- Mark a test tube **7b** and add 2 ml of distilled water, 2 ml of vegetable oil, and one spatula of bile salts.

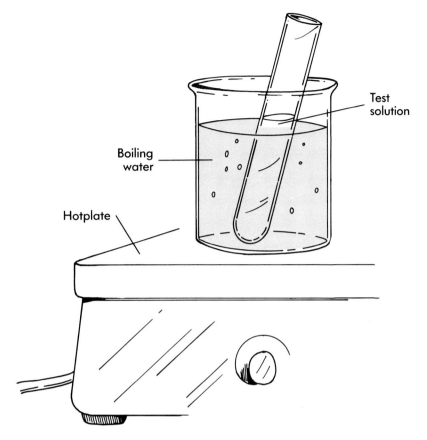

Figure 46-2 A boiling water bath, as used in Benedict's test for the presence of sugars.

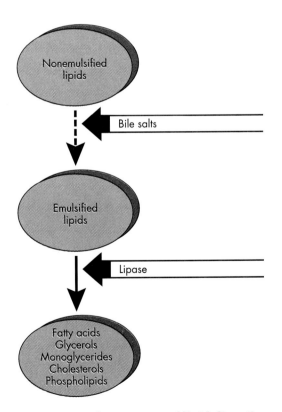

Figure 46-3 A summary of lipid digestion.

- Cap the tubes and shake them *simultaneously* for about 30 seconds.
- Compare the distribution of oil in both tubes. Continue watching them for about 10 minutes. How do you explain the difference between the two tubes?

☐ 2 To observe the results of lipid digestion by pancreatic lipase, follow this procedure:
- In a tube marked **8a** put 2 ml of litmus cream, 2 ml of distilled water, and one spatula of pancreatin.
- In a tube marked **8b** put 2 ml of litmus cream, 2 ml of distilled water, and one spatula each of pancreatin and bile salts.
- In a tube marked **8c** put 2 ml of litmus cream and 2 ml of distilled water.
- Place all three tubes in a 37°C water bath.
- Litmus turns from blue to red in the presence of acids. If fatty acids form as a result of lipid digestion, the litmus cream (cream with litmus added) will turn from its original bluish tint to a lavender-pink color and eventually to pink. Observe your tubes for about 1 hour. Do any of the tubes turn pink? How long does it take? How do you explain these results? Report your findings in Lab Report 46.

CLINICAL APPLICATION

Pancreatin is an extract of pancreatic secretions. It includes pancreatic amylase, lipase, and proteolytic (protein-digesting) enzymes. Wearers of contact lenses sometimes use pancreatin solutions, marketed as enzyme cleaners, to clean their lenses.

Contact lenses float on the tear fluid over the cornea. Over time, components of tear fluid build up on contact lenses. Tear fluid tends to form three layers, as shown in Figure 46-4. The top layer is mostly lipids from skin oil (sebum) and other sources. The middle layer is an aqueous solution of salts, proteins, and other substances. The bottom layer is composed mainly of *mucin,* which in turn is composed of combined polysaccharide-protein molecules called *glycoproteins.*

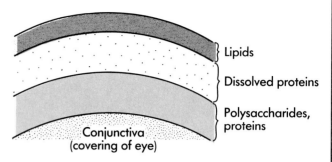

Figure 46-4 The three layers of tear fluid on the eye.

With this background and what you already know about enzymes and digestion, answer the following application questions.

1. On which components of the tear fluid can pancreatin act? On what do you base your answer?

2. How would digestion of tear fluid components help a wearer of contact lenses?

3. At what temperature would you recommend using enzymatic contact lens cleaning solutions? Explain.

4. Manufacturers of enzymatic contact lens cleaners recommend that the lenses be soaked in a nonenzyme solution for at least 4 hours after cleaning before wearing them again. What could happen if enzyme remains on a lens when it is put back on an eye?

LAB REPORT 46

Enzymes and Digestion

Tube	Beginning contents	Temp	Test	Result	Explanation
1a	Starch	22°C	Lugol's (starch)		
1b	Water	22°C	Lugol's (starch)		
2a	Maltose	22°C	Benedict's (sugar)		
2b	Water	22°C	Benedict's (sugar)		
3a	Starch, pancreatin	0°C	Lugol's (starch)		
3b	Starch, pancreatin	0°C	Benedict's (sugar)		
4a	Starch, pancreatin	22°C	Lugol's (starch)		
4b	Starch, pancreatin	22°C	Benedict's (sugar)		
5a	Starch, pancreatin	37°C	Lugol's (starch)		
5b	Starch, pancreatin	37°C	Benedict's (sugar)		
6a	Starch, pancreatin	100°C	Lugol's (starch)		
6b	Starch, pancreatin	100°C	Benedict's (sugar)		

Discuss the overall meaning of your results by answering the questions posed in step 3 of Activity A:

Tube	Beginning contents	Temp	Test	Result	Explanation
7a	Water, oil	22°C	Visual observation		
7b	Water, oil, bile salts	22°C	Visual observation		
8a	Litmus cream, water, pancreatin	37°C	Litmus (acid)		
8b	Litmus cream, water, pancreatin, bile salts	37°C	Litmus (acid)		
8c	Litmus cream, water	37°C	Litmus (acid)		

Discuss the overall meaning of your results by answering the questions posed in step 1 and 2 of Activity B:

Fill-in

_____ 1
_____ 2
_____ 3
_____ 4
_____ 5
_____ 6
_____ 7
_____ 8
_____ 9
_____ 10
_____ 11
_____ 12
_____ 13
_____ 14
_____ 15

Fill-in (complete each statement with the correct term)

1. Lipase is an enzyme that digests __?__ molecules.
2. Amylase is an enzyme that digests __?__ molecules.
3. Bile is helpful in preparing __?__ for digestion.
4. The __?__ secretes bile salts.
5. Polysaccharides are broken apart to yield __?__ molecules before being completely digested.
6. Carbohydrates are absorbed into the body in the form of __?__ molecules.
7. The substance __?__ contains several enzymes and is often used to clean contact lenses.
8. __?__ test is used to detect the presence of starch.
9. __?__ test is used to detect the presence of certain sugars.
10. __?__, here used in fatty cream, is a substance used to detect the presence of acids.
11. Enzymes that break apart proteins are termed __?__ enzymes.
12. Glucose is an example of a(n) __?__ type of carbohydrate molecule.
13. Bile __?__ globules of fat.
14. A(n) __?__ is a protein that catalyzes biochemical reactions.
15. Enzymes called __?__ digest sugars such as sucrose and maltose.

LAB EXERCISE 47

Urinary Structures

The **urinary system** is composed of the **kidneys, ureters, urinary bladder, urethra,** and associated structures. The functions of the urinary system include maintaining extracellular fluid balance, excreting wastes, and maintaining blood pH.

In this exercise, you are challenged to discover the basic anatomy of the urinary system on both the gross and microscopic levels. In Lab Exercise 49, you will analyze the components of the **urine** produced by this system.

Before you begin

❑ Read the appropriate chapter in your textbook.

❑ Set your learning goals. When you finish this exercise, you should be able to
  ■ identify the major organs of the urinary system and find them in models and charts
  ■ describe the gross anatomical features of the kidney and identify them in figures, models, and specimens
  ■ describe the features of the nephron and locate them in figures and models
  ■ identify the renal corpuscle in a prepared microscopic specimen

❑ Prepare your materials:
  ■ model of the human torso (dissectible)
  ■ model of the kidney (frontal section)
  ■ preserved sheep kidney (double or triple injected, if available)
  ■ dissection tools and trays
  ■ model of a nephron and associated structures
  ■ microscope
  ■ prepared microslide: kidney cortex c.s.

❑ Read the directions and safety tips for this exercise **carefully** before starting any procedure.

A. The urinary plan

Study the layout of the urinary system by locating these features in a dissectible model of the human torso:

❑ 1 The left and right **kidney** are located behind the peritoneum, along the posterior abdominal wall. Locate these kidney features:
  ■ **Renal fat pad**—Tissue that surrounds and protects each kidney (not shown in all models)
  ■ **Renal capsule**—The fibrous outer wall of the kidney
  ■ **Hilum**—An indentation on the medial side of each kidney where vessels and nerves enter or exit

❑ 2 The **ureter** is a muscular tube that exits each kidney at the hilum and extends posteriorly to the pelvic cavity.

❑ 3 The **urinary bladder** is a collapsible, muscular sac for the temporary storage of urine. The two ureters enter on each side of the posterior floor of the bladder.

❑ 4 The **urethra** is a muscular tube that extends from the anterior floor of the bladder to the outside of the body. In the female, it is a short tube that ends just anterior to the vagina. In the male, it is much longer, extending all the way through the penis. In the male, it conducts semen and urine.

Urine is formed in each kidney and is conducted through the ureters to the bladder. When the bladder is full and it is convenient, **urinary sphincters** that control flow through the urethra relax and allow urine to exit the body.

B. The kidney model

Locate these features of the human kidney on a model (or chart) of a frontal section:

❑ 1 Locate the **hilum** and **renal capsule** from this perspective.

❑ 2 Identify these structures located just within the hilum:
  ■ **Renal sinus**—A fat-filled cavity
  ■ **Renal pelvis**—A wide section of the urinary channel, distal to the ureter
  ■ **Calyces**—Branches of the pelvis that extend from the kidney tissue proper

❏ 3 The kidney tissue proper is divided into an outer **cortex** and an inner **medulla,** as is the tissue of many organs. **Renal pyramids** are cone-shaped sections of tissue lying mostly within the medulla. Each appears as a triangle in a frontal section. Each is composed of collecting ducts that conduct urine toward its tip (papilla), which is surrounded by the end of a calyx.

Urine formed by the kidney is conducted through the collecting ducts of the renal pyramids to the calyces. The calyces fuse to form the large renal pyramid. As urine is conducted out of the kidney through the hilum, the urinary channel narrows to form the ureter.

❏ 4 Identify these features of the renal arterial supply:
- **Renal artery**—A branch of the abdominal aorta that enters the hilum and extends through the renal sinus.
- **Interlobar arteries**—Branches of the renal arteries that extend outward through the tissue between the pyramids.
- **Arcuate arteries**—Branches of the interlobar arteries that turn to extend between the cortex and medulla.
- **Interlobular arteries**—Branches of the arcuate arteries that extend outward, into the cortex.
- **Afferent arterioles**—Small arteries that arise from branches of interlobular arteries, each one extending to a glomerulus.
- **Glomerular capillaries**—Capillaries that arise from an afferent arteriole and form a ball or small capillary bed called a **glomerulus.**

❏ 5 After a brief detour, the venous network of the kidney parallels the arterial network, as you may expect:
- **Efferent arterioles**—Each extending from a glomerulus.
- **Peritubular capillaries**—Small capillary beds that each arise from an efferent arteriole (a network of peritubular capillaries surrounds each *nephron,* the tubular, microscopic unit of the kidney).
- **Interlobular veins**
- **Arcuate veins**
- **Interlobar veins**
- **Renal veins**—Blood vessels that extend through the renal sinus, out the hilum, to drain into the inferior vena cava.

C. The sheep kidney

SAFETY FIRST! Observe the usual precautions when dissecting preserved specimens. Be sure to follow the safety advice that accompanies the preservative used in your specimen. Avoid injury with the dissection tools.

The sheep kidney is similar to the human kidney and makes an ideal specimen for study.

❏ 1 Examine the external aspect of your specimen. Identify as many parts of the kidney as you can. Refer to the aid given in Activities A and B if you have trouble. You may have to remove some of the renal fat pad.

❏ 2 Use a long knife or scalpel to cut a section, dividing the kidney into roughly equal dorsal and ventral portions. Try to identify as many features as you can.

HINT → Double-injected specimens have red latex injected into the arteries and blue latex injected into the veins. Triple-injected specimens also have yellow latex injected into the urinary channels and tubules. You may be able to see some of the finer detail visible in an injected specimen by using a hand lens or a dissection microscope.

The Urinary System

KIDNEY₁
URETER₂
URINARY BLADDER₃
URETHRA₄
RENAL CAPSULE₅
HILUM₆
RENAL SINUS₇
RENAL PELVIS₈
CALYCES₉
RENAL CORTEX₁₀
RENAL MEDULLA₁₁
RENAL PYRAMIDS₁₂
RENAL ARTERY₁₃
RENAL VEIN₁₄
NEPHRON₁₅

✎COLORING EXERCISE Using colored pens or pencils, shade in the figure and accompanying labels in contrasting colors of your choice as indicated by the red numerals.

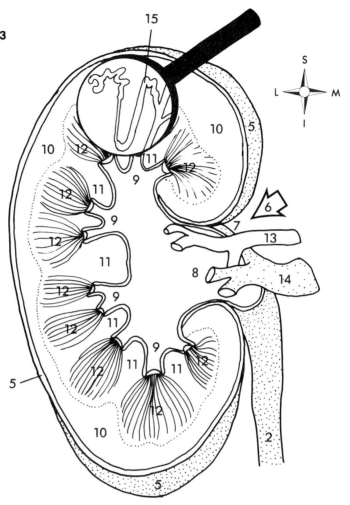

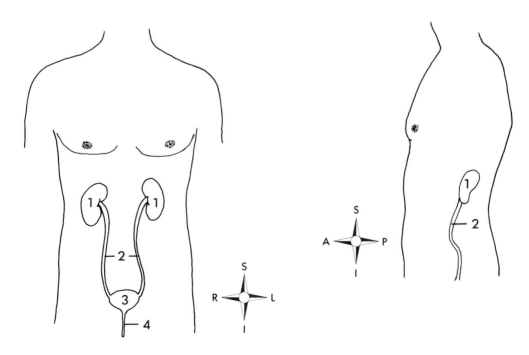

Figure 47-1

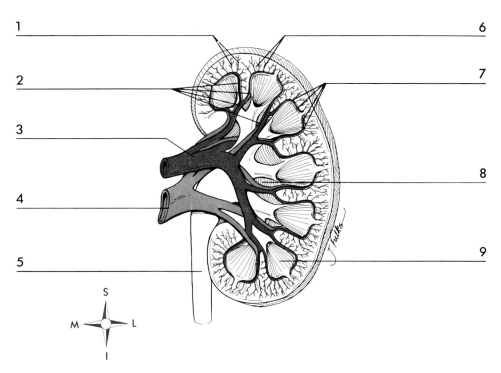

Figure 47-2 Identify the renal blood vessels indicated by label lines.

D. The nephron model

The functional units of the kidney are the tiny tubules called **nephrons.** The nephrons carry out the three basic processes that accomplish the kidney's function, forming *urine* as they do so: **filtration, tubular reabsorption,** and **tubular secretion.** Material from the blood is filtered into the beginning of the nephron. As the *filtrate* moves along the nephron tubule, some substances are reabsorbed into the blood and some additional substances are secreted from the blood into the filtrate. The urine thus produced is channeled to collecting ducts, which drain it from the kidney. Locate the main features of the nephron and associated structures in a model or chart.

❑ 1 The **renal corpuscle** is the roughly spherical structure at the beginning of the nephron. The inner portion of the renal corpuscle is the **glomerulus,** a ball of glomerular capillaries. Surrounding the glomerulus is the double-walled **Bowman's capsule.** Identify the afferent and efferent arterioles in the model. Can you trace their path into the renal corpuscle, out, then to the peritubular capillaries?

❑ 2 The **proximal convoluted tubule** is a narrow channel proceeding from the Bowman's capsule of the renal corpuscle. It is *convoluted,* meaning that it is coiled.

❑ 3 Filtrate formed in the Bowman's capsule flows through the proximal convoluted tubule and into the **loop of Henle.** Anatomically, the loop is a continuation of the proximal tubule. In many nephrons, the **descending limb** of the loop dips far down into the medulla, turns, and returns as the **ascending limb** to the cortex.

❑ 4 Filtrate flows from the ascending limb of the loop of Henle into the **distal convoluted tubule.** The filtrate is emptied into a **collecting duct,** a tubule that collects urine from many nephrons and conducts it through a renal pyramid to a calyx. Eventually, the urine is voided during **urination** or **micturition.**

❑ 5 Your model may represent a nephron whose distal convoluted tubule passes between the afferent arteriole and efferent arteriole, near the renal corpuscle. If so, you may notice that the afferent arteriole wall and distal tubule wall form a specialized structure where they meet. This structure is called the **juxtaglomerular apparatus.** *Juxtaglomerular* means "near the glomerulus." The juxtaglomerular apparatus secretes **renin,** an enzyme that catalyzes the conversion of *angiotensinogen* to *angiotensin I.* This begins a series of steps that help regulate blood pressure.

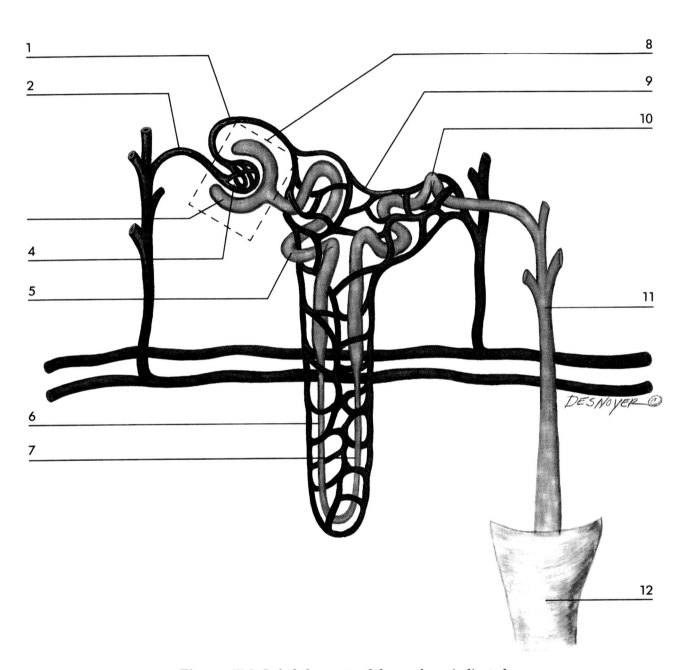

Figure 47-3 Label the parts of the nephron indicated.

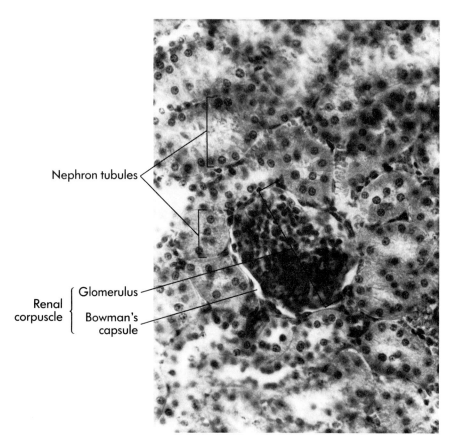

Figure 47-4 Micrograph of a renal cortex cross section. A renal corpuscle and surrounding tubules are visible.

E. Microscopic specimen

Obtain a prepared microscopic specimen of a cross section of renal cortical tissue. Scan it under low power first and locate one or more renal corpuscles, easily identifiable dark circles. Switch to high power and try to identify the features listed.

> **SAFETY FIRST!** Observe the usual precautions when using the microscope and a prepared slide.

❑ 1 The glomerulus is the dark region forming the center of the renal corpuscle. It is a network of glomerular capillaries.

❑ 2 The Bowman's capsule is seen as a very thin white, or light, area surrounding the glomerulus. Actually the *lumen* of the capsule appears white. The parietal wall of the Bowman's capsule is composed of simple squamous epithelium, which can sometimes be distinguished. The visceral wall of the Bowman's capsule is composed of specialized epithelial cells called **podocytes.** Podocytes have extensions that wrap around the capillary walls to form a filtration membrane. You will not be able to distinguish podocytes in your specimen.

> **HINT** → Plate 80 in the LABORATORY REFERENCE shows a color micrograph of the renal corpuscle.

❑ 3 The tissue surrounding each renal corpuscle is composed mostly of renal tubules. You will not be able to distinguish which are slices of proximal tubules or distal tubules, but you can appreciate the single-cell structure of their walls. What type of epithelium forms the tubule walls? How is this type of epithelial tissue specialized to perform the functions of the nephron tubule?

> **HINT** → If you are using double- or triple-injected sheep kidney specimens in the dissection, you may be able to see renal corpuscles and nephron tubules by examining a sectioned specimen with a hand lens or dissection microscope.

NAME _____ DATE _____ SECTION _____

LAB REPORT 47

Urinary Structures

Structure	Human	Sheep	Function(s)
Kidney	☐	☐	
Renal capsule	☐	☐	
Hilum	☐	☐	
Ureter	☐	☐	
Urinary bladder		☐	
Urethra		☐	
Renal sinus	☐	☐	
Renal pelvis	☐	☐	
Calyces	☐	☐	
Renal cortex	☐	☐	
Renal medulla	☐	☐	
Renal pyramids	☐	☐	
Renal artery, vein	☐	☐	
Interlobar arteries, veins	☐	☐	
Arcuate arteries, veins	☐	☐	
Interlobular arteries, veins	☐	☐	
Afferent arteriole	☐	☐	
Glomerulus	☐	☐	
Efferent arteriole	☐	☐	
Peritubular capillaries	☐	☐	
Renal corpuscle	☐	☐	
Bowman's capsule	☐	☐	
Proximal convoluted tubules	☐	☐	
Loop of Henle: descending limb, ascending limb	☐	☐	
Distal convoluted tubule	☐	☐	
Collecting duct	☐	☐	
Juxtaglomerular apparatus		☐	

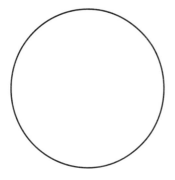

Specimen: *renal cortex c.s.*

Total Magnification: _____

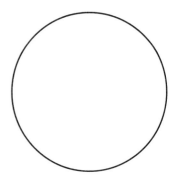

Specimen: *renal cortex c.s.*

Total Magnification: _____

Figure 47-2

_____ 1
_____ 2
_____ 3
_____ 4
_____ 5
_____ 6
_____ 7
_____ 8
_____ 9

Figure 47-3

_____ 1
_____ 2
_____ 3
_____ 4
_____ 5
_____ 6
_____ 7
_____ 8
_____ 9
_____ 10
_____ 11
_____ 12

Put in order

_____ 1
_____ 2
_____ 3
_____ 4
_____ 5
_____ 6
_____ 7
_____ 8
_____ 9
_____ 10
_____ 11

Fill-in

_____ 1
_____ 2
_____ 3
_____ 4
_____ 5
_____ 6
_____ 7
_____ 8
_____ 9
_____ 10

Put in order (arrange these structures in the order in which substances flow through the urinary system, beginning with the site of filtration)

    ascending limb of the loop of Henle
    Bowman's capsule
    calyces
    collecting duct
    descending limb of the loop of Henle
    distal convoluted tubule
    proximal convoluted tubule
    renal pelvis
    ureter
    urethra
    urinary bladder

Fill-in (complete each statement with the correct term)

1. The __?__ is a fat-filled cavity just inside the kidney's hilum.
2. The renal __?__ are branches of the renal pelvis.
3. The __?__ is a muscular tube extending from the renal pelvis to the urinary bladder.
4. The three basic processes observed in the nephron are: filtration, tubular reabsorption, and tubular __?__.
5. The __?__ conducts blood into the glomerular capillaries.
6. The __?__ conducts blood into the kidney.
7. The __?__ capillaries surround the tubules of the nephron.
8. The inner tissue of the kidney is termed the renal __?__ .
9. The __?__ apparatus secretes the enzyme renin.
10. The collecting ducts converge as they extend toward the calyces, forming the renal __?__.

LAB EXERCISE 48

Dissection: Urinary System

In this exercise, you will continue your study of urinary anatomy by dissecting a whole preserved specimen.

Activity A provides directions for studying the urinary anatomy of the cat. Activity B is an alternate activity, providing directions for studying the urinary anatomy of the fetal pig.

Before you begin

❑ Read the appropriate chapter in your textbook.

❑ Set your learning goals. When you finish this exercise, you should be able to
- dissect the urinary anatomy of a preserved cat or fetal pig
- identify the organs of the urinary tract in a dissected mammalian specimen

❑ Prepare your materials:
- preserved cat or fetal pig
- dissection tools and trays
- storage container (if specimen is to be reused)
- dissecting microscope (optional)

❑ Read the directions and safety tips for this exercise **carefully** before starting any procedure.

> **SAFETY FIRST!** Observe the usual precautions when working with a preserved specimen. Heed the safety advice accompanying preservatives used with your specimen. Use protective gloves while handling your specimen. Avoid injury with dissection tools. Dispose of or store your specimen as instructed.

A. Urinary anatomy of the cat

This activity challenges you to identify the major urinary structures of the cat. Assuming you have already removed the cat's skin (Exercise 10) and opened the abdominopelvic cavity (Exercise 40), you simply have to move the viscera out of the way—without cutting or removing them—to see the structures indicated in this activity.

> **HINT** → Figures 48-1 and 48-2 illustrate many of the structures of the cat listed in this activity. Useful information can also be found in the LABORATORY REFERENCE Plates 83 and 84.

❑ 1 Move the abdominal viscera out of the way so that you can see the left and right **kidneys** behind the parietal peritoneum. Because they are behind the parietal peritoneum, the kidneys' position is described as being *retroperitoneal*. Notice that each kidney is surrounded by a **renal fat pad** that supports and protects the underlying organ. Notice also that each kidney is capped by an **adrenal gland.** This gland is also known as the *suprarenal gland* in the human.

❑ 2 Expose each kidney by removing the peritoneum and renal fat pad. Identify the **hilum,** or medial notch, of each kidney. Locate these hollow organs that attach to each kidney at the hilum:
- **Ureter**
- **Renal artery**
- **Renal vein**

❑ 3 Remove one of the kidneys and cut it in a longitudinal section as you did with the sheep kidney in Exercise 47, Activity C. Try to identify all of the structures listed. You may wish to use a dissecting microscope to examine the detailed structure of the kidney section.
- **Renal capsule**
- **Renal cortex**
- **Renal medulla**
- **Renal pyramid**
- **Renal papilla**
- **Renal pelvis**

❑ 4 Follow the course of each ureter to the point at which it attaches to the **urinary bladder.** Cut the bladder open with a scalpel or scissors and locate the openings of the *left* and *right ureter* and the **urethra.**

ANATOMICAL ATLAS OF THE CAT
(URINARY STRUCTURES)

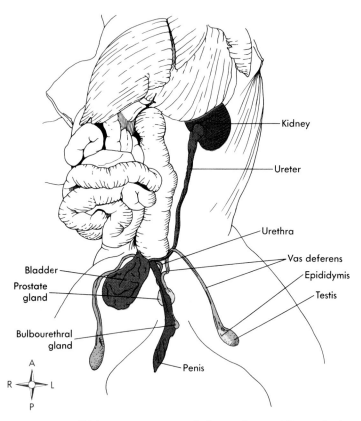

Figure 48-1 Urinary anatomy of the male cat. Ventral view.

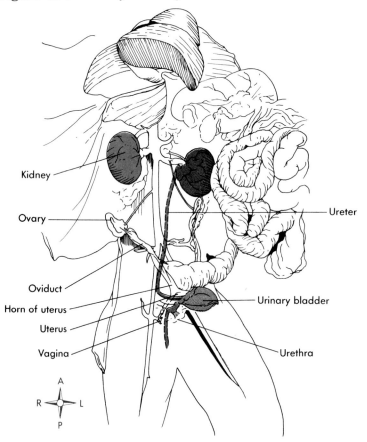

Figure 48-2 Urinary anatomy of the female cat. Ventral view.

❏ 5 Follow the course of the urethra to the outside of the body. How does the course of the urethra differ in the male and female cat?

B. Urinary anatomy of the fetal pig

This activity challenges you to identify the major urinary structures of the fetal pig. Assuming you have already removed the fetal pig's skin (Exercise 10) and opened the abdominopelvic cavity (Exercise 40), you simply have to move the viscera out of the way—without cutting or removing them—to see the structures indicated in this activity.

> **HINT** → Figures 48-3 and 48-4 illustrate many of the structures of the cat listed in this activity. Useful information can also be found in the LABORATORY REFERENCE Plates 85 and 86.

❏ 1 Move the abdominal viscera out of the way so that you can see the left and right **kidneys** behind the parietal peritoneum. Because they are behind the parietal peritoneum, the kidneys' position is described as being *retroperitoneal*. Notice that each kidney is surrounded by a **renal fat pad** that supports and protects the underlying organ. Notice also that each kidney is capped by an **adrenal gland.** This gland is also known as the *suprarenal gland* in the human.

❏ 2 Expose each kidney by removing the peritoneum and renal fat pad. Identify the **hilum,** or medial notch, of each kidney. Locate these hollow organs that attach to each kidney at the hilum:
- **Ureter**
- **Renal artery**
- **Renal vein**

❏ 3 Remove one of the kidneys and cut it in a longitudinal section as you did with the sheep kidney in Exercise 47, Activity C. Try to identify all of the structures listed. You may wish to use a dissecting microscope to examine the detailed structure of the kidney section.
- **Renal capsule**
- **Renal cortex**
- **Renal medulla**
- **Renal pyramid**
- **Renal papilla**
- **Renal pelvis**

❏ 4 Follow the course of each ureter to the point at which it attaches to the **urinary bladder.** Note that the *umbilical arteries* run alongside the bladder. Cut the bladder open with a scalpel or scissors and locate the openings of the *left* and *right ureter* and the **urethra.**

❏ 5 Follow the course of the urethra to the outside of the body. How does the course of the urethra differ in the male and female fetal pig?

ANATOMICAL ATLAS OF THE FETAL PIG
(URINARY ANATOMY)

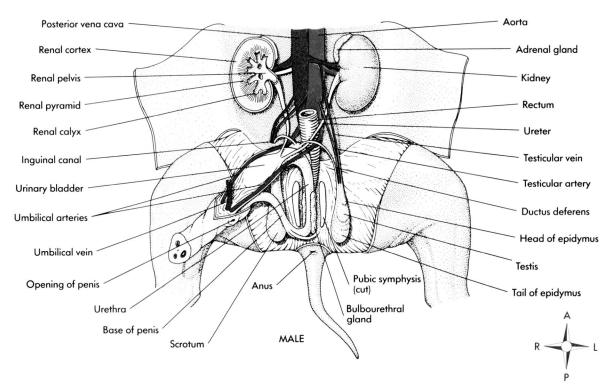

Figure 48-3 Urinary anatomy of the male fetal pig. Ventral view.

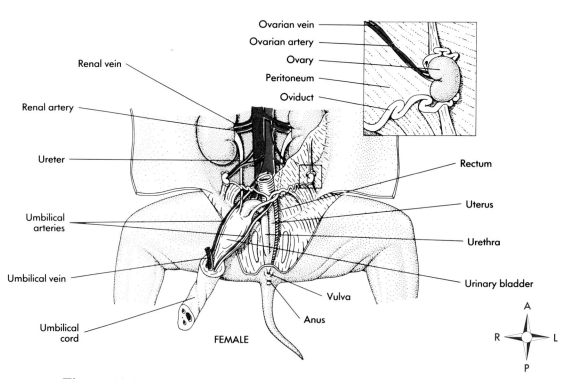

Figure 48-4 Urinary anatomy of the female fetal pig. Ventral view.

NAME_____ DATE_____ SECTION_____

LAB REPORT 48

Dissection: Urinary System

A. Cat dissection checklist
- ❏ kidney (left and right)
- ❏ parietal peritoneum
- ❏ renal fat pad
- ❏ adrenal gland
- ❏ hilum
- ❏ ureter
- ❏ renal artery
- ❏ renal vein
- ❏ renal capsule
- ❏ renal cortex
- ❏ renal medulla
- ❏ renal pyramid
- ❏ renal papilla
- ❏ renal pelvis
- ❏ urethra (male)
- ❏ urethra (female)

B. Fetal pig dissection checklist
- ❏ kidney (left and right)
- ❏ parietal peritoneum
- ❏ renal fat pad
- ❏ adrenal gland
- ❏ hilum
- ❏ ureter
- ❏ renal artery
- ❏ renal vein
- ❏ renal capsule
- ❏ renal cortex
- ❏ renal medulla
- ❏ renal pyramid
- ❏ renal papilla
- ❏ renal pelvis
- ❏ urethra (male)
- ❏ urethra (female)

LAB EXERCISE 49

Urinalysis

Urinalysis is the examination of urine and its contents. Urinalysis can be very comprehensive, testing for dozens of different physical and chemical characteristics. On the other hand, some clinical situations call for urinalysis that tests only one or two urine characteristics. For example, the pregnancy test and ovulation test performed in Lab Exercise 33 were types of urinalysis that each determined the presence or concentration of one particular hormone.

In this exercise, you will perform some of the more common clinical urinalysis tests, using the most current methods.

Before you begin

❑ Read the appropriate chapter in your textbook.

❑ Set your learning goals. When you finish this exercise, you should be able to
- state the normal characteristics of freshly voided urine
- demonstrate the use of dip-and-read clinical test strips for urinalysis
- explain the significance of common urine tests
- prepare and examine a stained urine sediment slide
- evaluate urinalysis results

❑ Prepare your materials:
- fresh urine specimen (your own or a packaged substitute) in a disposable container
- dip-and-read multiple test strips (Multistix 10SG are preferred)
- urine centrifuge and tapered tubes
- disposable 1 ml droppers
- Sedistain urine sediment stain
- paper towels and wipes
- microscope slides and coverslips
- microscope
- full-color urine sediment chart (optional)
- BIOHAZARD container
- unknown urine additives (optional)

❑ Read the directions and safety tips for this exercise **carefully** before starting any procedure.

Examination of urine

Urine reflects the overall status of extracellular fluid because it is derived from blood and its contents have been adjusted to some extent on the basis of homeostatic balance. Additionally, the health of the kidney and urinary tract in particular affect the characteristics of urine. Observe the urinalysis tests described or perform them yourself on *your own urine* or a packaged substitute provided by the instructor.

> **SAFETY FIRST!** Assume that all body fluids contain disease-causing agents. Protect yourself and others from contamination by wearing protective lab apparel, gloves, and eyewear. Disinfect all surfaces that have, or *could have,* come into contact with urine. Put all disposable urine containers, droppers, towels, wipes, slides, and coverslips in a BIOHAZARD container *immediately* after use. Your instructor may prefer to do this exercise as a demonstration.

❑ 1 Examine these physical characteristics of urine after placing some of the sample in a transparent container:
- **Transparency**—Normal urine is clear to slightly cloudy (especially after standing). Cloudy urine may contain fat globules, epithelial cells, mucus, microbes, or chemicals.
- **Color**—Normal urine is amber, straw, or transparent yellow resulting from the presence of **urochromes** such as *urobilinogen* (a bacterial product derived from *bilirubin*). Yellow-brown to greenish urine may occur when a high concentration of bile pigments are present. A red-to-dark-brown color may indicate the presence of blood. These or other abnormal colors may also result from the presence of food pigments (carotene, for example), drugs, or other chemicals.

❑ 2 Many characteristics of urine are determined by the use of paper that is impregnated with test reagents. The paper is dipped into a sample, and a color change indicates the presence (and sometimes, concentration) of a particular substance. You may have used pH test paper in your aquarium or in another lab course. In this step, you will use a plastic strip on which 9 or 10 test papers have been placed. Each paper is impregnated with different reagents.

> **HINT** → Carefully read the instructions provided by the distributor of your test strips. Follow the instructions exactly, especially those dealing with times at which test strips are to be read.

Shake or stir the sample so that any sediment on the bottom of the container becomes suspended. With gloved hands, dip the papered end of one test strip into your sample. Lift it out and tap the excess urine onto the inside rim of the container. You may need to transfer some urine to a test tube so that you can immerse all the paper pads. Read the results by comparing the test papers to standard charts as instructed. Usually, certain tests *must* be read at a certain time. A 10-test strip tests for these urine components:

- **Leukocytes**—Occasional WBCs are normal, but values increase in urinary infections.
- **Nitrite**—A positive nitrite result indicates the presence of large amounts of bacteria, as in an infection. This test is useful if one has a cloudy urine sample but no microscope to determine whether bacteria are causing the cloudiness.
- **Urobilinogen**—A derivative of bilirubin, high levels may indicate excessive RBC destruction or liver disease.
- **Protein**—Specifically *albumin,* a small protein molecule that is normally absent or present only in trace amounts; higher levels may indicate hypertension or kidney disease. Detectable levels are normal just after exercise.
- **pH**—The relative H$^+$ concentration, pH is a determinant of acidity. Normal urine pH is 4.6 to 8. Values are lower (more acid) in acidosis, starvation, and dehydration. pH is higher (more alkaline) in urinary infections and alkalosis.
- **Occult blood** (RBCs or free hemoglobin)—*Occult* means hidden. This strip tests for small amounts of hemoglobin that do not discolor the urine but are still clinically significant. Normally not present, hemoglobin may indicate kidney infection or the presence of stones in the kidney, ureter, or bladder.
- **Specific gravity**—This is the ratio of urine density to water density. If urine is pure water, the specific gravity is 1.000. Normal urine is 1.001 to 1.030. Lower values may indicate kidney disease. Higher values indicate high solute concentration and may occur during dehydration or diabetes mellitus. (This test is not present in 9-test strips.) An alternate method for determining specific gravity is to float a **hydrometer** in a cylinder containing urine, as in Figure 49-1. Read the scale on the hydrometer at the point that it meets the surface of the urine.
- **Ketone**—A by-product of fat metabolism, it may be present during fasting, diabetes mellitus, or a low-carbohydrate diet.
- **Bilirubin**—Normally not present, or present in trace amounts, this product of RBC destruction in the liver may indicate liver disease or bile tract obstruction if present in the urine.
- **Glucose**—Normally, there is no glucose in the urine. Trace amounts may be present after a meal high in carbohydrates. Continued high levels in the urine may indicate diabetes mellitus or pituitary problems.

> **SAFETY FIRST!** Do not forget to observe the usual precautions when using the microscope and stained, wet-mount slides.

❏ 3 Observe the components of urine sediment by performing this procedure:
- Shake or stir the urine sample to suspend any sediment that has settled.
- Transfer some of the urine to a tapered centrifuge tube.
- After balancing and securing the centrifuge as your instructor demonstrates, spin the sample for about 5 minutes.
- Without disturbing the sediment that has collected in the tip of the tube, squeeze the bulb of a disposable dropper and gently lower it into the urine. When the dropper reaches the bottom, release the bulb *slightly* to collect only a few drops of sediment-containing urine.
- Quickly place a drop of the sample on a clean microscope slide. Add one drop of Sedistain and cover it with a coverslip. Remove excess fluid from the edges with a lab wipe and immediately discard the wipe in the BIOHAZARD container.
- Examine the specimen under both low and high power. Move from field to field as you identify different types of sediment.

❑ 4 Your instructor may have a chart of sediment types for you to use in identifying urine solids. Figure 49-2 presents a minichart of common urine sediment components:
- **Cells**—Epithelial cells from the urinary tract lining, blood cells from injury or infection sites, or infectious microbes may be present in sediment. If more than trace amounts of blood or microbial cells are present, a urinary problem is indicated.
- **Artifacts**—Artifacts are materials that have accidentally gotten into the sample, including fabric fibers from underwear, powder used on the skin near the urethral opening, or skin oil droplets.
- **Crystals**—Very tiny crystals of normal urine components or drugs may be visible under high power. Small amounts of these crystals may be normal. A large number of crystals is seen in urinary *retention*, the inability to void urine from the bladder. Large masses of crystals are called *stones*, or *calculi*.
- **Casts**—Casts are chunks of material that have hardened somewhere in the urinary channel and sloughed off into the urine. They may be roughly cylindrical masses of cells, granules, or other substances.

❑ 5 Your instructor may opt to offer unknown urine additives. Add the additive to your sample as instructed and test the sample as if it were a new urine specimen. The additive will have altered the sample's characteristics in a way that mimics an abnormal condition. Your instructor may choose instead to give you a new sample, one that is abnormal in some respect. Test the new sample in the same way you tested the first sample. Report your results in Lab Report 49.

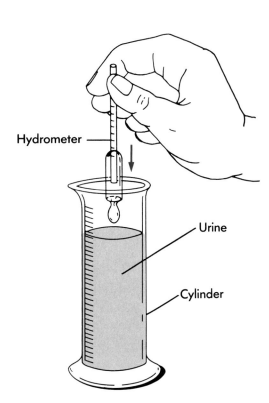

Figure 49-1 The specific gravity of urine may be determined by floating a hydrometer in room-temperature urine.

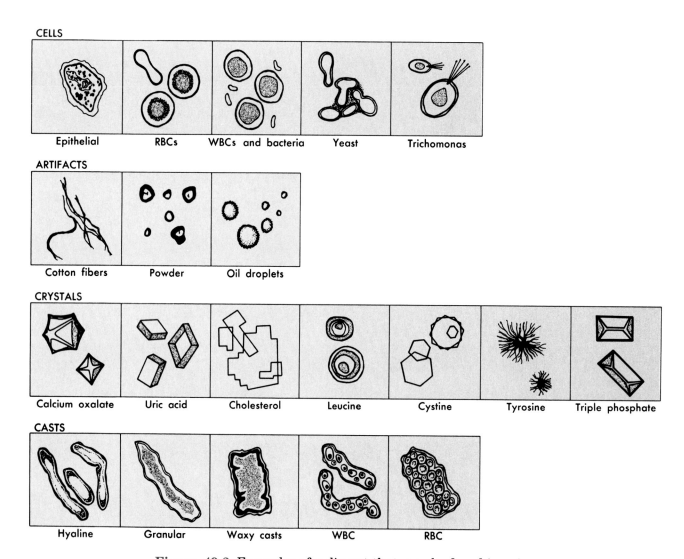

Figure 49-2 Examples of sediment that may be found in urine.

LAB REPORT 49

Urinalysis

Characteristic	Normal value	Observed value	Interpretation
Transparency			
Color			
Leukocytes			
Nitrite			
Urobilinogen			
Protein			
pH			
Occult blood			
Specific gravity			
Ketone			
Bilirubin			
Glucose			

Summary and evaluation of this sample:

Sketch (and label) examples of urine sediment found in your sample:

Use this table for results of the test on the "unknown" sample:

UNKNOWN CODE_____

Characteristic	Normal value	Observed value	Interpretation
Transparency			
Color			
Leukocytes			
Nitrite			
Urobilinogen			
Protein			
pH			
Occult blood			
Specific gravity			
Ketone			
Bilirubin			
Glucose			

Summary and evaluation of this sample:

LAB EXERCISE 50

The Male Reproductive System

Humans, as a species, have two **reproductive systems:** a male system and a female system. The primary function of both is sexual reproduction of offspring. Sexual reproduction in humans requires successful fusion of two **gametes,** or sex cells. Human gametes, **sperm** and **ova,** are formed by means of *meiosis* and so each has 23 chromosomes. When a sperm cell from the male parent unites with an ovum from the female parent, a cell with 46 chromosomes is formed. This offspring cell has a unique mix of DNA from both parents, providing the variation essential to the survival of the human species.

In this exercise, you are invited to explore the structure and function of the male reproductive system. In Lab Exercise 51, the female reproductive system is presented.

Before you begin

❑ Read the appropriate chapter in your textbook.

❑ Set your learning goals. When you finish this exercise, you should be able to
- describe the plan of the male reproductive tract
- describe the organs of the male reproductive system and their principal functions
- locate male reproductive organs in charts and models
- identify major features in a microscopic specimen of the testis
- identify the features of mature sperm in figures and in a prepared sperm smear

❑ Prepare your materials:
- models and charts of the male reproductive system
- microscope
- prepared microslide: *human testis c.s.*
- prepared microslide: *human sperm smear*

❑ Read the directions and safety tips for this exercise **carefully** before starting any procedure.

A. Human model

Find the major features of the male reproductive system in models and charts.

❑ 1 The primary sex organs of the male reproductive system are the two **testes.** The testes are located within the **scrotum,** a sac of skin and other tissues on the outer wall of the anterior trunk. A thin layer of smooth muscle called the **dartos muscle** forms part of the scrotum's wall. Find these structures associated with the testes:
- **Capsule**—A connective tissue outer wall, the capsule has inward extensions that divide each testis into **lobules.**
- **Seminiferous tubules**—These are long, coiled tubules in which sperm cells are produced. Tissue in between the tubules contains **interstitial cells (of Leydig)** that secrete *testosterone.*
- **Rete testis**—It is a network of tubules into which the seminiferous tubules empty. The rete testis, in turn, empties into 15 to 20 **efferent ductules** that leave the testis.

❑ 2 The **epididymis** is a set of coiled tubules on the outside of the testis (but still within the scrotum). Sperm cells formed in the testis move into the epididymis, where they continue to mature.

❑ 3 The **ductus deferens,** or **vas deferens,** conducts sperm cells from the epididymis out of the scrotum and into the pelvic cavity. Trace its path in your model. As the ductus deferens enters the abdominopelvic cavity through the *inguinal canal,* it is associated with nerves, vessels, and a connective tissue covering. Together, these structures form the **spermatic cord.** Near its end, each ductus deferens has an enlarged section called the **ampulla.**

❑ 4 Joining the ductus deferens at the ampulla is a duct from a **seminal vesicle.** Each seminal vesicle is a gland that produces fluid that becomes part of **semen.** Semen is the fluid medium containing sperm that is *ejaculated* during the male sexual response. The ductus deferens and seminal vesicle duct unite to form the **ejaculatory duct** that extends to the urethra, just inferior to the bladder.

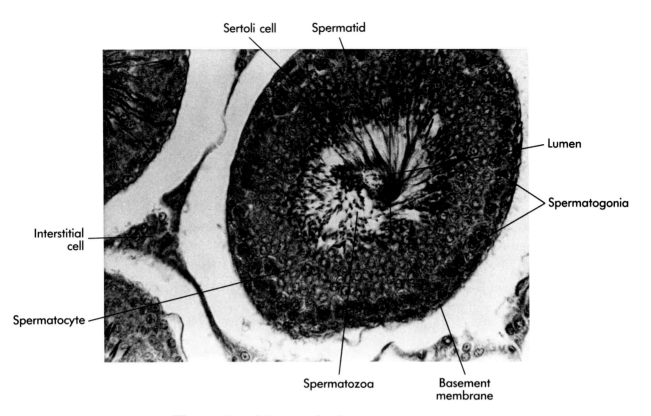

Figure 50-1 Micrograph of a testis cross section.

- ❏ 5 Surrounding the urethra and ejaculatory ducts below the bladder is the **prostate gland.** This gland, like the seminal vesicles, contributes to the seminal fluid.

- ❏ 6 Sperm-containing semen is conducted through the **urethra** during the male sexual response. Near the base of the penis, ducts from the **bulbourethral glands** join the urethra. These small glands contribute a small amount of fluid to the semen.

- ❏ 7 The **penis** is the erectile structure through which the urethra conducts semen out of the body during ejaculation. Erection of the penis is essential to insertion into the female tract and depositing the semen there. Three vascular bodies, the left and right **corpora cavernosa** and the **corpus spongiosum,** engorge with blood during the sexual response and stiffen the penis. The tip of the corpus spongiosum forms the head of the penis, or **glans penis.** The glans penis is covered by a fold of skin called the **foreskin, or prepuce,** unless it has been removed by means of *circumcision.*

> **SAFETY FIRST!** The remaining activities of this exercise call for the use of a microscope. Be sure to observe the usual precautions.

B. Microscopic structure of the testis

First in a model or chart, then in a prepared microscopic specimen, locate the following structures in a cross section of the testis:

- ❏ 1 The major feature of a testis cross section is the presence of numerous **seminiferous tubules.** Identify the **basement membrane,** lumen, and walls of a seminiferous tubule. Locate some of the **interstitial cells** between the tubules.

- ❏ 2 Examine the cells forming the wall of a seminiferous tubule. Large, pale cells with oval nuclei are supportive **Sertoli cells.** They extend all the way to the lumen, but their entire length is often difficult to distinguish. The majority of cells are **germ cells.** The outermost layer of germ cells are **spermatogonia.** Some daughter cells of the spermatogonia, nearer the lumen, become **spermatocytes** that divide by means of meiosis to form **spermatids.** Each spermatid develops to become a **spermatozoan,** many of which are seen in the lumen.

> HINT → LABORATORY REFERENCE Plate 82 shows a light micrograph of a cross section of the testis.

The Male Reproductive System

SCROTUM₁
TESTIS₂
EPIDIDYMIS₃
DUCTUS DEFERENS₄
SEMINAL VESICLE₅
EJACULATORY DUCT₆
PROSTATE GLAND₇
URETHRA₈
BULBOURETHRAL GLAND & DUCT₉
CORPUS SPONGIOSUM₁₀
 GLANS PENIS₁₁
CORPORA CAVERNOSA₁₂
FORESKIN₁₃
URINARY BLADDER₁₄
PUBIS₁₅
RECTUM₁₆

COLORING EXERCISE Using colored pens or pencils, shade in the figure and accompanying labels in contrasting colors of your choice as indicated by the red numerals.

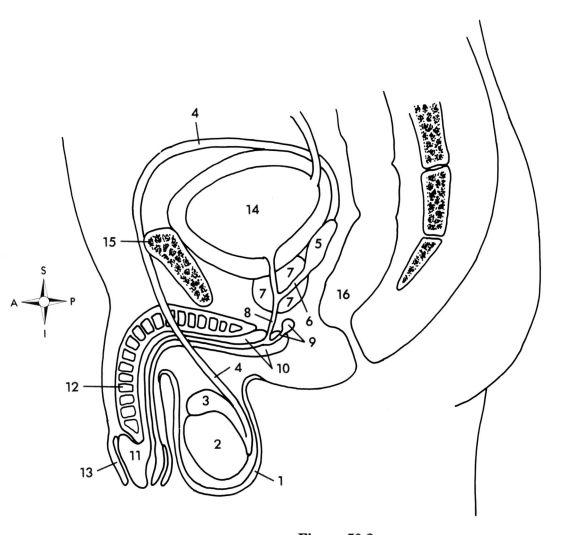

Figure 50-2

C. Sperm cells

Examine a prepared microscopic specimen of a sperm (semen) smear under high power. Notice that each normal sperm has a pear-shaped **head** containing a nucleus. A **midpiece** containing mitochondria is at the base of a **tail,** or *flagellum*.

Up to one third of the sperm cells in a normal specimen may be deformed. Look for sperm with multiple or misshapen heads or with tail deformities. High ratios of deformed sperm are associated with stress, infection, or high environmental temperatures.

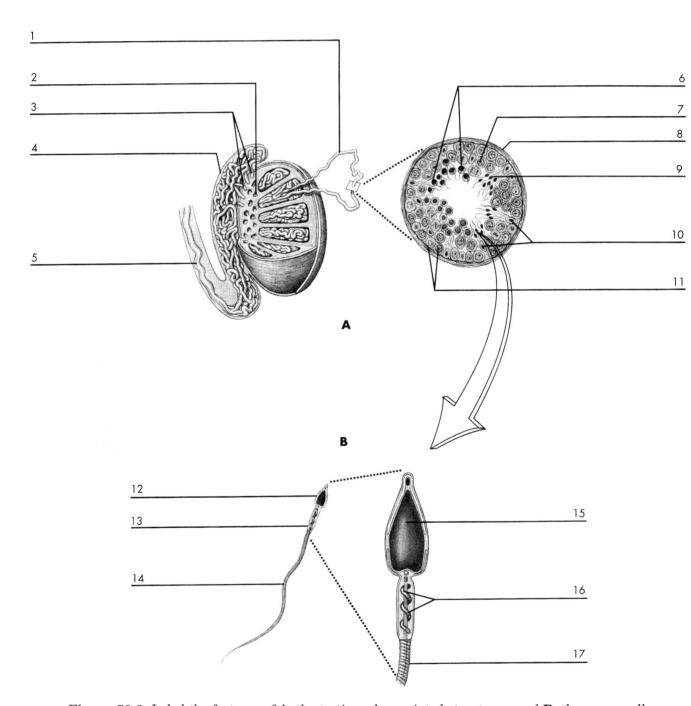

Figure 50-3 Label the features of **A,** the testis and associated structures, and **B,** the sperm cell.

NAME _____ DATE _____ SECTION _____

LAB REPORT 50

The Male Reproductive System

Structure	Model or chart	Function(s)
Testis	☐	
Scrotum	☐	
Capsule	☐	
Seminiferous tubule	☐	
Rete testis	☐	
Epididymis	☐	
Ductus deferens	☐	
Seminal vesicle	☐	
Ejaculatory duct	☐	
Prostate gland	☐	
Urethra	☐	
Bulbourethral gland	☐	
Penis	☐	
Corpora cavernosa	☐	
Corpus spongiosum	☐	
Glans penis	☐	
Foreskin	☐	
Seminiferous tubule	☐	
Interstitial cell	☐	
Serotoli cell	☐	
Germ cell	☐	
Spermatogonium	☐	
Spermatocyte	☐	
Spermatid	☐	
Spermatozoan: head, midpiece, tail	☐	

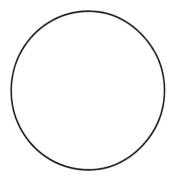

Specimen: *human testis c.s.*

Total Magnification: _____

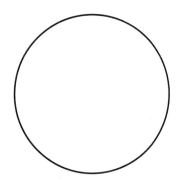

Specimen: *human sperm smear*

Total Magnification: _____

Figure 50-3

_____ 1
_____ 2
_____ 3
_____ 4
_____ 5
_____ 6
_____ 7
_____ 8
_____ 9
_____ 10
_____ 11
_____ 12
_____ 13
_____ 14
_____ 15
_____ 16
_____ 17

Multiple choice

____ 1
____ 2
____ 3
____ 4
____ 5
____ 6

Multiple choice (select the best response)

1. Semen travels through the male reproductive tract in this order
    a. ejaculatory duct, ductus deferens, epididymis, urethra
    b. epididymis, ductus deferens, ejaculatory duct, urethra
    c. urethra, ejaculatory duct, ductus deferens, epididymis
    d. ductus deferens, epididymis, ejaculatory duct, urethra

2. Sperm cells mature in phases that occur in this order:
    a. spermatogonium, spermatocyte, spermatozoan, spermatid
    b. spermatocyte, spermatozoan, spermatid, spermatogonium
    c. spermatogonium, spermatocyte, spermatid, spermatozoan

3. Erection of the penis is accomplished through increased blood volume in the
    a. corpus spongiosum
    b. foreskin
    c. corpora cavernosa
    d. a and c are correct
    e. none of the above

4. Mitochondria of the sperm cell are found in the
    a. head
    b. midpiece
    c. tail

5. The urethra of the male functions as a part of the
    a. urinary system
    b. digestive system
    c. reproductive system
    d. a and c are correct
    e. all of the above

6. The distal, widened portion of the ductus deferens is more specifically known as the
    a. ampulla
    b. vas deferens
    c. seminal vesicle
    d. prostate gland
    e. proximal tubule

LAB EXERCISE 51

The Female Reproductive Sytem

This exercise continues the study of human reproductive systems begun in Lab Exercise 50. In this exercise, you are invited to explore the essential structure and function of the female reproductive system.

Before you begin

❏ Read the appropriate chapter in your textbook.

❏ Set your learning goals. When you finish this exercise, you should be able to
  ■ describe the plan of the female reproductive tract
  ■ describe the organs of the female reproductive systems and their principal functions
  ■ locate female reproductive organs in charts and models
  ■ identify major features in a microscopic specimen of a mammalian ovary
  ■ identify structures of the mammary glands

❏ Prepare your materials:
  ■ models and charts of the female reproductive system and the breasts
  ■ microscope
  ■ prepared microslide: *mammalian ovary c.s.*

❏ Read the directions and safety tips for this exercise **carefully** before starting any procedure.

A. Human model

Find the major features of the female reproductive system in models and charts.

❏ 1 The primary sex organs of the female reproductive system are the two **ovaries.** They are small, rounded organs suspended in the pelvic cavity by ligaments. The ovarian mesentery, or **mesovarium,** attaches the ovaries to the **broad ligament.** The **suspensory ligaments** and **ovarian ligaments** also support the ovaries.

❏ 2 A **uterine tube** is associated with each ovary. Also called a **Fallopian tube,** or **oviduct,** this narrow tube with a ciliated lining conducts a mature **ovum** from the area of the ovary toward the uterus. It is the usual site of fertilization. Long processes called **fimbriae** surround the tube's margin at its ovarian end.

❏ 3 The **uterus,** or womb, is a single muscular sac at the midline in the pelvic cavity. Identify these parts of the uterus:
  ■ **Body of the uterus**—The larger, rounded portion in which is located the main part of the **uterine cavity**
  ■ **Cervix**—Narrow, inferior portion of the uterus (the **cervical canal** passes through this area to join the vagina)
  ■ **Serous layer of the uterus**—The portion of the peritoneum that forms the outer uterine wall
  ■ **Myometrium**—The muscular middle layer of the uterine wall
  ■ **Endometrium**—The inner layer of the uterine wall

❏ 4 The **vagina** is a muscular canal that receives the male penis during *sexual intercourse.* It functions as the **birth canal** during *delivery* and as an exit for sloughed-off endometrium during **menses.** The opening of the vagina to the outside may be covered partially or entirely by a thin, membranous **hymen.**

❏ 5 The external genitals of the female are collectively known as the **vulva,** or **pudendum.** Identify these features associated with the vulva:
  ■ **Vestibule**—The central space of the vulva, into which the vagina and urethra open (openings of **vestibular gland** ducts conduct lubricating fluid into the vestibule)
  ■ **Clitoris**—The female glans, a small erectile structure in the anterior corner of the vestibule (the clitoris is covered with a fold of skin called the **foreskin,** or **prepuce**)
  ■ **Labia minora**—A pair of thin, longitudinal folds on each side of the vestibule (sing. *labium minus*)
  ■ **Labia majora**—A pair of thick, longitudinal folds lateral to the labia minora (the space between the labia majora is called the **pudendal cleft**) (sing. *labium majus*)
  ■ **Mons pubis**—A fatty mound, covered with hair, over the pubic bone

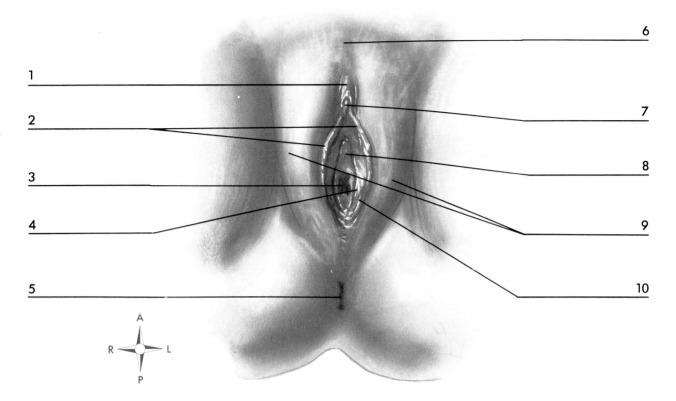

Figure 51-1 Label the features of the vulva.

B. Microscopic structure of the ovary

First, in a model or chart, then in a prepared microscopic specimen, locate important structures associated with the ovary.

> **SAFETY FIRST!** Be careful of electrical and other hazards when examining your specimen with a microscope.

❑ 1 Each ovary is covered with a portion of the peritoneum. The outer region of each ovary is composed of dense connective tissue with numerous **ovarian follicles.** Each follicle contains an **oocyte,** the germ cell of the female.

❑ 2 **Primary oocytes** are oocytes that have developed to a point early in meiosis and then temporarily stopped developing. Each primary oocyte is surrounded by **granulosa cells.** The oocyte and granulosa cells together constitute a **primary follicle.**

❑ 3 When stimulated by hormones, a primary follicle develops into a **secondary follicle.** Each secondary follicle has a central fluid-filled **antrum** and a mass of cells off to the side called the **cumulus mass.** The oocyte is embedded in the cumulus mass.

❑ 4 The follicle eventually develops into a **mature follicle** when the antrum fills with more fluid and the follicle becomes a bump on the ovary's surface. The tissue surrounding the follicle forms a layer called the **theca.** A clear layer, or **zona pellucida,** surrounds the enlarged oocyte. The developing follicle secretes the hormone **estrogen.**

❑ 5 As more fluid fills the follicle, it bursts open and spills its contents into the peritoneal cavity near the fimbriae of the uterine tube. This process is called **ovulation.** The ruptured follicle then develops into the glandular **corpus luteum,** which secretes **progesterone** and estrogen.

> **HINT** → Figure 51-3 illustrates the structures discussed in this activity. Several ovarian follicles, each in different stages of development, may be found in a single specimen. However, their distribution is rather random. Plate 81 in the LABORATORY REFERENCE shows a light micrograph of a mature (Graafian) follicle.

The Female Reproductive System

COLORING EXERCISE Using colored pens or pencils, shade in the figure and accompanying labels in contrasting colors of your choice as indicated by the red numerals.

OVARY₁
UTERINE TUBE₂
UTERUS
  BODY₃
  CERVIX₄
VAGINA₅
CLITORIS₆
LABIUM MINUS₇
LABIUM MAJUS₈
URINARY BLADDER₉
URETHRA₁₀
PUBIS₁₁
RECTUM₁₂

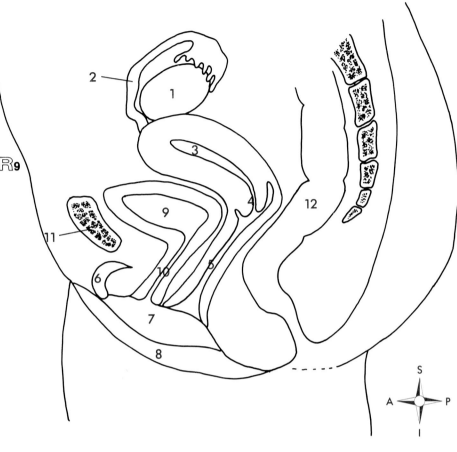

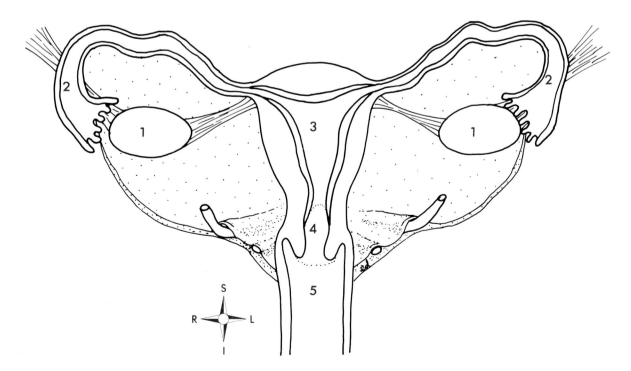

Figure 51-2

433

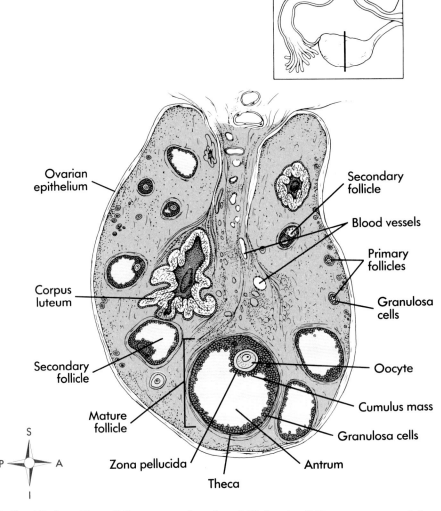

Figure 51-3 Sagittal section of the ovary, showing follicles in different stages of development.

C. The mammary glands

Locate the female breast in a model or chart. The **mammary glands** are modified sweat glands within the skin of the chest. Although structurally they are components of the integument, they are considered here because of their function of providing nutrition for offspring produced by the reproductive system. Find these features of the breasts:

❑ 1 Both the male and female breast have a pigmented area of skin called the **areola.** The areola surrounds a raised **nipple.**

❑ 2 Normally, only the mammary glands of the female develop. At puberty, adipose tissue develops under the skin of the breast. This tissue supports 15 to 20 glandular **lobes** of the mammary gland. Each lobe has a single duct that opens directly onto the surface of the nipple, giving the nipple a total of 15 to 20 openings.

❑ 3 Each lobe of the mammary gland is composed of several **lobules.** During *lactation* (milk production), the ends of the lobule ducts form secretory sacs called **alveoli.**

LAB REPORT 51

The Female Reproductive System

Structure	Model or chart	Function(s)
Ovary	☐	
Mesovarium	☐	
Broad ligament	☐	
Suspensory ligament	☐	
Ovarian ligament	☐	
Uterine tube	☐	
Fimbriae	☐	
Body of uterus, uterine cavity	☐	
Cervix, cervical canal	☐	
Serous layer of uterus	☐	
Myometrium	☐	
Endometrium	☐	
Vagina	☐	
Vestibule (of vulva)	☐	
Clitoris	☐	
Prepuce	☐	
Labium minus	☐	
Labium majus	☐	
Mons pubis	☐	
Ovarian follicles	☐	
Areola (of breast)	☐	
Nipple	☐	
Lobes (of mammary gland)	☐	
Lobules (of mammary gland)	☐	
Alveoli (of mammary gland)	☐	

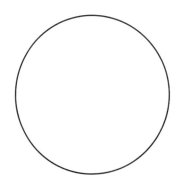

Specimen: *mammalian ovary c.s.*

Total Magnification: _____

Figure 51-1

_____ 1
_____ 2
_____ 3
_____ 4
_____ 5
_____ 6
_____ 7
_____ 8
_____ 9
_____ 10

Fill-in

_____ 1
_____ 2
_____ 3
_____ 4
_____ 5
_____ 6
_____ 7
_____ 8
_____ 9
_____ 10
_____ 11
_____ 12

Fill-in (complete each statement with the correct term)

1. The fluid-filled space with a follicle is called the __?__.
2. The tissue surrounding a mature follicle is called the __?__.
3. A follicle secretes the hormone __?__.
4. The external genitals of the female are known by the term __?__.
5. The __?__ secrete lubricating fluid into the vestibule.
6. The release of an oocyte from a follicle is termed __?__.
7. The normal site of fertilization of an egg is in the __?__.
8. The narrow portion, or neck, of the uterus is called the __?__.
9. The oocyte in a secondary follicle is embedded in the __?__ mass.
10. The thin membrane covering all or part of the vaginal opening is called the __?__.
11. The __?__ is a structure of the ovary that secretes progesterone.
12. The birth canal is also called the __?__.

LAB EXERCISE 52

Dissection: Reproductive Systems

In this exercise, you will continue your study of reproductive anatomy by dissecting a whole preserved specimen.

Activity A provides directions for studying the reproductive anatomy of the male cat. Activity B provides directions for studying the reproductive anatomy of the female cat. Activities C and D are alternate activities, providing directions for studying the (immature) reproductive anatomy of the male and female fetal pig.

Before you begin

❑ Read the appropriate chapter in your textbook.

❑ Set your learning goals. When you finish this exercise, you should be able to
- dissect the reproductive anatomy of a preserved cat or fetal pig
- identify the major organs of both the male and female reproductive systems in a dissected mammalian specimen

❑ Prepare your materials:
- preserved cat or fetal pig
- dissection tools and trays
- storage container (if specimen is to be reused)

❑ Read the directions and safety tips for this exercise **carefully** before starting any procedure.

> **SAFETY FIRST!** Observe the usual precautions when working with a preserved specimen. Heed the safety advice accompanying preservatives used with your specimen. Use protective gloves while handling your specimen. Avoid injury with dissection tools. Dispose of or store your specimen as instructed.

A. Reproductive anatomy of the male cat

This activity challenges you to identify the major reproductive structures of the male cat. Assuming you have already removed the cat's skin (Exercise 10) and opened the abdominopelvic cavity (Exercise 40), you simply have to move the viscera out of the way—without cutting or removing them—to see many of the structures indicated in this activity. You may have to cut some additional skin or muscle tissue to see a few of the structures more clearly.

> **HINT** → Figure 52-1 illustrates many of the structures of the cat listed in this activity. Useful information can also be found in the LABORATORY REFERENCE Plate 83.

❑ 1 Locate the **scrotum,** a sac of skin between the hindlimbs of the animal. Cut the scrotum open with your scissors or scalpel, taking care not to cut the structures inside the scrotum.

❑ 2 Identify the **testes,** the male *gonads* within the scrotum. Remove the connective tissue covering each testis to reveal the coiled **epididymis.** Note that the **vas (ductus) deferens** leads from the epididymis, through the *inguinal canal,* into the abdominopelvic cavity.

❑ 3 Trace the course of one of the vas deferens into the abdominopelvic cavity to where it meets its partner from the other side. Note that they join together where they meet the **urethra.** At this location a mass of glandular tissue called the **prostate gland** can be found.

❑ 4 Follow the course of the urethra toward the base of the **penis,** where a pair of small **bulbourethral glands** can be found.

❑ 5 Identify the **glans penis** at the distal tip of the penis.

ANATOMICAL ATLAS OF THE CAT (REPRODUCTIVE ORGANS)

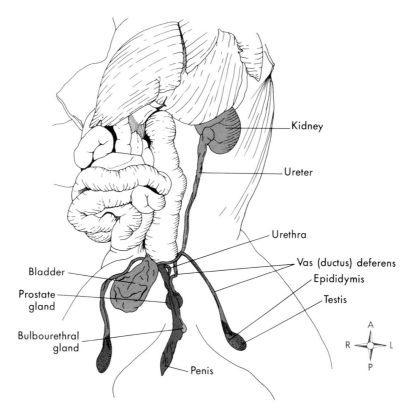

Figure 52-1 Reproductive anatomy of the male cat. Ventral view.

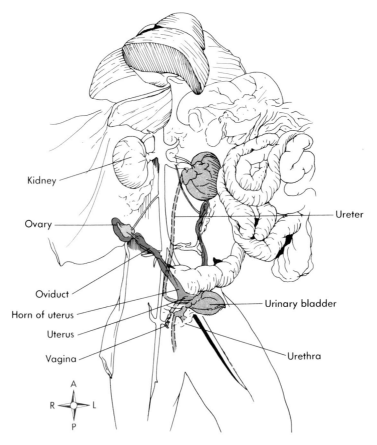

Figure 52-2 Reproductive anatomy of the female cat. Ventral view.

B. Reproductive anatomy of the female cat

This activity challenges you to identify the major reproductive structures of the female cat. Assuming you have already removed the cat's skin (Exercise 10) and opened the abdominopelvic cavity (Exercise 40), you simply have to move the viscera out of the way—without cutting or removing them—to see many of the structures indicated in this activity. You may have to cut some additional skin or muscle tissue to see a few of the structures more clearly.

> HINT → Figure 52-2 illustrates many of the structures of the cat listed in this activity. Useful information can also be found in the LABORATORY REFERENCE Plate 84.

❑ 1 Locate the tiny lightly colored **ovaries,** the female *gonads,* located just posterior to the kidneys.

❑ 2 Identify the **oviduct,** with its distal *fimbriae,* that leads from each ovary toward a **uterine horn.** Each uterine horn is a lateral extension of the superior portion of the uterus.

❑ 3 Locate the main portion of the **uterus** along the midline of the body. The female reproductive tract continues posteriorly from the uterus as the **vagina** toward the outside of the body. The vagina ends at the *urogenital sinus.*

C. Reproductive anatomy of the male fetal pig

This activity challenges you to identify the major reproductive structures of the male fetal pig. Assuming you have already removed the fetal pig's skin (Exercise 40), you simply have to move the viscera out of the way—without cutting or removing them—to see many of the structures indicated in this activity. You may have to cut some additional skin or muscle tissue to see a few of the structures more clearly.

> HINT → Figure 52-3 illustrates many of the structures of the fetal pig listed in this activity. Useful information can also be found in the LABORATORY REFERENCE Plate 85.

❑ 1 Locate the **scrotum,** a sac of skin between the hindlimbs of the animal. Cut the scrotum open with your scissors or scalpel, taking care not to cut the structures inside the scrotum.

❑ 2 Identify the immature **testes,** the male *gonads* within the scrotum. They should have descended into the scrotum late in fetal development. Remove the connective tissue covering each testis to reveal the coiled **epididymis.** Note that the **vas (ductus) deferens** leads from the epididymis, through the *inguinal canal,* into the abdominopelvic cavity.

❑ 3 Trace the course of one of the vas deferens into the abdominopelvic cavity to where it meets its partner from the other side. Note that they join together where they meet the **urethra.** A small, glandular **seminal vesicle** may be observed at the end of each vas deferens. Also at this location is a single mass of glandular tissue called the **prostate gland.**

❑ 4 Follow the course of the urethra toward the base of the **penis,** where a pair of small, elongated **bulbourethral glands** can be located.

D. Reproductive anatomy of the female fetal pig

This activity challenges you to identify the major reproductive structures of the female fetal pig. Assuming you have already removed the fetal pig's skin (Exercise 10) and opened the abdominopelvic cavity (Exercise 40), you simply have to move the viscera out of the way—without cutting or removing them—to see many of the structures indicated in this activity. You may have to cut some additional skin or muscle tissue to see a few of the structures more clearly.

> HINT → Figure 52-4 illustrates many of the structures of the fetal pig listed in this activity. Useful information can also be found in the LABORATORY REFERENCE Plate 86.

❑ 1 Locate the tiny lightly-colored **ovaries,** the female *gonads,* located just posterior to the kidneys.

❑ 2 Identify the **oviduct,** with its distal *fimbriae,* that leads from each ovary toward a **uterine horn.** Each uterine horn is a lateral extension of the superior portion of the uterus.

❑ 3 Locate the main portion of the **uterus** along the midline of the body. The female reproductive tract continues posteriorly from the uterus as the **vagina** toward the outside of the body. The vagina ends at the *urogenital sinus.*

ANATOMICAL ATLAS OF THE FETAL PIG
(REPRODUCTIVE STRUCTURES)

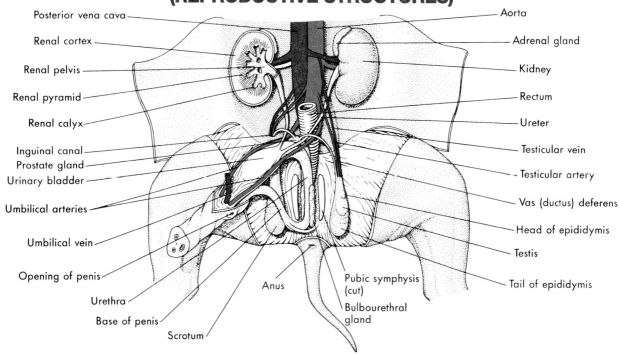

Figure 52-3 Reproductive anatomy of the male fetal pig. Ventral view.

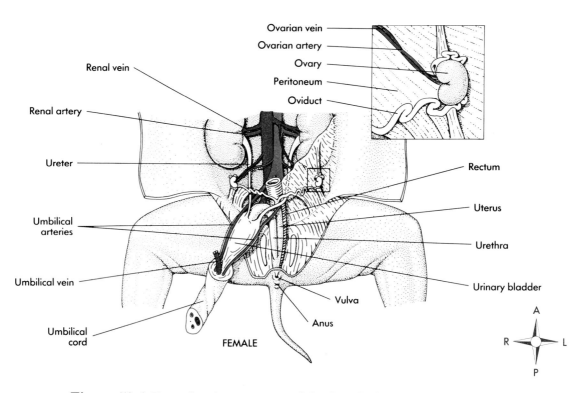

Figure 52-4 Reproductive anatomy of the female fetal pig. Ventral view.

NAME _____ DATE _____ SECTION _____

LAB REPORT 52

Dissection: Reproductive Systems

A. Male cat dissection checklist
- ❏ scrotum
- ❏ testis
- ❏ epididymis
- ❏ vas (ductus) deferens
- ❏ urethra
- ❏ prostate gland
- ❏ penis
- ❏ bulbourethral glands
- ❏ glans penis

B. Female cat dissection checklist
- ❏ ovaries
- ❏ oviduct
- ❏ uterine horn
- ❏ uterus
- ❏ vagina

C. Male fetal pig dissection checklist
- ❏ scrotum
- ❏ testis
- ❏ epididymis
- ❏ vas (ductus) deferens
- ❏ urethra
- ❏ seminal vesicle
- ❏ prostate gland
- ❏ penis
- ❏ bulbourethral glands

D. Female fetal pig dissection checklist
- ❏ ovaries
- ❏ oviduct
- ❏ uterine horn
- ❏ uterus
- ❏ vagina

LAB EXERCISE 53

Development

Human development begins from the moment of fertilization in the female reproductive tract. A sperm cell from the male parent and an oocyte from the female parent unite to form the first cell of the offspring. This cell divides (using mitosis) again and again, forming a mass of cells that will eventually become a mature human. This exercise presents a brief exploration of the stages of human prenatal development.

Before you begin

❑ Read the appropriate chapter in your textbook.

❑ Set your learning goals. When you finish this exercise, you should be able to
  ■ describe the basic plan of prenatal development
  ■ identify major features of successive developmental stages in models or charts

❑ Prepare your materials:
  ■ models or charts of early human development
  ■ videotaped film: MIRACLE OF LIFE (Nova/PBS)

❑ Read the directions and safety tips for this exercise **carefully** before starting any procedure.

A. Early human development

Obtain a set of models or charts that depict the major stages of early human development. Use the guidance given here and in your textbook to identify the major events of early development and the landmark structures associated with each stage.

❑ 1 The original cell of an offspring is termed the **zygote.** About 18 to 36 hours after being formed by fusion of an egg and sperm, this cell divides. The two daughter cells divide, then their daughter cells divide, and so on, until a ball of cells is formed. Locate a zygote in Figure 53-1 and in your model.

❑ 2 When the ball has about 32 cells and a fluid-filled cavity called the **blastocele** forms, the offspring is called a **blastocyst.** Most of the blastocyst is a single layer of cells called the **trophoblast,** but one portion also has an **inner cell mass** several cells thick. The inner cell mass develops into the embryo. The trophoblast develops into the **chorion,** which later becomes the placenta and the membranes that surround the embryo. Locate this stage in Figure 53-1 and in your model.

❑ 3 Around 7 days after fertilization, the blastocyst begins **implantation** by digesting its way into the endometrium of the mother's uterus. During this stage, the chorion (formerly the trophoblast) develops projections (**chorionic villi**) into the maternal blood supply. This close association of the embryonic and maternal tissue is called the **placenta.** Later in development, the placenta allows diffusion of substances between the embryo's blood and the mother's blood. Identify this stage in Figure 53-1, then in your model.

❑ 4 By day 11, several features have become visible. Identify these features in Figure 53-1, then in a model or chart:
  ■ **Connecting stalk**—Narrow piece of tissue that connects the inner cell mass to the developing placenta (it will eventually develop into the **umbilical cord**).
  ■ **Amniotic cavity**—Fluid-filled space within the cell mass, within a layer of cells called the **ectoderm** (this cavity will eventually develop into the **amniotic sac** surrounding the embryo).
  ■ **Yolk sac**—Another fluid-filled space within the cell mass, this one within a layer of cells called the **endoderm.**
  ■ **Embryonic disk**—Flat sheet of tissue formed by adjacent layers of endoderm and ectoderm.

❑ 5 By day 14, more features of the developing offspring become apparent, as seen in Figure 53-1:
  ■ **Primitive streak**—A thickened line in the endoderm, at the center of the elongated embryonic disk (the embryo will eventually form around the streak; the **notochord** extends from the cephalic end of the streak).
  ■ **Mesoderm**—A new *germ layer* that arises between the ectoderm and mesoderm.

❑ 6 Follow the developmental stages represented in your model or chart, noting that the three germ layers (endoderm, mesoderm, and ectoderm) develop into different tissues and organs. Using your textbook or a reference book, determine which major tissues or organs are derived from each of the three germ layers. Report your findings in Lab Report 53.

❑ 7 After about 60 days, the cartilage and membranous tissue of the embryonic skeleton begin to ossify. From this point onward, the offspring is termed a **fetus.** Examine a model or chart of a full-term fetus. Because its organs and systems have developed sufficiently, the full-term fetus is able to survive outside the uterus. Locate these structures associated with the fetus:

- **Amniotic sac**—A fluid-filled space derived from the amniotic cavity, it cushions the developing offspring. At **parturition,** the sac breaks and spills its contents through the birth canal (vagina).
- **Placenta**—The organ originally formed by the chorionic villi's implantation into the endometrial lining of the uterus, the mature placenta allows exchange of materials between the fetal circulatory system and the maternal circulatory system. Examine the anatomical relationship of these two sets of vessels.
- **Umbilical cord**—Derived from the connecting stalk observed in earlier stages, the umbilical cord includes **umbilical arteries** and an **umbilical vein.** Exercise 38 describes how these vessels fit into the fetal circulatory plan. The umbilical cord serves as a connection between the developing offspring and the mother's body.

❑ 8 When your exploration of the models and charts is complete, view the film MIRACLE OF LIFE, if available. This film follows the course of human development from fertilization onward and is an outstanding summary activity for your study of human reproduction and development.

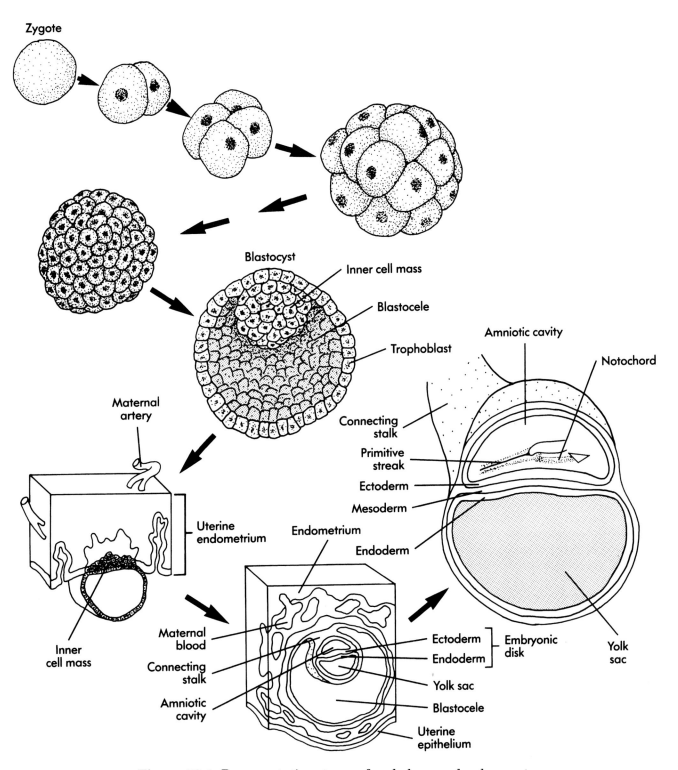

Figure 53-1 Representative stages of early human development.

ULTRASONOGRAPHY

Ultrasonography is a method of imaging body parts by using high-frequency sound (ultrasound) waves. Usually, a handheld wand that emits ultrasonic waves is placed against the patient's skin. The waves pass through different tissues at different speeds, and some waves are reflected back to the wand. Sensors in the wand detect the reflected ultrasound and relay the information to a computer capable of constructing an image based on the reflections. The image produced by this technique is called a *sonogram*.

An advantage of the technique is that ultrasound energy is not as dangerous as the radiation used in CT scans and regular radiographs. This is one reason that it has become popular among obstetric physicians. Developing fetal tissue that could be harmed by x-rays is not harmed by ultrasound waves. Therefore, the development of a particular embryo or fetus can be monitored by ultrasonography.

Sonograms can be used to determine whether there is a single fetus or multiple fetuses. Abnormally developed structures are sometimes visible. Late in development, male or female genitals can sometimes be seen.

Occasionally, measurements of a fetus at various stages are made. The normal rate of development of organs can then be verified. Usually, the age of a fetus is given as the clinical age rather than the developmental age used earlier in this exercise. The clinical age is the time since the mother's last menstrual period, whereas the developmental age is the time since fertilization.

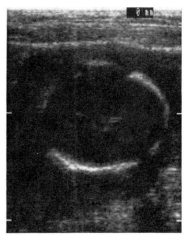
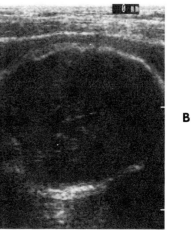

Figure 53-2 A, shows the outline of a fetal skull at 153 days (clinical age). Compare it to the skull of the same fetus at 206 days, shown in Figure 53-2, **B.**

1. The size of the skull has increased by what percentage during the 53 days between sonograms?

2. Which structures in particular do you think are likely to show up well in a sonogram?

3. What other applications for ultrasonography can you think of?

4. What advantages do CT scans, MR images, and regular radiographs have over sonograms?

LAB REPORT 53

Development

Identify (Identify the germ layer from which each of these organs or tissues is derived. Write "ENDO" for endoderm, "ECTO" for ectoderm, and "MESO" for mesoderm.)

1. Adrenal medulla
2. Anterior pituitary gland
3. Bones of the face
4. Bones (except those of the face)
5. Brain
6. Cardiovascular organs
7. Dermis
8. Epidermis
9. Gonads
10. Kidney ducts and bladder
11. Kidneys
12. Lens and cornea of eye
13. Lining of GI tract
14. Linings of hepatic and pancreatic ducts
15. Lining of lungs
16. Melanocytes
17. Muscle
18. Nasal cavity
19. Outer ear
20. Parathyroid gland
21. Skeletal muscles of the head
22. Spinal cord
23. Thymus gland
24. Thyroid gland
25. Tonsils
26. Tooth dentin and pulp
27. Tooth enamel

Multiple choice

_____ 1
_____ 2
_____ 3
_____ 4
_____ 5
_____ 6
_____ 7

Multiple choice (choose the best response)

1. The cell formed by the union of a sperm cell and an oocyte is called a
    a. blastula
    b. morula
    c. zygote
    d. blastocele
    e. b and c are correct

2. The connecting stalk seen early in human development later becomes the
    a. placenta
    b. amniotic sac
    c. umbilical cord
    d. yolk sac

3. The earliest developmental stage at which one can see a fluid-filled extracellular cavity is
    a. morula
    b. blastula
    c. zygote
    d. fetus
    e. 4-cell stage

4. The organ in which maternal and fetal blood exchange material is the
    a. umbilical cord
    b. umbilical vein
    c. amniotic sac
    d. amniotic cavity
    e. placenta

5. The process by which a baby is born is called
    a. gestation
    b. pregnancy
    c. differentiation
    d. parturition
    e. lactation

6. The chorion is derived from the
    a. trophoblast
    b. blastocele
    c. inner cell mass

7. The embryo develops around a thickened line in the endoderm called the
    a. trophoblast
    b. inner cell mass
    c. notochord
    d. primitive streak

LAB EXERCISE 54

Genetics and Heredity

Heredity, or the inheritance of traits, falls into the realm of **genetics.** Genetic information, in the form of **genes** found in the 23 DNA molecules that you inherit from your mother and the 23 that you inherit from your father, is passed from generation to generation. Twenty-two pairs of your DNA molecules, or *chromosomes,* are called **autosomes.** The remaining pair is called the **sex chromosomes.** One member of each pair is inherited from one parent, the other member of the pair from the other parent.

Each autosome in a pair is homologous to the other member of the pair. With this arrangement, a person has two genes for every inherited characteristic. If one gene is always expressed, whether or not its mate is the same gene, geneticists call that gene **dominant.** A gene that is not expressed when its mate is different is termed a **recessive gene.**

Before you begin

❑ Read the appropriate chapter in your textbook.

❑ Set your learning goals. When you finish this exercise, you should be able to
- identify the phenotype and possible genotypes of selected human characteristics
- distinguish between the concepts of *dominant* and *recessive* as applied to inherited traits
- predict probabilities of phenotypes and genotypes in offspring given the parents' genotypes or phenotypes

❑ Prepare your materials:
- taste papers: control, sodium benzoate, PTC, thiourea
- BIOHAZARD container
- blood typing materials (if not done previously in Lab Exercise 34)

❑ Read the directions and safety tips for this exercise **carefully** before starting any procedure.

A. Human phenotypes and genotypes

For any particular inherited characteristic, a **phenotype** can be identified. The phenotype is the characteristic actually expressed in an individual. A person's **genotype** is a statement of both genes that influence a particular trait. For example, *albinism* is a recessive genetic condition in which a person lacks skin pigmentation. By convention, a gene for normal skin pigmentation, which is dominant, is represented as *A*. The recessive gene is represented as *a*. A person with normal skin pigmentation may have the genotype *AA,* meaning that both genes are normal. Geneticists may state that the person's genotype is **homozygous,** meaning that both genes are the same. A person with normal skin color could also have the *Aa* genotype. Because *A* is dominant, the abnormal *a* is not expressed. This person has a **heterozygous** genotype, meaning that the two genes are not the same. A person with albinism must have the genotype *aa*, because it is the only genotype that allows the recessive abnormal trait to be expressed. Thus, the normal phenotype (normal skin color) may be associated with the genotype *AA* or the genotype *Aa*. The abnormal phenotype (albinism) is always associated with the genotype *aa*.

> **HINT** → Skin color is determined by a number of factors, both environmental and genetic. A number of different *A* genes exist in the human gene pool, and genes at other locations in the DNA may influence skin color. The example of albinism is simple when first observed, but the effect of multiple factors complicates things. This is typical of inherited characteristics. Often, many genes influence one particular trait. Also, many environmental factors can affect inherited characteristics.

This section instructs you to determine your phenotype for a variety of easily observed characteristics. Once you know your phenotype, you can determine your possible genotypes. For example, if you are an albino, you know that your genotype is *aa*. If you do not have albinism, your genotype is either *AA* or *Aa*. For each characteristic described, record your phenotype and possible genotypes in Lab Report 54.

> **SAFETY FIRST!** The first step of this activity involves tasting test papers. Be sure to use clean procedures in handling the papers before and after you taste them. Used papers are to be placed in a BIOHAZARD container immediately after use.

☐ 1 The first set of characteristics to be determined involves the sense of taste. It is known that the ability to taste certain compounds depends on the presence of certain genes. You will taste papers that have each been impregnated with a different compound. Before starting, put a piece of CONTROL test paper on your tongue and chew it. If you taste something, you must be sensitive to a compound in the paper itself. Ignore that particular taste in the tests that remain or try to sense *differences* in taste.
  - **Sodium benzoate test**—Ability to taste something sweet, salty, or bitter in the paper is dominant.
  - **PTC (phenylthiocarbamide) test**—Ability to sense a bitter taste is dominant.
  - **Thiourea test**—Ability to taste something bitter is dominant.

☐ 2 The next set of determinations involves anatomical characteristics of your hand:
  - **Bent little finger**—Place your relaxed hand flat on the lab table. If the distal phalanx of the little (fifth) finger bends toward the fourth finger, you have the dominant trait.
  - **Middigital hair**—Dorsal hair on the skin over the middle phalanges of the hand is dominant.
  - **Hitchhiker's thumb**—If you can hyperextend the distal joint of the thumb noticeably, you have the recessive trait.

☐ 3 Determine your phenotype for these facial features:
  - **Pigmented anterior of the iris**—If you have pigment on the anterior *and* posterior of the iris, your eyes are green, brown, black, or hazel. If you lack pigment on the anterior aspect of the iris, your eyes are blue or gray. Anterior pigmentation is dominant.
  - **Attached earlobes**—If the inferior, fatty lobe of the ear is attached rather than free, you have the recessive trait.
  - **Widow's peak**—Assuming you have a hairline, if it is straight across the forehead, you have the recessive trait. If it forms a downward point near the midline, you have the dominant "widow's peak."
  - **Tongue roll**—Try to curl your tongue as you extend it from your mouth. If you can't curl it, you have the recessive trait.
  - **Freckles**—If your face has a scattering of freckles, you have the dominant form of this characteristic. If your face is free of freckles, you have the recessive condition.

☐ 4 There are two dominant genes for ABO blood types. One is I^A, which signifies the presence of the A antigen. The other is I^B, signifying the B antigen. The recessive gene is i, signifying neither ABO antigen. A person with the phenotype TYPE A has the genotype $I^A I^A$ or $I^A i$. A person with the phenotype TYPE B has the genotype $I^B I^B$ or $I^B i$. A person with phenotype TYPE AB has the genotype $I^A I^B$. Phenotype TYPE O requires the genotype ii. Record your phenotype (your ABO blood type). What is your genotype?

> **HINT →** If you have not already typed your blood (as instructed in Lab Exercise 34), consult your health records or perform the typing now. Refer to Lab Exercise 34 and heed the safety advice given there.

☐ 5 The Rh blood type is also determined by genetics. Presence of the Rh antigen is dominant. What is your Rh blood type? What is your genotype?

B. Probabilities of inheritance

Now that you have a grasp of the concept of phenotype and genotype, you can move on to the concept of **probability**. Probability is the likelihood of a certain outcome in a particular event. As applied to human genetics, probability refers to the likelihood that the offspring of a particular set of parents will have a certain inherited condition. **Genetic counselors** work with prospective parents to determine their possible genotypes for a variety of traits. Then they predict the probability of their children having those traits. In this way, parents can anticipate possible abnormal genetic conditions.

In this exercise, we will use a simple method developed by the English geneticist Punnett. He devised a simple grid, or **Punnett square,** with which one can easily predict simple ratios of genetic probability.

First, an example:

*The father has freckles; the mother does not. The gene for freckles is **F**, and the gene for no freckles is **f**. The father's genotype is either **FF** or **Ff**. The mother's genotype must be **ff**. We must consider two different outcomes in offspring from this couple because the father has two possible genotypes for this trait. First, let's assume that the father's genotype is **FF**. Any sperm cell contributed by the father will have the **F** gene. Any oocyte contributed by the mother will have the **f** gene. We place the possible gene for the father in the left margin of the Punnett square and the mother's in the top margin:*

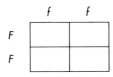

To use the Punnett square, start in the top left square of the grid. Combine the gene from the left margin with the gene from the top margin:

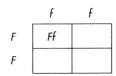

*We now have one possible offspring genotype: **Ff**. We then do the same for the other three squares of the grid:*

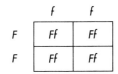

*Our Punnett square predicts a 100% probability that any one offspring will have genotype **Ff** and therefore have the freckled phenotype. The second possibility is that the father's genotype is **Ff**. We can set up a Punnett square for this possibility:*

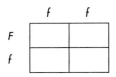

We can now fill in the grid with possible offspring genotypes, combining the gene from the left margin with the gene from the top margin for each square:

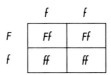

*Because half the squares have genotype **Ff**, there is a 50% probability that any offspring of this couple will have the **Ff** genotype (and therefore the freckled phenotype). There is also a 50% probability that the genotype will be **ff**, expressed as the nonfreckled phenotype. If the phenotypes or genotypes of the clients' parents and grandparents are known, a genetic counselor might construct a **pedigree,** or family tree, to determine the father's likely genotype. If that information is unavailable, the parents will have to deal with the two different probabilities.*

Now that you have seen an example, try your hand at the following problems. Draw your Punnett squares and record your results in Lab Report 54.

❑ 1 Huntington's chorea is a degenerative nerve disorder with a genetic basis that becomes apparent after about the age 40. The Huntington's chorea gene, *H,* is dominant. The normal, recessive gene is *h*. One of Heather's parents has Huntington's chorea, and the other does not. Can you predict the highest probability that Heather will develop Huntington's chorea later in her life?

❑ 2 Kevin has Rh positive blood. Christine has Rh negative blood. Their first child, Andrew, has Rh positive blood. Both of Kevin's parents have Rh positive blood. What is the probability that the second child Kevin and Christine are expecting will be Rh negative?

❑ 3 Leo's father has albinism, but Leo does not. Cleo's father has albinism, but she does not. If Leo and Cleo have a child, what is the probability that it will have albinism? What is the probability that their second child will have albinism? Their third child?

❑ 4 In the ABO blood typing system, Mario is type O. Ana is type AB. What ABO blood types might their children have?

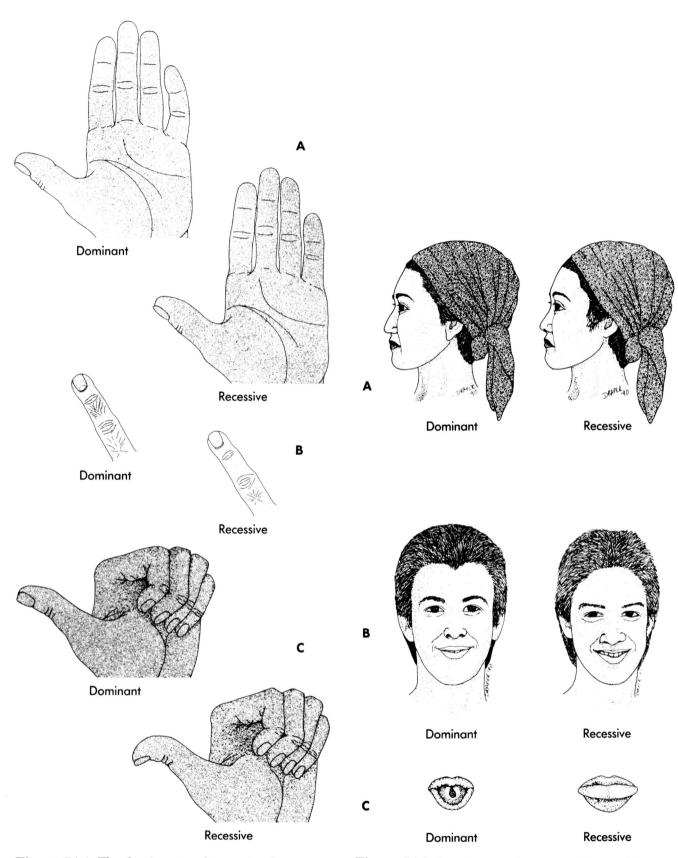

Figure 54-1 The dominant and recessive forms of **A,** bent little finger, **B,** middigital hair, and **C,** hitchhiker's thumb.

Figure 54-2 Dominant and recessive forms of **A,** attached earlobes, **B,** widow's peak, and **C,** tongue roll.

LAB REPORT 54

Genetics and Heredity

Trait	Dominant gene(s)	Recessive gene	Your phenotype	Possible genotypes
Sodium benzoate taste	S	s		
PTC taste	P	p		
Thiourea taste	T	t		
Bent little finger	L	l		
Middigital hair	M	m		
Hitchhiker's thumb	H	h		
Pigmented anterior of iris	I	i		
Attached earlobes	A	a		
Widow's peak	W	w		
Tongue roll	R	r		
Freckles	F	f		
ABO blood type	I^A, I^B	i		
Rh blood type	D	d		

1. What is the highest probability that Heather will develop Huntington's chorea? _____%

2. What is the probability that Kevin and Christine's second child will have Rh negative blood? _____%

3. What is the probability that Leo and Cleo's first child will be albino? _____%
   Their second child? _____%
   Their third child? _____%

4. What are the possible ABO blood types of Mario and Ana's children? _____

LAB EXERCISE 55

The Whole Body

This exercise is intended as a possible summary or synthesis activity for the laboratory course. Whether your instructor will invite you to complete this exercise will depend on the course schedule and the availability of the required materials.

The first activity of this exercise suggests that you observe a live or videotaped demonstration of human anatomy using a human body. The second activity presents a special focus on transverse (horizontal) sectional anatomy, which is becoming increasingly important in practical applications of human anatomy.

> **SAFETY FIRST!** Do not touch the cadaver unless invited to do so by the demonstrator. If you are invited to touch it, use gloved hands. Cadavers are normally fixed in formalin, which is a toxic substance. Do not inhale the fumes that are immediately above the specimen. Excuse yourself from the demonstration if you feel ill, or if you may be sensitive to formalin. Pregnant women should avoid contact with formalin fumes.

Before you begin

❑ Set your learning goals. When you finish this exercise, you should be able to
- describe the overall body plan of the human
- identify the major organs of the human in a previously dissected cadaver or in a human body model or chart
- identify structures in selected transverse sections of the human body

❑ Prepare your materials:
- prosected human cadaver (for demonstration, if available) or human dissection videotaped film
- plastinated preparation (or chart): *human thorax, horizontal section (just above heart)*
- plastinated preparation (or chart): *human abdomen, horizontal section (pancreas level)*
- demonstration pointers
- hand lens

❑ Read the directions and safety tips for this exercise **carefully** before starting any procedure.

A. The human cadaver

Advanced laboratory courses in human anatomy use the human cadaver as the basic dissection specimen. A **cadaver** is an embalmed corpse. If available, view a live or videotaped demonstration using a prosected human cadaver. A prosected specimen is one that has been previously dissected and prepared for anatomical demonstrations.

B. Transverse sectional anatomy

Clinical and research institutions are seeing an increased use of computed tomography (CT) scans, magnetic resonance imaging (MRI), ultrasonography, and other advanced imaging techniques. Each of these techniques requires a basic knowledge of human anatomy as seen in a transverse, or horizontal, section.

Your studies so far have emphasized the sagittal and frontal aspects of anatomy. Now that you are familiar with the essentials of human anatomy from those perspectives, it is time to apply that knowledge to locating structures in transverse preparations.

> **SAFETY FIRST!** Plastinated specimens are natural tissues that have been embedded in plastic. Even though they long outlast normal embalmed specimens, they are very fragile. If you handle them at all, be careful not to damage them. Always use a demonstration pointer in pointing to structures, never a pen or pencil. Do not touch the specimen with your pointer, just point at it.

❑ 1 Obtain a plastinated transverse section of a human torso just superior to the heart (at about T4). Try to locate these structures:
- **Aorta**—If the aorta is present, what section is it (ascending, arch, or descending)?

- **Superior vena cava**—Can you identify this vein by its position and the thinness of its wall?
- **Esophagus**—This is a small (and perhaps flattened) muscular tube anterior to a vertebra.
- **Trachea**—Are any of the C-shaped cartilage rings visible in the tracheal wall?
- **Thymus**—It is posterior to the sternum.
- **Lungs**—Can you distinguish any portions of the respiratory tract within the lung tissue? Identify the parietal and visceral pleurae and the pleural cavity.
- **Sternum**—This structure is in the anterior wall of the thoracic cavity.
- **Ribs**
- **Thoracic vertebra**—What features of the vertebra are visible in your specimen?
- **Spinal cord**—It is inside the vertebral foramen. What features of the spinal cord can you identify?
- **Skeletal muscles**—They form the wall of the thoracic cavity.
- **Subcutaneous tissue**—Identify the areolar and/or adipose tissue under the skin.
- **Integument**—Can you distinguish between the dermis and epidermis?

Are there any other organs or structures visible in your specimen? If the upper arms are present in your specimen, identify the humerus and other arm structures.

> **HINT** → Use a hand lens to examine the detail of the structures you have found. Refer to Plates 88 through 90 in the LABORATORY REFERENCE, which show color photographs of transverse sections of human cadavers.

☐ 2 Obtain a plastinated transverse section of the human abdomen at the level of the pancreas (about T12). Try to locate these structures:
- **Vertebra**—What features are visible in this preparation?
- **Spinal cord**—Can you distinguish white and gray matter?
- **Kidneys**—Are they the same size? Why or why not? Identify the renal fat pad around each kidney. Can you distinguish between the renal cortex and the renal medulla?
- **Aorta**—The abdominal aorta may be present just anterior to the body of the vertebra.
- **Inferior vena cava**—This large vein is normally next to the aorta.
- **Pancreas**—This glandular tissue is near the stomach.
- **Stomach**—This large muscular organ should be present in the left portion of the specimen. Can you identify the rugae (large folds) of the gastric mucosa?
- **Liver**—This is the dominant feature of your specimen, a large mass of tissue to the right and anterior of the abdominal cavity.
- **Colon**—Is any portion of the colon visible in your specimen? If so, which section is it?
- **Peritoneum**—Can you distinguish between the parietal and visceral portions of the peritoneum? Identify the peritoneal cavity. Are any mesenteries distinguishable in your specimen? The lesser omentum?
- **Ribs**
- **Abdominal muscles**—Which abdominal muscles could be present in your section?
- **Subcutaneous tissue**—Is the adipose tissue in this layer of the same thickness all the way around the abdominal wall?
- **Integument**—Are the dermis and epidermis distinguishable in this section?

> **HINT** → Each specimen is unique, just as every person's body is unique. Therefore, some structures listed may not be present (or identifiable) in your specimen. Refer to Plates 99 and 100 in the LABORATORY REFERENCE. These plates show transverse sections of a cadaver's abdomen.

Transverse Thoracic Section

AORTA 1
SUPERIOR VENA CAVA 2
ESOPHAGUS 3
TRACHEA 4
THYMUS 5
LUNG 6
PLEURAL CAVITY 7
STERNUM 8
RIB 9
VERTEBRA 10
SPINAL CORD 11
SKELETAL MUSCLE 12
SUBCUTANEOUS TISSUE 13
INTEGUMENT 14

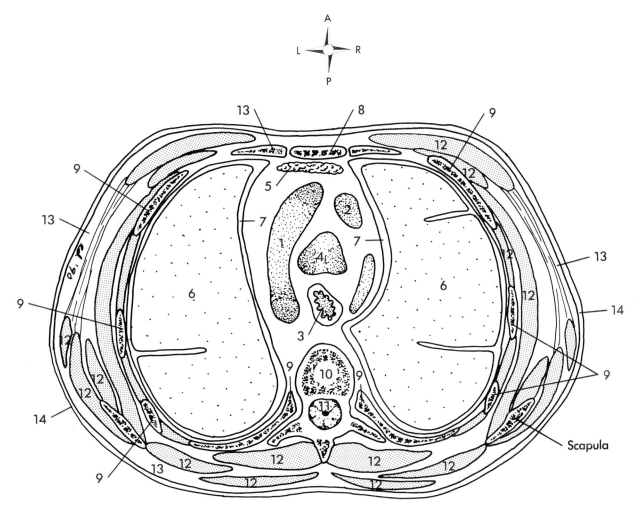

Figure 55-1

COLORING EXERCISE Using colored pens or pencils, shade in the figure and accompanying labels in contrasting colors of your choice as indicated by the red numerals.

VERTEBRA 1
SPINAL CORD 2
KIDNEY 3
AORTA 4
INFERIOR VENA CAVA 5
PANCREAS 6
STOMACH 7

LIVER 8
COLON 9
PERITONEAL CAVITY 10
RIB 11
ABDOMINAL MUSCLE 12
SUBCUTANEOUS TISSUE 13
INTEGUMENT 14

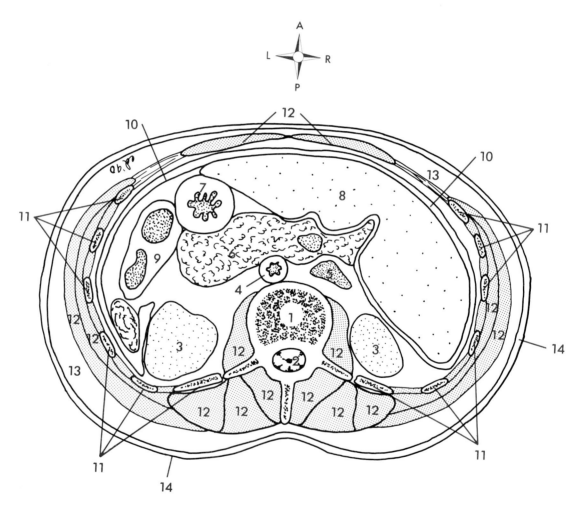

Figure 55-2

NAME _____ DATE _____ SECTION _____

LAB REPORT 55

The Whole Body

Sketch and label your thorax (transverse section) specimen:

Sketch and label your abdomen (transverse section) specimen:

NAME _____ DATE _____ SECTION _____

LAB REPORT SUPPLEMENT

Microscopic Observations

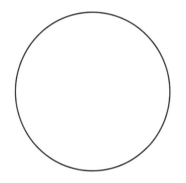

Specimen: _____

Total Magnification: _____

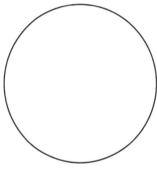

Specimen: _____

Total Magnification: _____

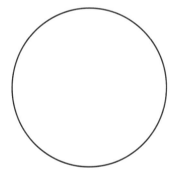

Specimen: _____

Total Magnification: _____

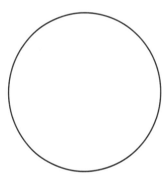

Specimen: _____

Total Magnification: _____

NOTES:

NAME _____ DATE _____ SECTION _____

LAB REPORT SUPPLEMENT

Microscopic Observations

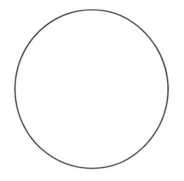

Specimen: _____

Total Magnification: _____

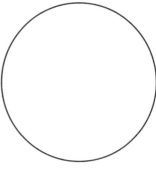

Specimen: _____

Total Magnification: _____

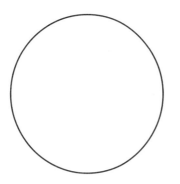

Specimen: _____

Total Magnification: _____

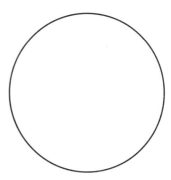

Specimen: _____

Total Magnification: _____

NOTES:

NAME_____ DATE_____ SECTION_____

LAB REPORT SUPPLEMENT

Microscopic Observations

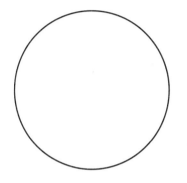

Specimen: _____

Total Magnification: _____

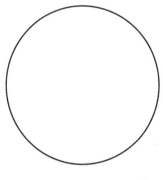

Specimen: _____

Total Magnification: _____

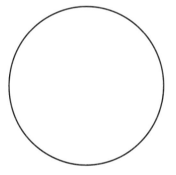

Specimen: _____

Total Magnification: _____

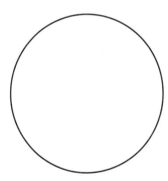

Specimen: _____

Total Magnification: _____

NOTES:

NAME _____ DATE _____ SECTION _____

LAB REPORT SUPPLEMENT

Microscopic Observations

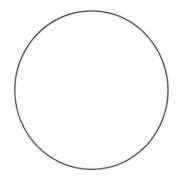

Specimen: _____

Total Magnification: _____

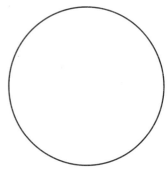

Specimen: _____

Total Magnification: _____

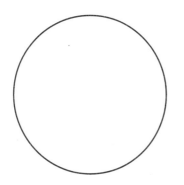

Specimen: _____

Total Magnification: _____

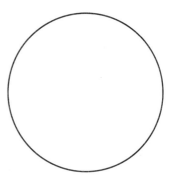

Specimen: _____

Total Magnification: _____

NOTES:

LABORATORY REFERENCE

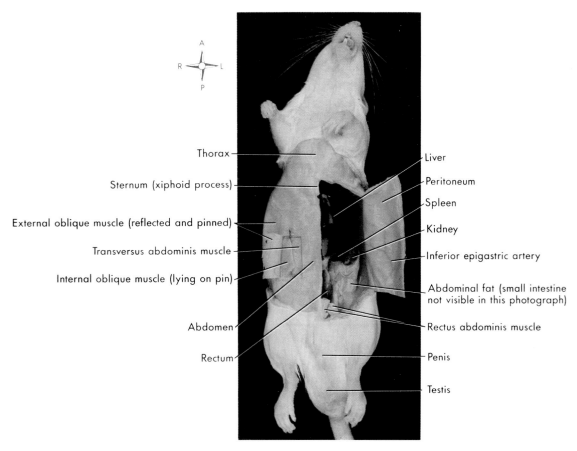

Plate 1 Abdomen of the male rat. Ventral view, left portion of abdominal cavity exposed.

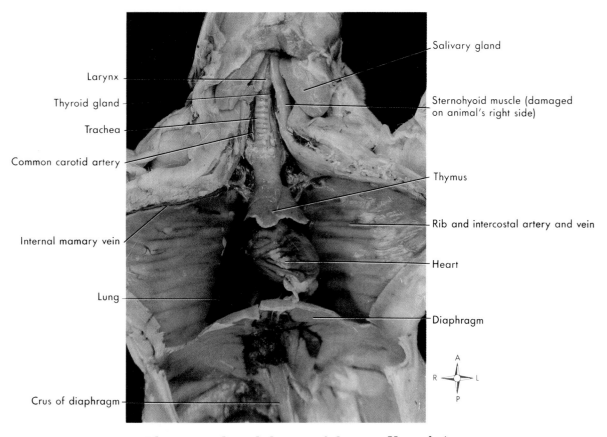

Plate 2 Neck and thorax of the rat. Ventral view.

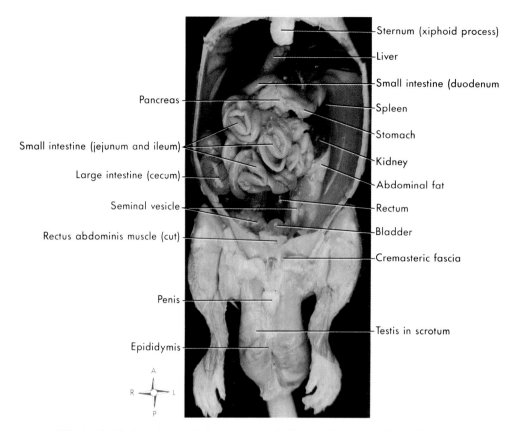

Plate 3 Abdominopelvic cavity of the male rat. Ventral view.

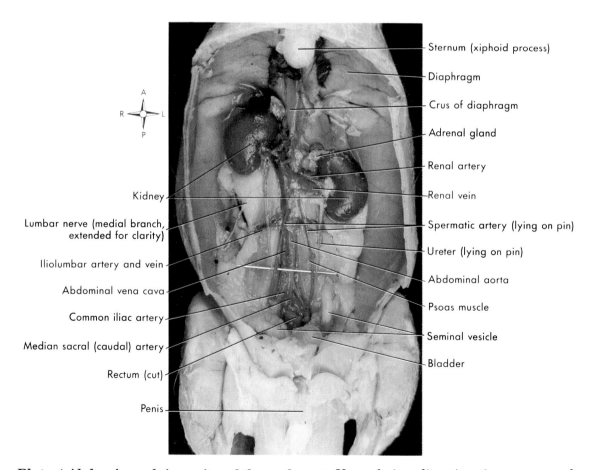

Plate 4 Abdominopelvic cavity of the male rat. Ventral view, digestive viscera removed.

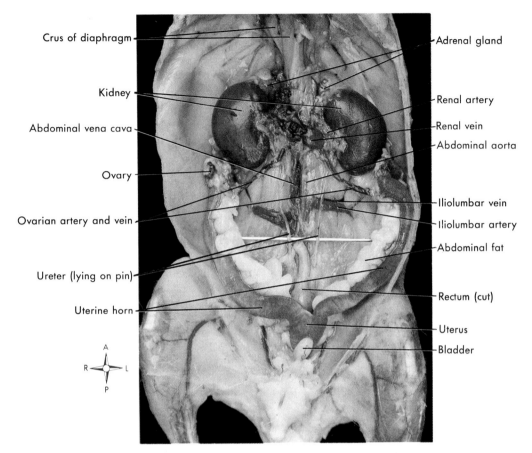

Plate 5 Abdominopelvic cavity of the female rat. Ventral view, digestive viscera removed.

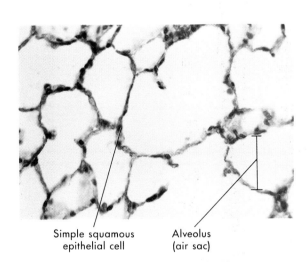

Plate 6 Simple squamous epithelium. Cross section of lung shows this thin tissue (viewed from side) lining the air sacs.

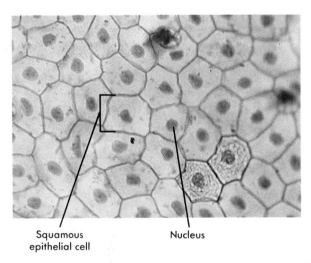

Plate 7 Simple squamous epithelium. Sheet of this tissue viewed from above.

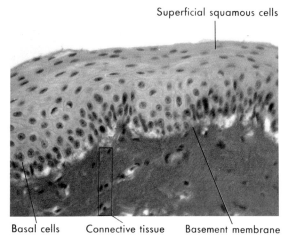

Plate 8 Stratified squamous epithelium (nonkeratinized). Vaginal tissue. Each cell in the top layer is flattened and most have visible nuclei.

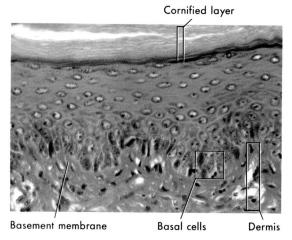

Plate 9 Stratified squamous epithelium (keratinized). Note cells are progressively flattened and scale-like toward the surface. Outer surface contains many layers of extremely flattened dead cells with no visible nuclei.

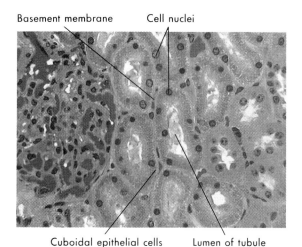

Plate 10 Simple cuboidal epithelium. Kidney tubules. Note that the single layer of cuboidal cells encloses the tubule opening (lumen).

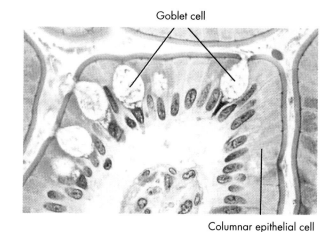

Plate 11 Simple columnar epithelium. Note the mucus-producing goblet cells.

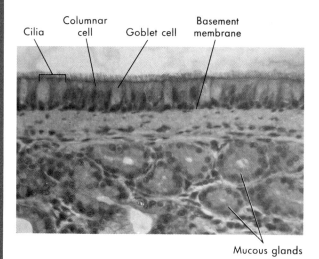

Plate 12 Pseudostratified ciliated epithelium. Trachea. Each irregularly shaped columnar cell touches the underlying basement membrane. Nuclei at irregular levels in the cells gives a false (pseudo) impression of stratification.

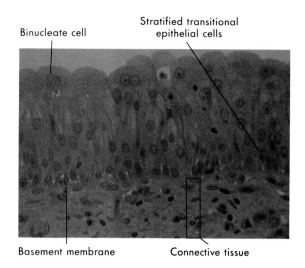

Plate 13 Transitional epithelium. Urinary bladder lining. Note variable cell shapes among the several layers present. Intermediate and surface cells do not touch the basement membrane.

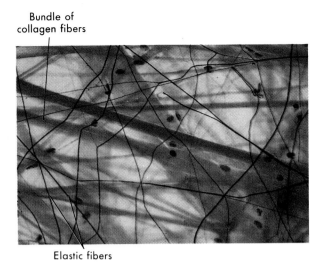

Plate 14 Loose, ordinary (areolar) connective tissue. Note loose arrangement of lightly stained bundles of collagenous fibers and thick, darkly stained elastic fibers.

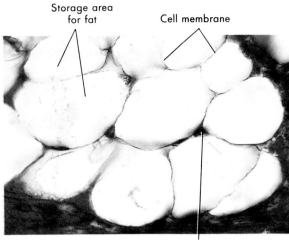

Plate 15 Adipose tissue. Note the large storage spaces for fat inside the adipose tissue cells.

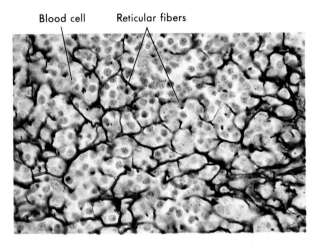

Plate 16 Reticular connective tissue. The supporting framework of reticular fibers is stained black in this section of spleen tissue.

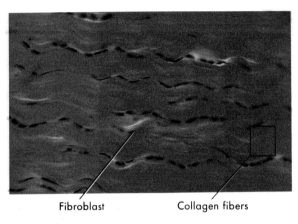

Plate 17 Dense fibrous connective tissue (regular). Photomicrograph of tissue in a tendon. Note the multiple, regular bundles of collagenous fibers arranged in parallel rows.

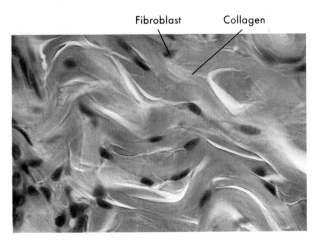

Plate 18 Dense fibrous connective tissue (irregular). Section of dermis showing irregular arrangements of collagenous fibers (pink) and purple-staining fibroblast cell nuclei.

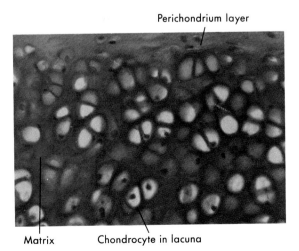

Plate 19 Hyaline cartilage. Trachea. Note the many cell spaces, or lacunae, in the gel-like matrix.

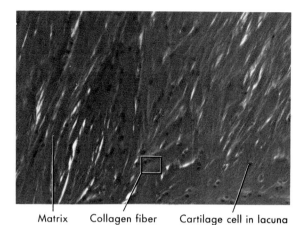

Plate 20 Fibrocartilage. Pubic symphysis joint. Strong dense fibers filling the matrix convey shock-absorbing qualities.

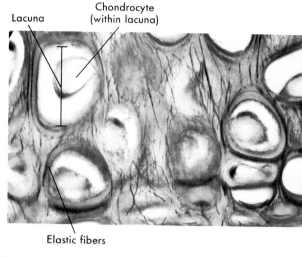

Plate 21 Elastic cartilage. Note the cartilage cells in the lacunae surrounded by matrix and dark-staining elastic fibers. (Very high power.)

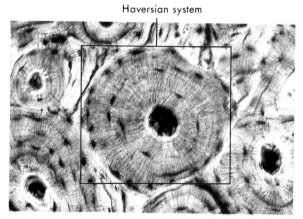

Plate 22 Bone tissue (compact). Dried, ground bone. Many wheel-like *osteons* are apparent in this section.

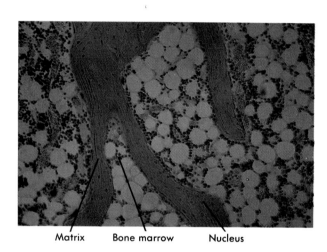

Plate 23 Bone tissue (cancellous). Note branching trabeculae surrounded by marrow (myeloid tissue) with a reticular meshwork.

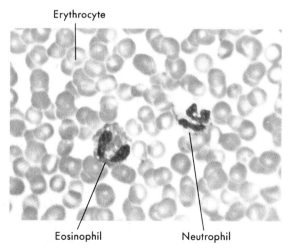

Plate 24 Blood. Blood smear shows two white blood cells (leukocytes) surrounded by many smaller red blood cells (erythrocytes).

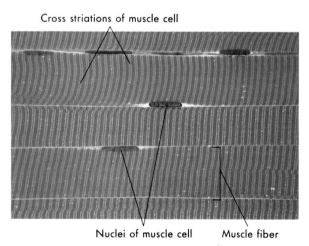

Plate 25 Skeletal muscle. Note striations of muscle cell fibers in longitudinal section.

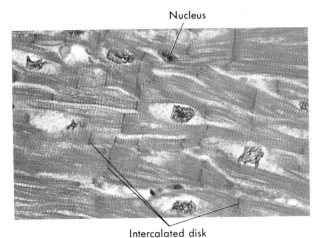

Plate 26 Cardiac muscle. Darker bands are the edges of *intercalated discs,* which join cardiac muscle cells end to end.

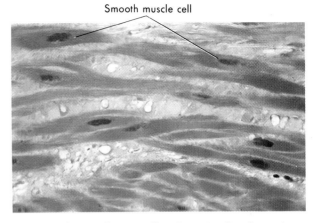

Plate 27 Smooth muscle. Longitudinal section. Note the central placement of nuclei in the spindle-shaped smooth muscle fibers.

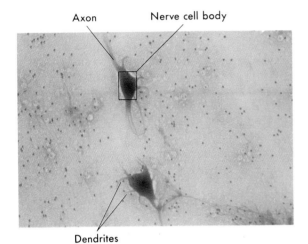

Plate 28 Nervous tissue. Multipolar neurons in smear of spinal cord, showing characteristic cell bodies and multiple cell processes.

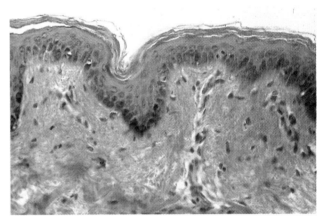

Plate 29 Thin skin. From a dark-skinned individual. Dark brown melanin in lower epidermis. (From Erlandsen SL, Magney JM: *Atlas of histology,* St Louis, 1992, Mosby.)

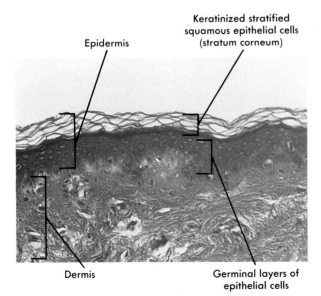

Plate 30 Thin skin. Note that the keratinized layer has split apart during slide preparation. (Low power.)

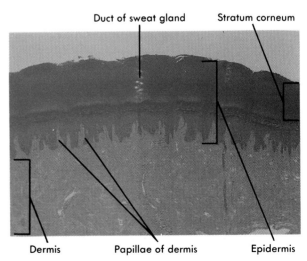

Plate 31 Thick skin. Note the coiled duct of a sweat gland. (Low power.)

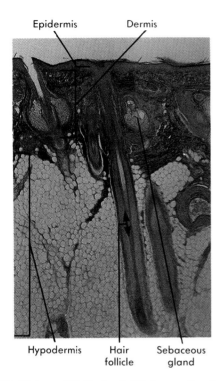

Plate 32 Skin with hair follicle and sebaceous gland. A hair follicle, root of the hair, and a sebaceous gland are all visible. (Low power.)

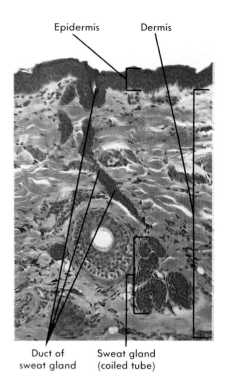

Plate 33 Skin with sweat glands. Coiled sweat ducts and darkly stained secretory portions of sweat glands are visible.

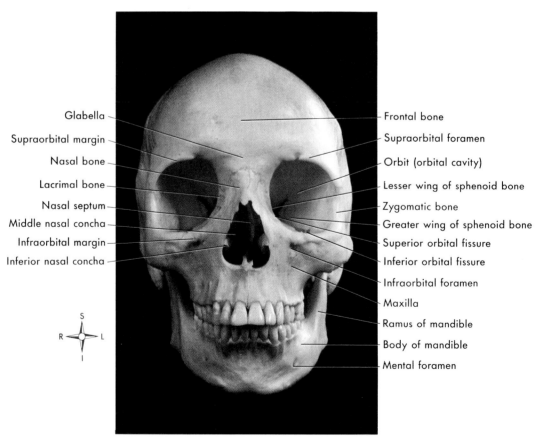

Plate 34 Skull. Anterior view.

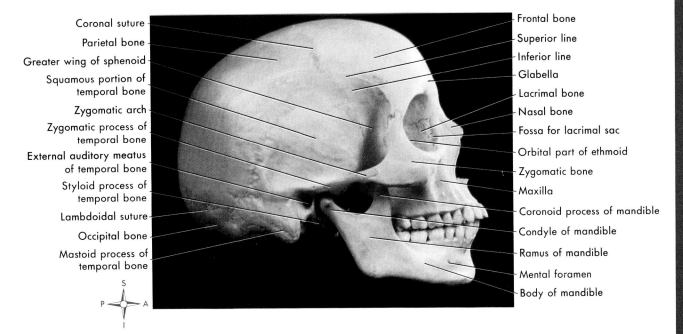

Plate 35 Skull. Lateral view (from subject's right).

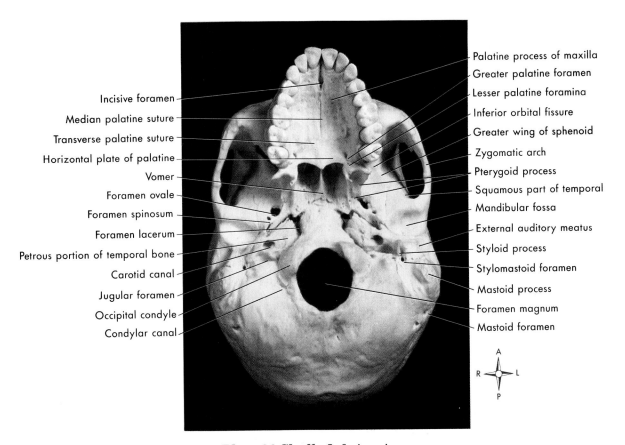

Plate 36 Skull. Inferior view.

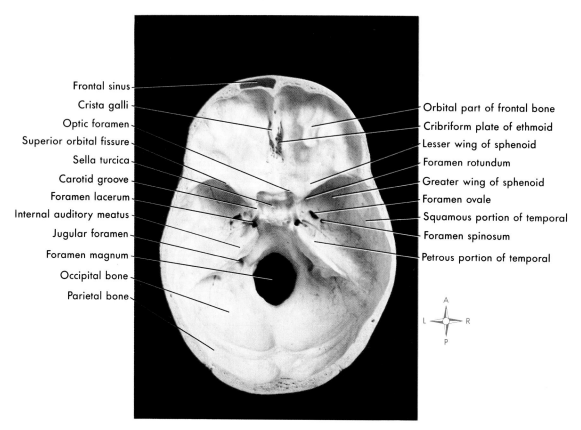

Plate 37 Skull. Superior view (with top, or calvarium, removed).

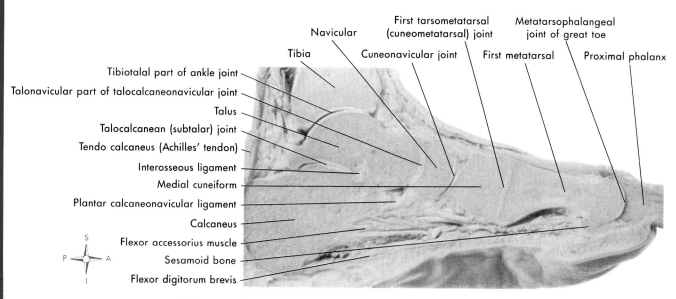

Plate 38 Sagittal section of the ankle and foot.

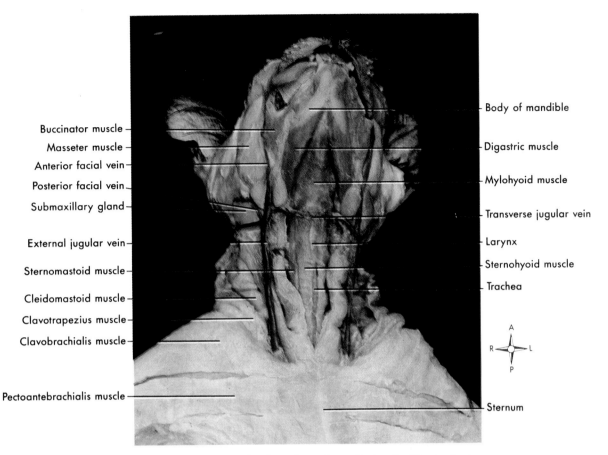

Plate 39 Muscles of the head and neck (cat). Ventral view.

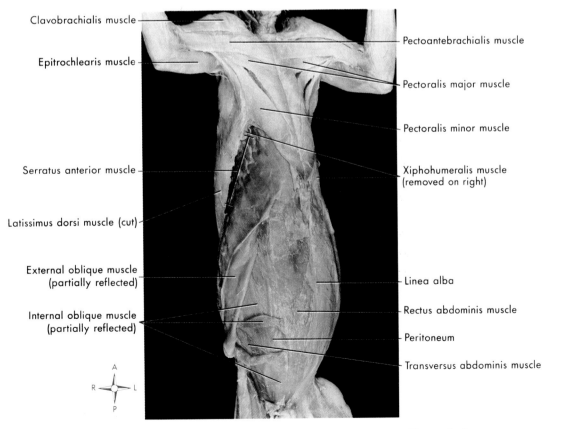

Plate 40 Muscles of the trunk and shoulders (cat). Ventral view.

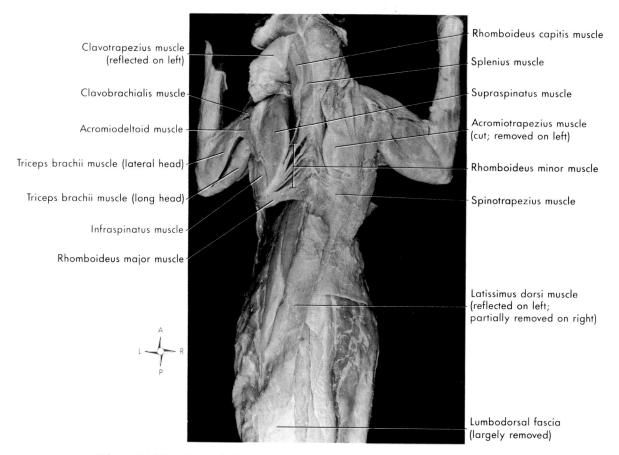

Plate 41 Muscles of the trunk and shoulders (cat). Dorsal view.

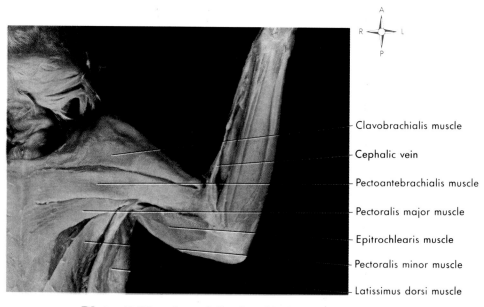

Plate 42 Muscles of the forelimb (cat). Medial view.

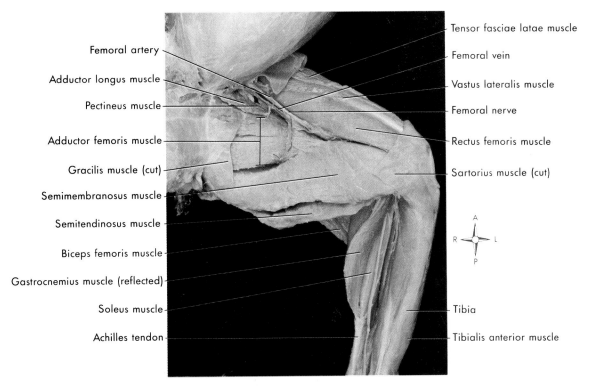

Plate 43 Muscles of the hindlimb (cat). Medial view.

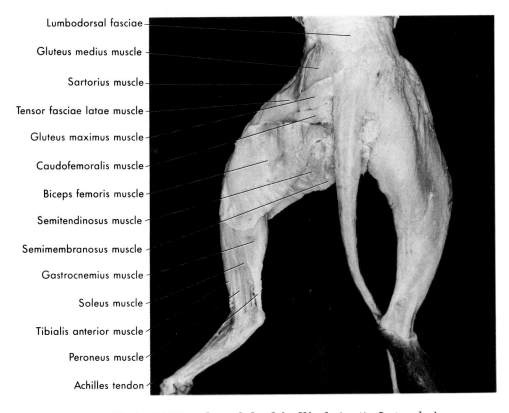

Plate 44 Muscles of the hindlimb (cat). Lateral view.

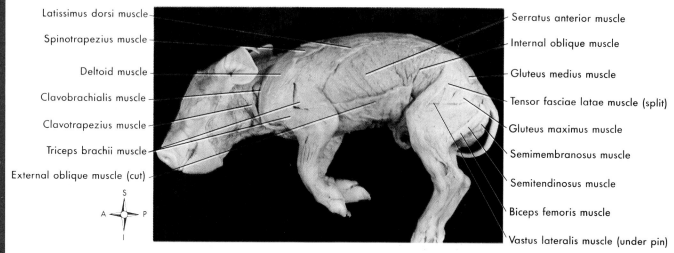

Plate 45 Lateral view of fetal pig muscles.

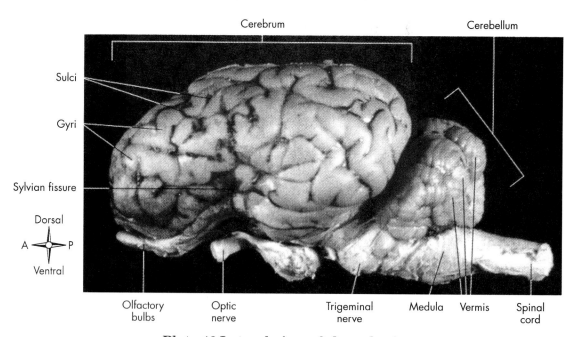

Plate 46 Lateral view of sheep brain.

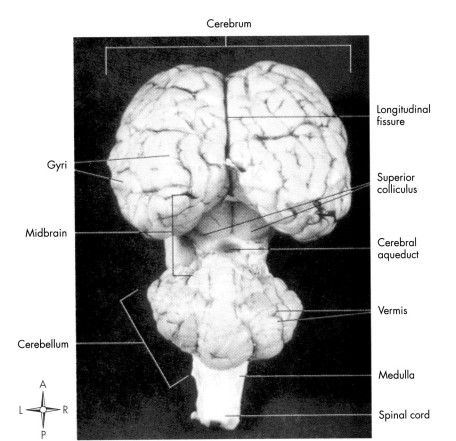

Plate 47 Dorsal view of sheep brain. The cerebellum has been pulled away from the cerebrum to show the dorsal aspect of the midbrain.

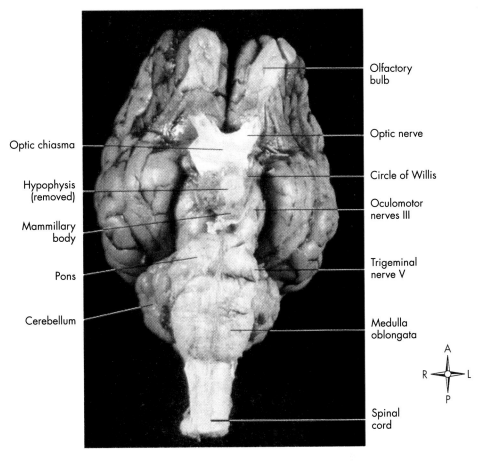

Plate 48 Ventral view of sheep brain.

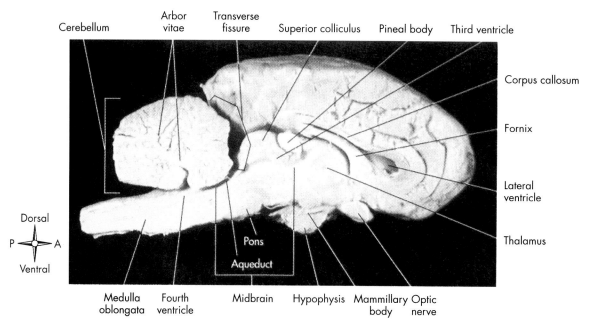

Plate 49 Midsagittal section of sheep brain.

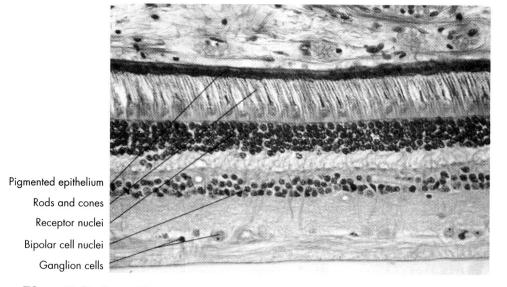

Plate 50 Retina. Photomicrograph of a cross section of the human retina.

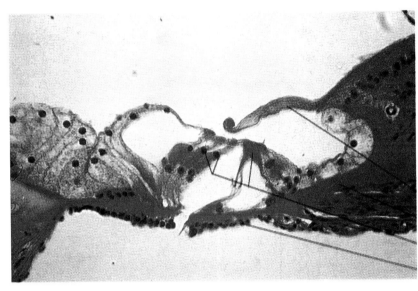

Plate 51 Organ of Corti. Photomicrograph showing details of the organ of Corti within the cochlea of the ear.

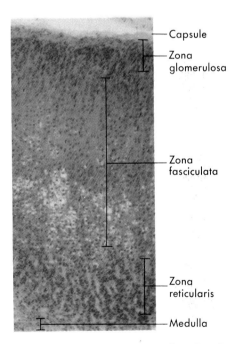

Plate 52 Adrenal gland tissue. Cross section showing layers (zona) of the cortex. A portion of the medulla is also visible. (×35.)

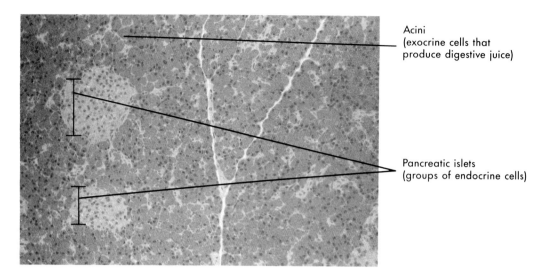

Plate 53 Pancreas tissue. The hormone-producing pancreatic islets are evident among acinar cells that produce pancreatic digestive juice.

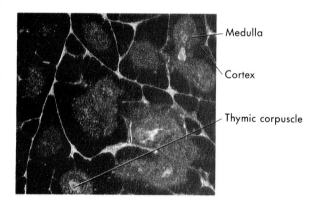

Plate 54 Thymus tissue. Photomicrograph of thymic tissue showing several lobules, each with a cortex and a medulla.

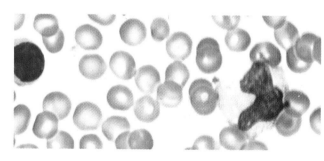

Plate 55 Monocyte. Compare the large monocyte (*right*) with the smaller lymphocyte (*left*).

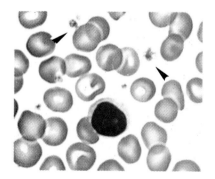

Plate 56 Lymphocyte. Arrowheads point to platelets.

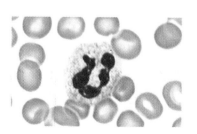

Plate 57 Neutrophil.

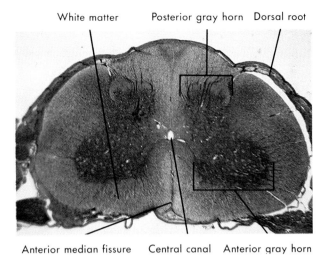

Plate 58 Mammalian spinal cord. Cross section of cord in the cat. (Low power.)

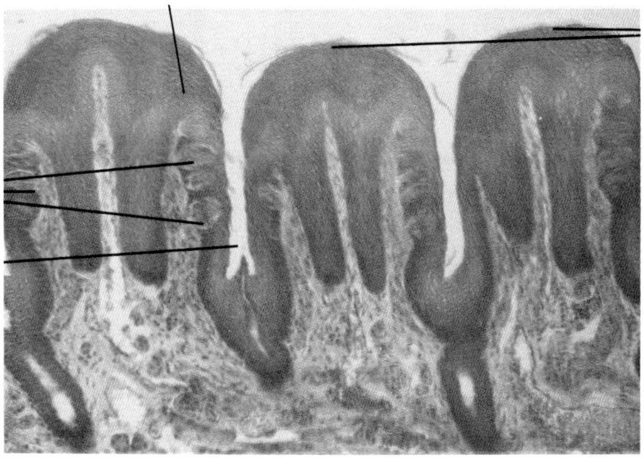

Plate 59 Vallate papillae on surface of tongue. Taste buds are located on lateral surfaces of papillae. Several taste buds can be seen facing into the moat from the sides of the papillae. (×35.)

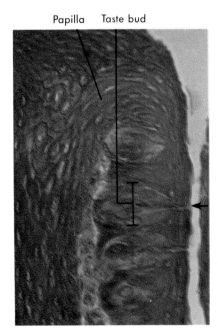

Plate 60 Taste buds. Enlargement of photomicrograph of taste buds. Arrow points to pore in outer surface of taste bud. (×140.)

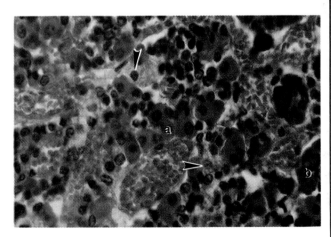

Plate 61 Histology of the adenohypophysis. In this light micrograph, nonstaining chromophobes are indicated by arrowheads. Examples of hormone-secreting cells are labeled a (acidophil) and b (basophil).

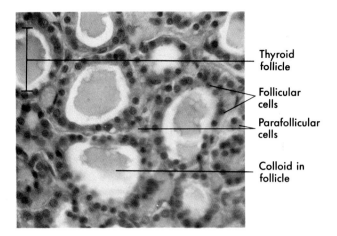

Plate 62 Thyroid gland tissue. Note that each of the follicles is filled with colloid. (×140.)

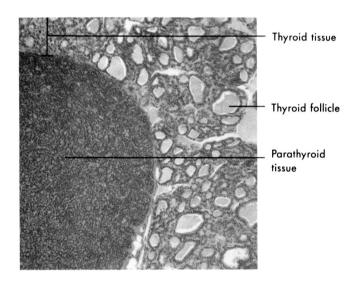

Plate 63 Parathyroid tissue. Parathyroid tissue is shown surrounded by thyroid tissue, in which it is embedded. (×35.)

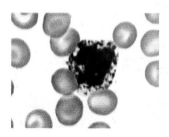

Plate 64 Eosinophil.

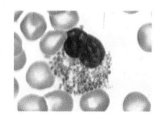

Plate 65 Basophil.

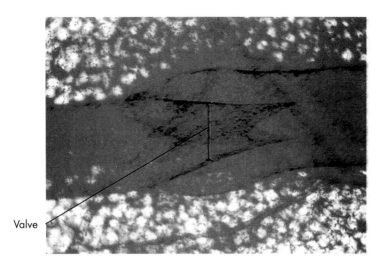

Plate 66 Lymphatic vessel. Photomicrograph shows valve that ensures one-way flow of lymphatic fluid. (×25.)

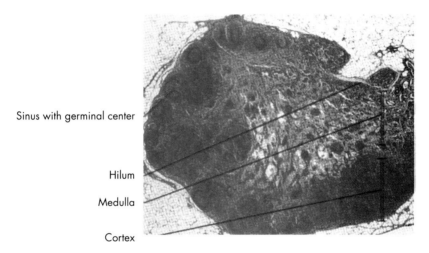

Plate 67 Lymph node. Photomicrograph of cross section of lymph node shows germinal centers in sinuses of cortex region. Note also medullary region with narrow, dark cords. Hilum is a notch through which blood and lymphatic vessels pass. (×5.)

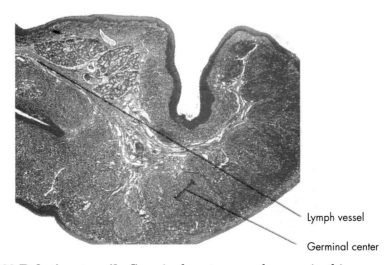

Plate 68 Palatine tonsil. Germinal centers can be seen in this cross section.

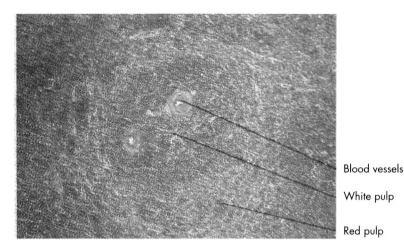

Plate 69 Spleen tissue. Photomicrograph of spleen cross section shows the white pulp surrounding splenic arterial vessels and the peripheral red pulp.

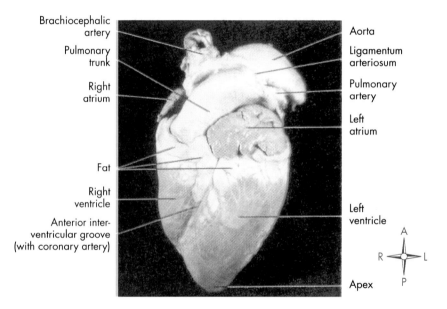

Plate 70 Sheep heart. Ventral view.

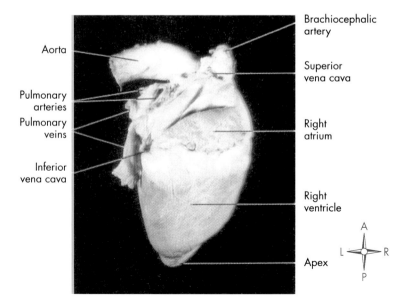

Plate 71 Sheep heart. Dorsal view.

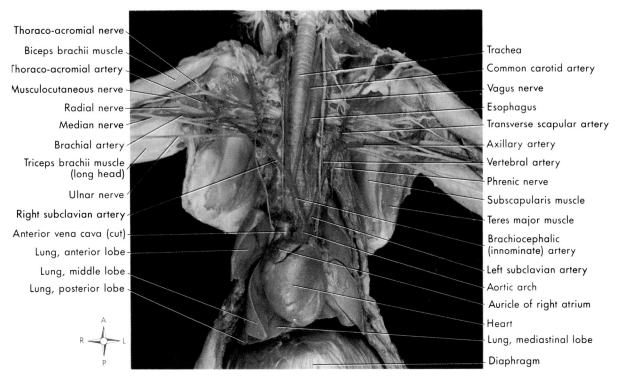

Plate 72 Arteries of the cat. Neck and thorax, ventral view.

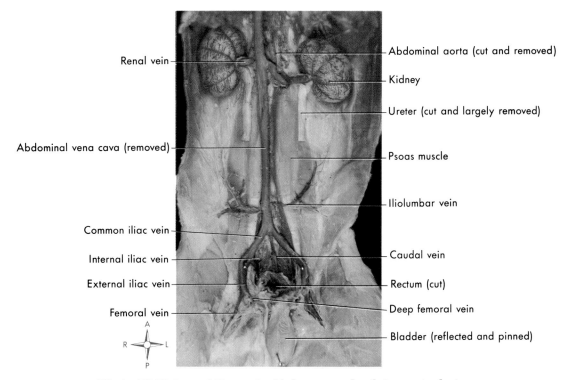

Plate 73 Veins of the cat. Abdomen and pelvis, ventral view.

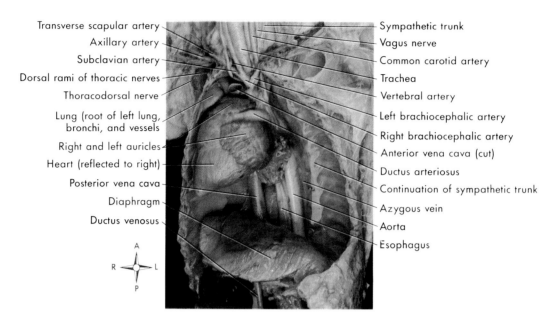

Plate 74 Vessels of the fetal pig. Neck and thorax, ventral view.

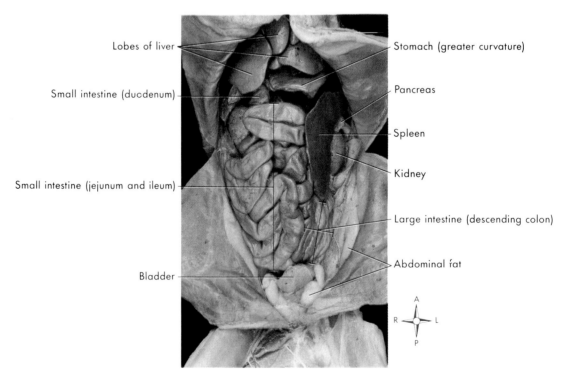

Plate 75 Digestive organs of the cat. Abdominopelvic cavity, ventral view.

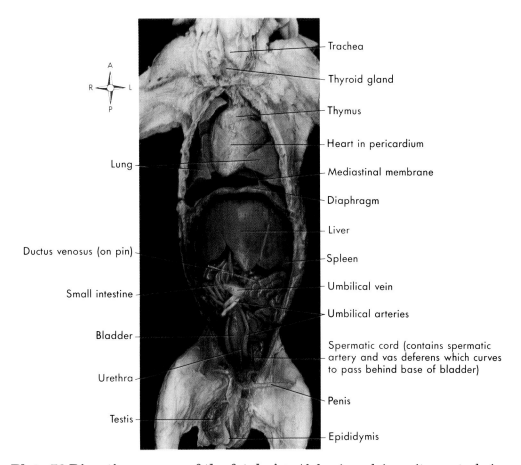

Plate 76 Digestive organs of the fetal pig. Abdominopelvic cavity, ventral view.

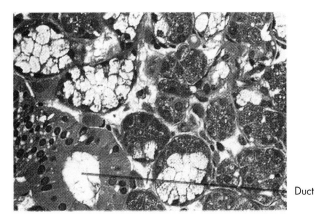

Plate 77 Salivary gland tissue. Photomicrograph of submandibular salivary gland. Note mucus-filled glandular cells and rings of duct cells.

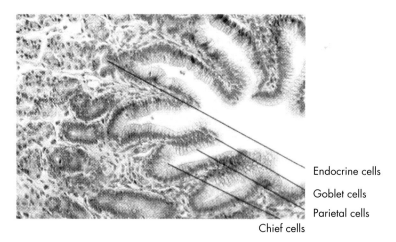

Plate 78 Epithelial tissue of stomach wall. Note different cell types within the lining of a gastric gland from the wall of the stomach.

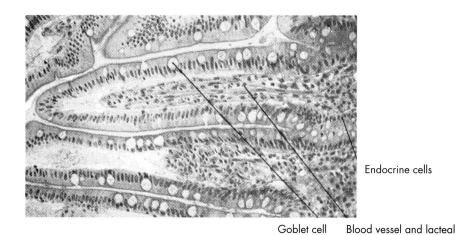

Plate 79 Epithelial tissue of the intestinal wall. Note goblet cells among the columnar cells of the intestinal lining.

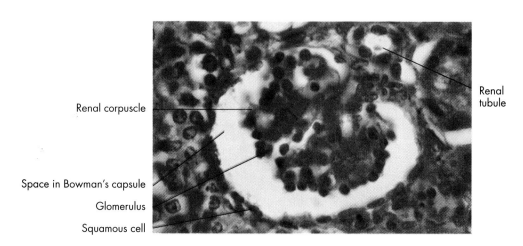

Plate 80 Renal corpuscle. Photomicrograph of kidney tissue shows a Bowman's capsule surrounding a glomerulus.

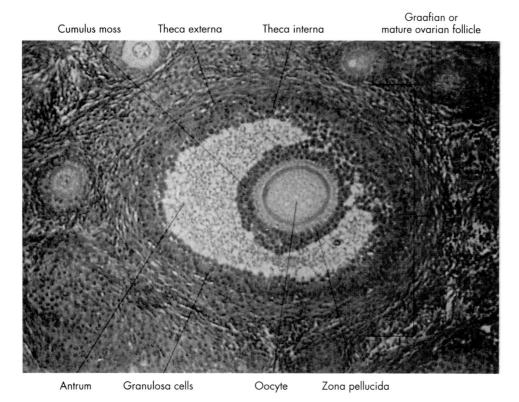

Plate 81 Graafian follicle. Photomicrograph of ovarian cortex, showing a mature (Graafian) ovarian follicle.

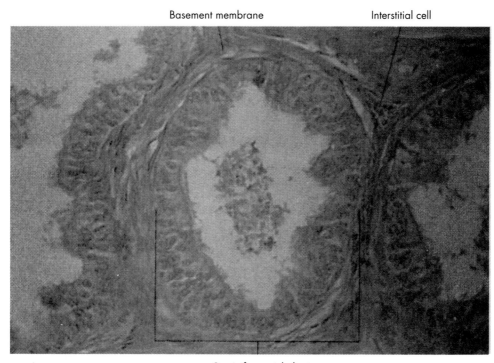

Plate 82 Testis tissue. Photomicrograph of testicular tissue shows seminiferous tubules in cross section. Cells between tubules are call *interstitial cells*.

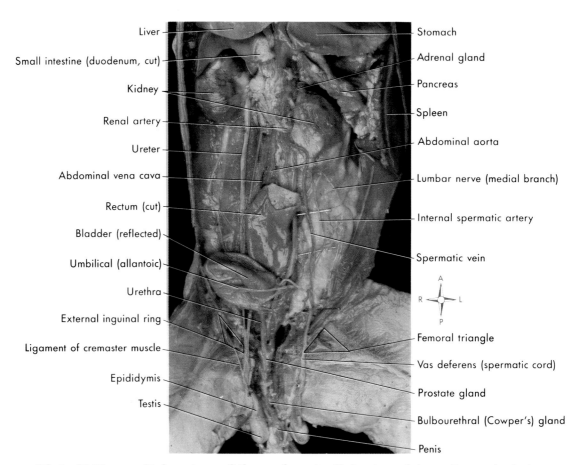

Plate 83 Urogenital system of the male cat. Abdominopelvic cavity, ventral view.

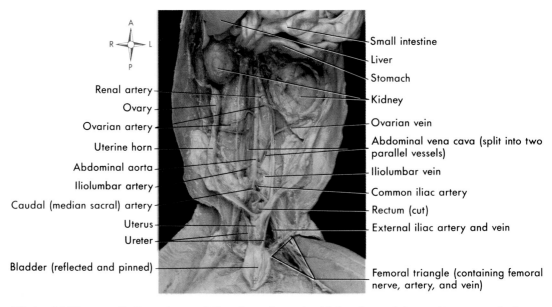

Plate 84 Urogenital system of the female cat. Abdominopelvic cavity, ventral view.

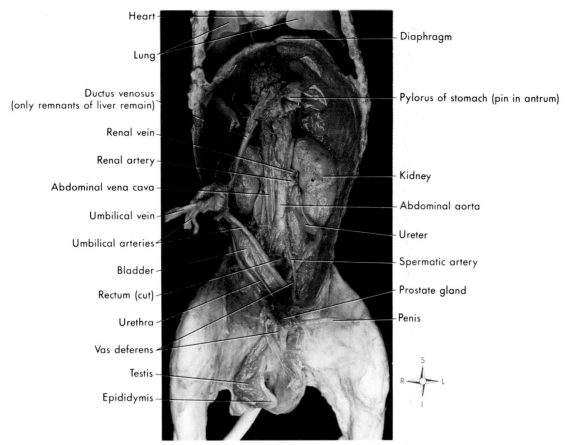

Plate 85 Urogenital system of the male fetal pig. Abdominopelvic cavity, ventral view.

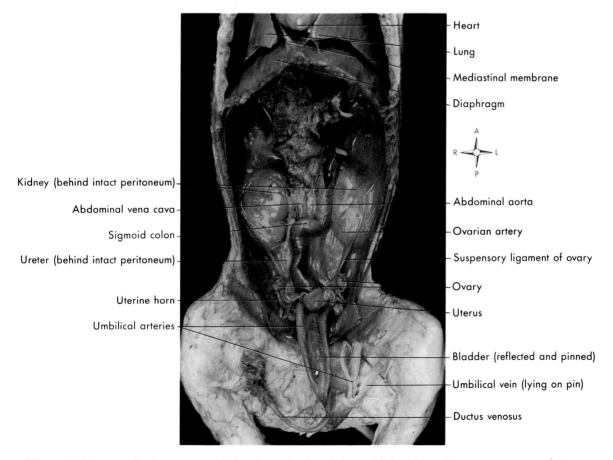

Plate 86 Urogenital system of the female fetal pig. Abdominopelvic cavity, ventral view.

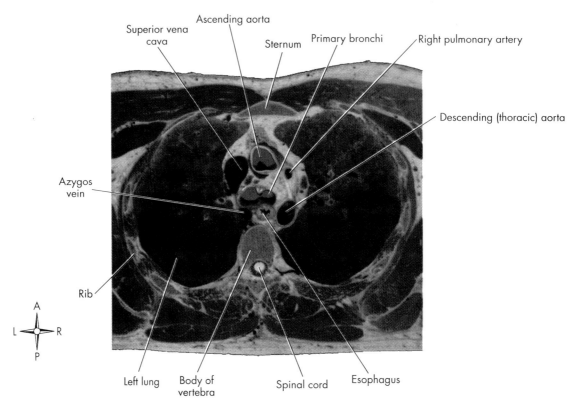

Plate 87 Transverse (horizontal) section of human thorax. Section at level of vertebra T4 and the bifurcation of the trachea.

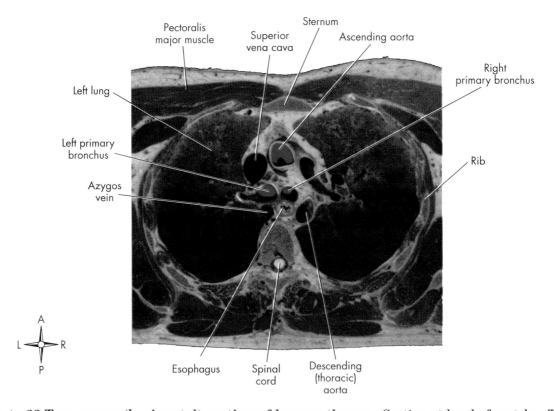

Plate 88 Transverse (horizontal) section of human thorax. Section at level of vertebra T5.

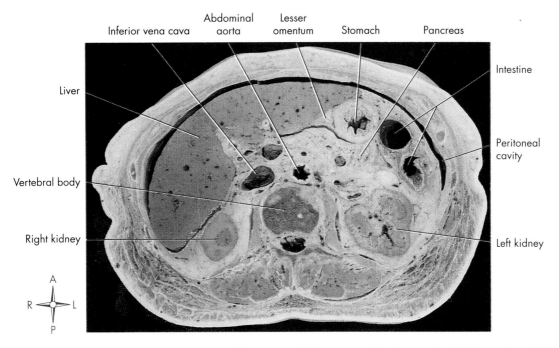

Plate 89 Transverse (horizontal) section of human abdomen. Section at level of vertebra T12, upper face of section, viewed from below.

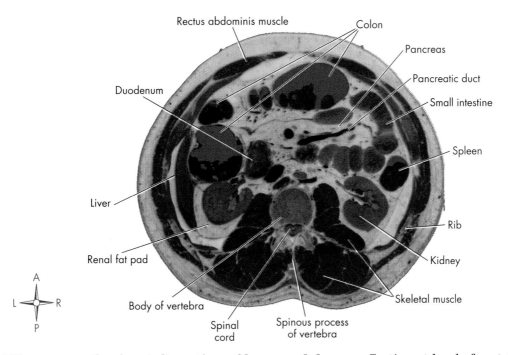

Plate 90 Transverse (horizontal) section of human abdomen. Section at level of vertebra L2.